TABLE OF CONTENTS

Reduction and Dekomposition of Modulo Optimization Problems

A. Bachem

Institut für Operations Research
Universität Bonn, Germany

1. Introduction

Let G_1, G_2 be subgroups of the additive group $(Z^m,+)$, let $S \subset Z^m$, Ψ, f be maps from S into G_2, $\mathbb{R}$ resp. and let $\Phi: G_1 \rightarrow G_2$ be a homomorphism. We consider the problem

$$\begin{array}{llll} & & \min & f(x) \\ (1.1) & \text{s.t.} & & \Psi(x) \in \text{kernel } \Phi \\ & & & x \in S \end{array}$$

where kernel Φ denotes the annulator of Φ. Since "$\Psi(x) \in$ kernel Φ" is usually a congruence relation, we look at (1.1) as a generalized "modulo problem " and call it a modulo optimization problem. Semigroup optimization (1), problems in coding theory (3) and other problems (4) may be reformulated as modulo optimization problems, though there will be no advantage from an algorithmic point of view.

Here we are going to investigate a special homomorphism $\Phi(x) := x - A \lfloor A^+ x \rfloor$, where A^+ denotes the Moore-Penrose inverse of an (m,n) integer matrix A and " $\lfloor x \rfloor$" denotes the integer part of x. The homomorphism Φ generalizes in some sense a congruence relation as: "$\Phi(x) := x (\equiv \text{mod } A)$".
If we denote by G(A) the group $\Phi(Z^m)$ and by $G(\Psi/A) = \{\Phi(x) \mid x = \psi(\lambda), \lambda \in Z^n\}$ the subgroup generated by ψ we are interesting in the following three questions:

1) Characterize isomorphic groups G(A).

2) Find a reduced representation of $G(\psi/A)$ in case $G(\psi/A)$ is not isomorphic to G(A)

3) Decompose (1.1) in subproblems $G(\psi_i/A)$, and link the optima of the subproblems to an optimum of problem (1.1)

2. Reduction of Modulo Optimization Problems

If A is an (m,n) integer matrix and rank of A equals m, we denote by H(A) the uniquely determined matrix B, such that for an unimodular matrix K, AK = (B,0) is in Hermite normal form.

Definition (2.1)

Let A be an (m,n) integer matrix, rank A ε {m,n}

$$\phi_A(x) := \begin{cases} x - A\,[A^+x] & \text{if rank } A = n \\ x - H(A)\,H(A)^+x & \text{otherwise} \end{cases}$$

where A^+ denotes the Moore-Penrose inverse of A; and "$[x]$" the integer part of x.

Proposition (2.2)

Let G be an additiv subgroup of Z^m. The map $\phi_A: G \to \phi_A(G)$ is a homomorphism onto $(\phi_A(G), \oplus)$, where $x \oplus y := \phi_A(x+y)$.
kernel $\phi_A = \{x \varepsilon G / x = A\lambda,\ \lambda \varepsilon Z^n\}$

Proof:

Let's first take the case "rank A = n". Since $A^+A = (A'A)^+A'A = I_n$ and A'A regular, Φ_A is clearly a homomorphism. Let x be an element of kernel ϕ_A, that is $x = A[A^+x]$. Since A'A is regular, $A^+x = (A'A)^{-1}A'x = [A^+x] = \lambda$ and we have $x = A\lambda$.
Let now x be an element of the set $\{x\varepsilon G/x = A\lambda,\ \lambda\varepsilon Z^n\}$.
The rank of A equals n, so we have $A^+ = (A'A)^{-1}A'$ which means that the system $A^+A\lambda = \lambda$ is consistent for all integers λ and $x = A\lambda$ is a solution of $A^+z = \lambda$. Since λ is integer, it follows $A^+x = [A^+x]$ or $A(A^+x - [A^+x]) = o$ and because of $AA^+x = A\lambda = x$ we get the result $\phi_A(x) = x - A[A^+x] = o$, which completes the proof. The case "rank A=m" is almost the same as the first case and we don't want to treat it here.

Definition (2.3)

Let A be an (m,n) integer matrix such that the row or column rank is maximal. Because of proposition (2.2), $(\phi_A(Z^m), \oplus)$ is an additive group and we will denote this group by G(A). We call the product of the invariant factors of B the determinant of B (det B). If $d = \prod_{j=1}^{k} \pm p_j^{\varepsilon_j}$ is a representation of d = det B as a product of prime factors, and p a function from $\mathbb{N}^2$ into $\mathbb{N}$ defined recursively as

$$\underline{p(n,m):} = \begin{cases} p(n,m) & 1 \le n \le m \\ p(n,m-1) + p(n-m,m) & n \ge m \ge 1 \end{cases}$$

$p(o,m): = 1$, $p(n,o):=o$ $n,m\varepsilon\mathbb{N}$

we define

$$\underline{K(d)} : = \sup_{m\varepsilon\mathbb{N}} \prod_{j=1}^{k} p(\varepsilon_j,m)$$

Theorem (2.4)

The number of nonisomorphic groups G(A), where A is an (m,n) integer matrix with maximal row rank, equals K(d). (K(d) is finite, d=det A).

For a proof of the theorem we refer to (2). Additionally in (2) a list of some **of** the numbers K(d) up to determinants 100 000 is given. We have restricted ourselves to affine maps ψ, so let $\psi(x) = Nx - b$, where N is an (m,r) integer matrix and $b\varepsilon Z^m$. We denote the underlying group of problem (1.1) by

$$\underline{G(N/B)} : = \{\phi_B(x) \ / \ x = N\lambda, \ \lambda\varepsilon Z^n\}$$

To get a canonical representation of G(N/B) we need

Definition (2.5)

Let $\rho\varepsilon\mathbb{N}_+^m$ and rank B = n. We call $\delta\varepsilon\mathbb{N}^m$ the reduction of B under ρ if $G(\text{diag}(\rho)^{-1}B)$ is isomorphic to $G(\text{diag}(\delta))$.

In this case we denote $\delta: = \text{Red}\ (B/\rho)$

Proposition (2.6)

For every (m,n)-integer matrix B with maximal column rank and every $\rho\varepsilon\mathbb{N}^m$ *there exists a* $\delta\varepsilon\mathbb{N}^m$*, such that*

$$\delta = \text{Red}\ (B/\rho)$$

For a proof we refer to (2).

Theorem (2.7) (Reduction Procedure)

Let N be a (m,r) integer matrix and $WPNQ = (\text{diag}\ (\rho_1,\ldots,\rho_{m_1}, 0_{n-m_1})$ *the Smith mormal form of N up to zero rows. Let B be an (m,n) integer matrix, rank* $B\varepsilon\{m,n\}$*. If* δ *is the reduction of H(WPB) under* ρ *we have*

$$G(N/B) \cong G(\text{diag}(\delta))$$

Proof:

Let us denote by $\{N\}: =\{x\varepsilon Z^m/x=N\lambda,\ \lambda\varepsilon Z^n\}$.

Since Φ_B is a homomorphism (proposition 2.2), G(N/B) is isomorphic to the factor group $\{N\}/_{\text{Kern } \phi_B}$.

The map WP : $\{N\} \to \text{Kern } \phi_{\text{diag}(\rho)}$ is a surjective homomorphism, which yields an isomorphism between the factor groups $\{N\}/_{\text{Kern } \phi_B}$ and $\text{Kern } \phi_{\text{diag}(\rho)}/_{\text{Kern } \phi_{WPB}}$. Let A : = H(WPB) and $F := \text{diag}\ (\rho)^{-1} A$ be matrices, then we have shown above the isomorphism

$$G(N/B) \cong Z^{m_1}/_{\text{Kern } F},$$

and defining $\delta := \text{Red}\ (H(WPB)/P)$
we get the desired result applying proposition (2.6).

Corollary (2.8)
Let us denote by $g_1,\dots,g_k$ the group elements of G(N/B) which are generated by the matrix N. The corresponding group elements in the isomorphic representation G(diag(δ)) have the form

$$\phi_{\text{diag}(\delta)}\ (Ag_i) \qquad i = 1,\dots,k$$

where $A := U\ \text{diag}\ (\rho)^{-1}\ WP$ and U denotes the left hand side unimodular matrix, which transforms E in Smith normal form.

Example (2.9)
Consider the problem

$$\min\ 5x_1 + 6x_2 + x_3$$

$$\text{s.t.} \quad \begin{pmatrix} 4 & 1 & 3 \\ 5 & 3 & 2 \end{pmatrix} \begin{pmatrix} x_1 \\ x_2 \\ x_3 \end{pmatrix} \equiv \begin{pmatrix} 6 \\ 4 \end{pmatrix} \bmod \begin{pmatrix} 7 \\ 7 \end{pmatrix}$$

$$x_1,\ x_2,\ x_3 \geq 0 \text{ integer}$$

Theorem (2.7) yields

$$\delta = \begin{pmatrix} 1 \\ 7 \end{pmatrix},\quad U\ \text{diag}\ (\rho)^{-1}\ P = \begin{pmatrix} -3/7 & 1/7 \\ 1 & 0 \end{pmatrix}$$

$$\phi_{\text{diag}(\delta)} \begin{pmatrix} -1 \\ 4 \end{pmatrix} = \begin{pmatrix} 0 \\ 4 \end{pmatrix} \qquad \phi_{\text{diag}(\delta)} \begin{pmatrix} -2 \\ 6 \end{pmatrix} = \begin{pmatrix} 0 \\ 6 \end{pmatrix}$$

$$\phi_{\text{diag}(\delta)} \begin{pmatrix} 0 \\ 1 \end{pmatrix} = \begin{pmatrix} 0 \\ 1 \end{pmatrix}$$

$$\phi_{\text{diag}(\delta)} \begin{pmatrix} -1 \\ 3 \end{pmatrix} = \begin{pmatrix} 0 \\ 3 \end{pmatrix}$$

and we get the equivalent problem

$$\min \quad 5x_1 + 6x_2 + x_3$$

$$\text{s.t.} \qquad 4x_1 + x_2 + 3x_3 \equiv 6 \bmod 7$$

$$x_1, x_2, x_3 \geq 0 \text{ integer}$$

It is well known that the condition "(B,N) contains an (m,m) unimodular submatrix" is sufficient but not necessary for G(N/B) and G(B) to be isomorphic.

For instance the matrices $N = \begin{pmatrix} 3 & 0 \\ 0 & 6 \end{pmatrix}$ and $B = \begin{pmatrix} 5 & 0 \\ 0 & 10 \end{pmatrix}$ yields G(N/B)=G(B), but (B,N) contains no unimodular submatrix.

If G(N/B) is isomorphic to G(B), we obtain that the ith invariant factor of B and the matrix elements $N_{i_1}, \ldots, N_{i_n}$ are relatively prime for all $i=1,\ldots,m$ as a necessary condition.

3. Partitioning of Modulo Optimization Problems

Because of theorem (2.7) we now need only consider problems where the matrix B is reduced in such a way that G(N/B) is isomorphic to G(B). Let us denote the set of feasible solutions of problem (1.1) by

$$LG(N,b/B) := \{x \in S \;/\; Nx - b \in \text{Kern } \phi_B\}$$

Proposition (3.1)

Let $i_o, j_o \in \{1,\ldots,n\} =: L$ such that $N_{j_o} \in G(N_{i_o}/B)$, where N_{j_o} denotes the j_o^{th} column of the matrix N.

x is an optimal solution to

$$(P1) \quad \left| \begin{array}{l} \min f(x) \\ x \in LG(N_J, b/B) \end{array} \right.$$

if and only if $L_{j_o}(x)$ is an optimal solution to

$$(P2) \quad \left| \begin{array}{l} \min \sum_{i \in L} c_i' x_i \\ x \in LG(N, b/B) \end{array} \right.$$

where

$\lambda \in \mathbb{N}_+$ and $N_{j_o} = \phi_B(\lambda N_{i_o})$, $J = L \{i_o\}$ and

$$\delta := \begin{cases} 1 & c_{j_o} \leq \lambda c_{i_o} \\ 0 & \textit{otherwise} \end{cases} \quad \textit{and}$$

$$f(x) = \sum_{i \in J} c_i' x_i - \delta \left[\frac{x_{i_0}}{\lambda} \right] (\lambda c_{i_0} - c_{j_0})$$

and

$$L_{j_0}(x) = \begin{cases} x_i - \delta \cdot \lambda \left[\dfrac{x_{i_0}}{\lambda} \right] & i = i_o \\ \delta \left[\dfrac{x_{i_0}}{\lambda} \right] & i = j_o \\ x_i & i \in J \; \{i_0\} \end{cases}$$

Proposition (3.1) reduces the matrix N of problem (1.1) to the minimal set of generators of the group G(B) in case that an algorithm for piecewise linear objective function is available. For a proof see (2).

The computational effort to solve problem (1.1) usually grows rapidly according to the determinant of B. It is therefore sometimes advantageous to decompose the problem into smaller subproblems and to link the optima of the subproblem to a solution of the master-problem. We give now two examples of decomposing problem (1.1) in case the matrix N is of the form

$$(3.2) \qquad N = \left(\begin{array}{c|cccc} & N_2 & & & 0 \\ & & \cdot & & \\ N_1 & & & \cdot & \\ & 0 & & \cdot & N_r \end{array} \right)$$

or

$$(3.3) \qquad N = \begin{pmatrix} A_1, \ldots, A_r & & \\ N_1 & \cdot & 0 \\ & \cdot & \\ 0 & & \cdot N_r \end{pmatrix} \qquad b = \begin{pmatrix} b_0 \\ \cdot \\ \cdot \\ b_r \end{pmatrix}$$

Let N be an (m,r) integer matrix of form (3.2), let $b_i(x) := \Phi_B(b - N_1 x)_{I_i}$ where I_i corresponds to the row indices of the submatrix N_i and let denote by

$$z(b_i(y)) := \begin{cases} \infty & \text{if } b_i(y) \notin G(N_i/B) \\ \min c_i' X & \\ x \in LG(N_i, b_i(y)/B) & \end{cases}$$

the optimal value of the subproblems.

Theorem (3.4)

The programs

$$\min_{x \in LG(N,b/B)} c'x \qquad (3.5)$$

and

$$\min_{y \in \mathbb{N}} c_1 y + \sum_{i=2}^{r} z(b_i(y)) \qquad (3.6)$$

are equivalent.

Let again N be an (m,r) integer matrix which has form (3.3) and define

$$z_1(x_2,\dots,x_n) := \min_{x_1 \in LG\left(\binom{A_1}{N_1}, \begin{pmatrix} b_0 - \sum_{i=2}^{r} A_i x_i \\ b_1 \end{pmatrix} / B \right)} c_1 x_1$$

$$z_i(x_i,\dots,x_n) := \min_{x_i \in LG(N_i,\, b_i/B)} c_i x_i + z_{i-1}(x_{i-1},\dots,x_n) \qquad i=2,\dots,r$$

as the optimal value of the subproblems.

Theorem (3.7)

The programs

$$\min_{x \in LG(N,b/B)} c'x \qquad (3.8)$$

and

$$\min_{x_r \in LG(N_r, b_r/B)} c_r x_r \; 0 \; z_{r-1}(x_{r-1}, x_r) \qquad (3.9)$$

are equivalent.

For a proof of theorem (3.4) and (3.7) see (2).

References:

(1) Araoz-Durand, J., "Polyhedral Neopolarities, Thesis, Faculty of Mathematics, Department of Computer Sciences and applied Analysis, University of Waterloo, Waterloo, Ontario, November 1973

(2) Bachem, A., "Algorithmen zur Modulo-Optimierung strukturierter Matrizen", Report 7539-OR, Institut für Ökonometrie und Operations Research, Universität Bonn, 1975

(3) Berlekamp, E. R., "Algebraic Coding Theory", McGraw Hill, New York, 1970

(4) Bradley, G.H., "Modulo Optimization Problems and Integer Linear Programming", in: Application of Number Theory to Numerical Analysis", (S.K. Zaremba, ed.) Academic Press New York 1972

OPTIMIZATION OF ELASTIC STRUCTURES BY MATHEMATICAL PROGRAMMING TECHNIQUES

H. Baier

Fachgebiet Leichtbau der TH

61 Darmstadt, Petersenstr. 18

Summary

Structural optimization as a technical field of application of mathematical programming is presented. Since the behaviour of the system to be optimised can be examined only by a rather complex algorithm, a major effort lies in solving the optimization problem efficiently. Penalty function techniques, sequential linear programming and optimality criteria approaches are examined and compared for that purpose.

Introduction

One of the disciplines concerned with design of civil, mechanical or aeronautical constructions is structural design, where the optimal distribution of structural material is sought while performance constraints must be satisfied. This can be solved by mathematical programming techniques . Though other methods of mathematical economics and operations research are useful here, the paper concentrates on three different types of optimization algoritms.

1. Formulation of the problem

The structural optimization problem leads to

$$\text{minimize} \quad \{f(x) \mid g(x) \geq 0 \, , \, h(x) = 0 \, , \, x \in R^n\} \tag{1}$$

An example is presented in chapter 3. The objective function f, the inequality and equality constraints g and h are nonlinear, real valued, continuously differentiable functions from R^n to R, R^m and R^p respectively. The variables x are named design variables. As the problem is not necessarily convex, only local solutions x_s can be expected. Sometimes the specific problem is solved with different starting vectors and the best solution is used. For simplicity in most cases the cost of the structure is represented by its amount of material whereas the constraints say that the systemresponse, such as displacements or stresses caused by the external loading, must be less or equal to some prescribed value. Thicknesses or crosssections are usually taken as design variables and n typically ranges within $5 \lesssim n \lesssim 50$ whereas m is often much larger and p=0 .

Though other possibilities exist [1] , the most natural approach for the solution of the optimal design problem is the following: the variables x are iteratively changed by an optimization procedure and if the values of constraints or their gradient with respect to the design variables are needed, a finite element method is used which allows the analysis, i.e. determination of the systemresponse, for rather complex structures. Strictly speaking, in this method the structure is divided into a lot of geometrical simple elements for which the underlying variational problem of elastomechanics is solved similar a Ritz procedure. An important part of that method are the matrix relations

$$K(x) \cdot U = F \tag{2a}$$

$$\sigma = B(x) \cdot U \tag{2b}$$

$$U, F \in R^N \, , \sigma \in R^M$$

Here it is assumed that the structure behaves elastic and responds linear to the external forces. K is the stiffnes matrix, U the vector of discrete displacements of certain nodal points of the structure, F is one of possibly several force vectors defined by the structur's environment and is applied at that points and B connects U with stresses σ in the elements. More details are found in Whiteman [2], Zienkiewicz [3].

The gradients are obtained by

$$K(x) \cdot \frac{\partial}{\partial x_i} U = -\left(\frac{\partial}{\partial x_i} K(x)\right) \cdot U \tag{3a}$$

$$i = 1,\dots n$$

$$\frac{\partial}{\partial x_i} \sigma = \left(\frac{\partial}{\partial x_i} B(x)\right) \cdot U + B(x) \cdot \frac{\partial}{\partial x_i} U \tag{3b}$$

For complex problems N and M are very large, say several hundred or even thousand so that a major computational burden lies in solving the equations (2) and (3) at each iteration step. It would be far to expensive to compute gradients of higher orders.

2. Solution procedures

Though analytical procedures, see Sheu and Prager [4], which give some theoretical and physical reasoning for the numerical techniques, exist, these techniques can be applied only for rather small and idealised systems, so that for more complex problems only numerical solutions are possible. Three classes of such techniques, the penalty functions, sequential linear programming and optimality criteria are discussed and measured with the performance criteria efficiency and reliability.

2.1 Penalty Functions

By means of penalty functions, as suggested by Caroll [5] and Fiacco, McCormick [6], the constrained problem (1) is transformed to a sequence of free ones by

$$T(x, r_k, r_k^*) = f(x) + r_k \sum_{j=1}^{m} \Phi_j(g_j) + r_k^* \sum_{j=1}^{p} \Psi_j(h_j) \tag{4}$$

where k means the number of iterationstep, r_k must strictly decrease to zero and r_k^* strictly increase to infinity. For k large enough the solution of (4) will be sufficiently close to that of (1). Besides some continuity requirements Φ_j and Ψ_j must satisfy

$$\Phi_j \to \infty \quad \text{if} \quad g_j \to 0 \tag{5a}$$

$$\Psi_j \to 0 \quad \text{if} \quad h_j \to 0 \tag{5b}$$

and simple and often used versions are

$$\Phi_j = \frac{1}{g_j} \tag{6a}$$

$$\Psi_j = h_j^2 \tag{6b}$$

Then if p=0, (4) is an interior-point version of penalty functions, i.e. beginning with a feasible starting point x^0 satisfying $g(x) > 0$, (which to find causes no serious problem here), all subsequent points theoretically lie within the feasible region. But since on a computer one is doing finite steps, f and g_j might be computed in the infeasible region and because of (6a) Φ_j is negative. So this step is assumed as productive for minimizing T and no attempts are made to return to the feasible region, see figure 1 . Some modifications of (6a) have been tried [7] and one of the most promising is due to Kavlie [8] :

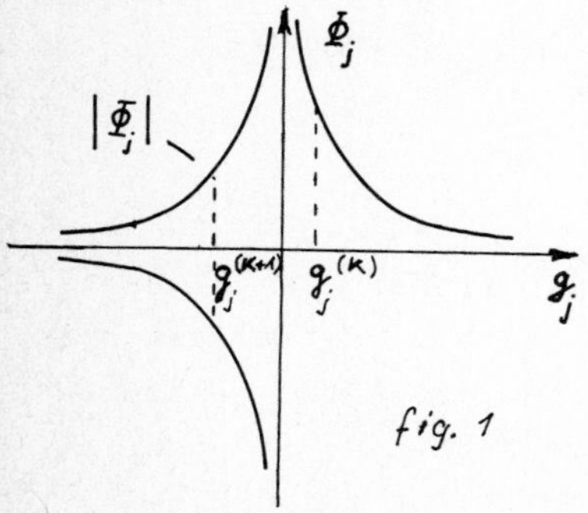

fig. 1

$$\Phi_j = \begin{cases} \dfrac{1}{g_j} & \text{, if } g_j \geq \varepsilon \\ \dfrac{2\varepsilon - g_j}{\varepsilon^2} & \text{, if } g_j < \varepsilon \end{cases} \tag{7a}$$

where $\varepsilon = r_k/\delta$ (7b)

and δ is a positive constant that defines the transition between the two types of penalty terms. With (7a) one renounces on strict feasibility at least during the first iterationsteps. Another problem is the reasonable determination of r_k and r_k^*, and though some more or less heuristical proposals exist [6, 7], very often the proper choice is known only after a specific problem has been computed several times. From the Hession matrix H of (4)

$$H = \nabla^2 f + r_k \sum_{j=1}^{m} \frac{2}{g_j^3} \nabla g_j \nabla g_j^T - r_k \sum_{j=1}^{m} \frac{1}{g_j^2} \nabla^2 g_j + 2r_k^* \sum_{j=1}^{p} \nabla h_j \nabla h_j^T + 2r_k^* \sum_{j=1}^{p} h_j \nabla^2 h_j \tag{8}$$

can be seen that near a solution H becomes ill conditioned and this may even lead to nonoptimal solutions if the r's are not properly chosen.

The advantages of the technique are quite obvious: its simplicity, original convexity is maintained by (4) and algorithms for free problems can be used. From derivative free search procedures we have found [7] these of Hooke and Jeeves [9], Powell [10] and Davies and Swann [11] as rather efficient. From gradient techniques the variable metric method of Fletcher and Powell [12] is one of the best. The onedimensional minimization procedures have a great influence on the total efficiency of these algorithms, and especially for gradient methods that of Lasdon [13] can be recommended because it takes into account the special nature of the penalty functions.

As the number of functionevaluations of (4) or computing gradients drastically increases with the number of design variables, which is especially true for the gradient free techniques, these methods should be used only for smaller problems, say $n \lesssim 10$.

2.2 Sequential Linear Programming

Substituting nonlinear problems by a sequence of linear ones always has had a certain appeal to engineers. Such an approach in optimization is that of Griffith and Stewart [14], which is also usable for nonconvex problems because after every step a total relinearization is undertaken. For the purpose of Structural Optimization it has the following form (p = 0):

step 1: r:=0, k:=0; $x^{(k)}$ and a scalar $0 < \alpha^{(r)} \leq 1$ are given

step 2: Analyse the structure with x^k, equations (2)

step 3: Scaling: $x^{(k)} := \gamma^{(k)} \cdot x^{(k)}$

step 4: Determine the gradients at $x^{(k)}$, equations (3)

step 5: Determine a new approximation $x^{(k+1)}$ for the solution x_s by

$$\text{minimize} \quad f^{(k)} + \sum_{i=1}^{m} \frac{\partial f}{\partial x_i}\bigg|_{x^k} \cdot (x_i - x_i^{(k)}) \tag{9a}$$

$$g_j^{(k)} + \sum_{i=1}^{n} \frac{\partial g_j}{\partial x_i}\bigg|_{x^k} \cdot (x_i - x_i^{(k)}) \geq 0 \text{ , } j = 1,\ldots m \tag{9b}$$

$$x_i^{(k)}(\alpha - 1) + x_i \geq 0 \tag{9c}$$

$$i = 1,\ldots n$$

$$x_i^{(k)}(1 + \alpha) - x_i \geq 0 \tag{9d}$$

$$x_i \geq 0 \tag{9e}$$

step 6: Analyse the structure with $x^{(k+1)}$

step 7: Scaling: $x^{(k+1)} := \beta^{(k)} \cdot x^{(k+1)}$

step 8: Termination criterion

$$\left| \frac{f^{(k+1)} - f^{(k)}}{f^{(k)}} \right| \quad \begin{array}{ll} > \varepsilon & \text{go to step 9} \\ \leq \varepsilon & \text{stop} \end{array}$$

where ε is some prescribed small value.

step 9:

$$f^{(k+1)} \begin{cases} < f^{(k)} & : k := k+1, \quad \text{go to step 4} \\ > f^{(k)} & : \text{go to step 10} \end{cases}$$

step 10:

$$x^{(k+1)} := 0.5 \cdot (x^{(k+1)} + x^{(k)}) \tag{10}$$

$$r := r+1$$

decrease α, go to step 7

The scaling factors γ and β are obtained from the behaviour of the structure by physical reasoning and they make at least one constraint active while the others are inactive. So in step 8 only feasible designs are compared. Since in the linear problem (9) the coefficients of the objective function are positive and the number of constraints m is often much larger than n, the dual form of the simplex routine should be used. Constraints (9c), (9d) can be handled very efficiently by the lower- and upper bounding technique. These constraints are incorporated for two reasons: first, the linear approximation does hold only for a certain region around $x^{(k)}$, and second they should prevent the well known cycling between two corners of the linear problem which are not solutions of the nonlinear one. Step 10 plays a similar role: if the objective function has increased because of an ill representation around $x^{(k)}$ of (1) by (9), the 'steplength' is halfed until $f^{(k+1)}$ becomes smaller than $f^{(k)}$ or the termination criteria is active.

The total number of iterations depends on the starting vector $x^{(0)}$ and a reasonable choice of $\alpha^{(r)}$. Heuristically we have found [15]

$$\alpha^0 = 0.75 \tag{11a}$$

$$\alpha^{(r+1)} = \frac{\alpha^{(r)}}{1+\alpha^{(r)2}} \tag{11b}$$

as quite satisfactorily. This guarantees that changes in $x^{(k)}$ can be large enough at the beginning, reduced if side iterations (10) must be taken and are not to small at the end.

Our experience shows that this approach is rather efficient and very reliable, at least compared with penalty function techniques.

If one wants to renounce the move-limits of equations (9c), (9d) one can use a modification of the objective function discussed by Beale [16], with the geometrical interpretation shown in figure 2. A similar thing can be done in structural optimization. Though we do not know the

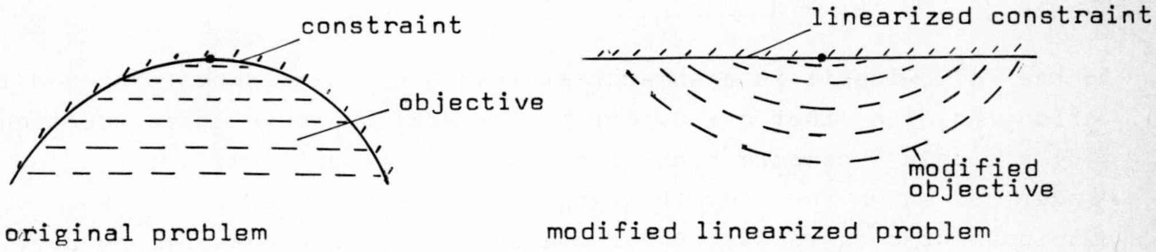

fig. 2

functions g_j explicitly, from elastomechanics it is known that terms $x_i^{-q_i}$, $q_i > 0$, will dominate in the stress constraints. So a Taylor expansion with neglected terms of second and higher orders in the variables $x_i^{-q_i}$ will substitute the original problem much better than the linear one. With $\bar{x}_i = x_i^{-q_i}$ in the above mentioned algorithm step 5 is substituted by

$$\text{minimize } f(\bar{x}) \tag{12a}$$

$$g_j^{(k)} + \sum_{i=1}^{n} \frac{\partial g_j}{\partial \bar{x}_i}\bigg|_{\bar{x}^{(k)}} \cdot (\bar{x}_i - \bar{x}_i^{(k)}) \geq 0 \quad , \quad j = 1,\ldots m \tag{12b}$$

$$\bar{x}_i^{(k)}(\bar{\alpha} - 1) + \bar{x}_i \geq 0 \tag{12c}$$

$$\bar{x}_i^{(k)}(1 + \bar{\alpha}) - \bar{x}_i \geq 0 \tag{12d}$$

$$\bar{x}_i \geq 0 \tag{12e}$$

Problem (12) can be solved by any routine for nonlinear minimization problems with linear constraints. Now the hypercube defined by (12c), (12d) can be made greater than in (9c), (9d) or even quite neglected.

3.3 Iteration Formulae derived from the Kuhn-Tucker-Equations

Intuitively it is assumed that if one uses a criteria which is satisfied by an optimal solution to obtain that solution, the process will converge faster than the aforementioned comparatively 'blind' techniques. Such a criterion are the Kuhn-Tucker-equations [17]

$$\frac{\partial f}{\partial x_i} = \sum_{j=1}^{m} \frac{\partial g_j}{\partial x_i} \lambda_j + \sum_{l=1}^{p} \frac{\partial h_l}{\partial x_i} \nu_l \tag{13a}$$

$$\Sigma\, g_j\, \lambda_j = 0 \qquad i = 1,\ldots n \tag{13b}$$

$$\lambda_j \geq 0 \qquad j = 1,\ldots m \tag{13c}$$

$$g_j \geq 0 \tag{13d}$$

$$l = 1,\ldots p$$

$$h_l = 0 \tag{13e}$$

In the following it is assumed that the Kuhn-Tucker-constraint-qualifications hold and that $p = 0$. For the general nonconvex case equations (13) are only necessary conditions for a local optimum, but in [18] Prager has shown that for the case of stiffness constraints, such as displacement constraints, the above condition is necessary and sufficient for a global minimum. Such a constraint is defined by the work W_j of a force vector F_j acting along the displacements mentioned in (2a).

$$W_j = F_j^T\, U \tag{14}$$

$$g_j = C_j - W_j \tag{15}$$

where C_j is some prescribed value. After some manipulations using equations (2a) and (3a) one obtains for (13a)

$$\frac{\partial f}{\partial x_i} = \sum_j U_j^T \left(\frac{\partial}{\partial x_i} K\right) \cdot U \cdot \lambda_j \tag{16}$$

where U_j is a displacement vector caused by F_j. Note that no gradient of the system response U must be determined.

Only simple recursion formulae should be used, and one possibility is obtained by multiplying (16) with x_i and forming

$$x_i^{(k+1)} = x_i^{(k)} \cdot (\sum_j U_j^T (\frac{\partial}{\partial x_i} K) \cdot U \cdot \lambda_j)^{(K)} / (\frac{\partial f}{\partial x_i})^{(K)} \tag{17}$$

with $\frac{\partial f}{\partial x_i} \neq 0$. Initial values $\lambda^{(0)}$ are obtained by considering all the constraints seperately and solving (13b), (16) for each $\lambda_j^{(0)}$. The next approximation is obtained by

$$\lambda_j^{(k+1)} = \lambda_j^{(k)} \left(\frac{W_j^{(k)}}{C_j}\right)^Z \tag{18}$$

with $Z \approx 2$. This satisfies the requirement of eliminating the λ_j corresponding to noncritical constraints. Though this scheme is somewhat intuitive, results published so far [19] show that it fails in very few problems and because of its efficiency it is up to now the only procedure for great problems (say $n \geq 50$) used in industrial environment [20].

An algorithm handling stress constraints, where the gradients of the system response σ do not vanish, is proposed by the anthor in [21]. The design problem is comprehended as a problem of optimum allocation of resources, i. e. the strength of the structural material and equations (13) are solved by alternating sensitivity analysis by use of the multipliers λ and recursion formula for x depending on that sensitivity.

4. Examples

As examples for testing the performance of optimization routines trusses are used, because they are simpel yet not trivial, all important parts of the analysis program are needed, system modifications can be easily undertaken etc. Technically they are a system of straight bars connected at joints where the loads are applied. For the three bar truss of figure 3 with two not simultaneously acting loading conditions the volume minimization problem under stress constraints is

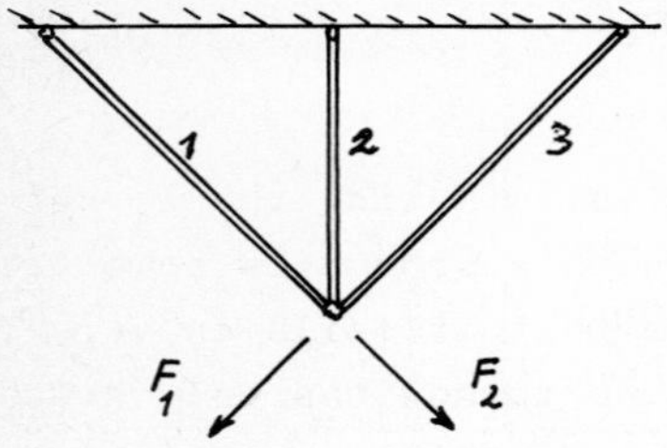

fig. 3

$$f = \sum_i l_i \, A_i \qquad (19a)$$

$$1 - \sigma_{ij}(A)/\sigma_{izul} \geq 0 \qquad i = 1,2,3 \quad j = 1,2 \qquad (19b)$$

$$A_i \geq 0 \qquad (19c)$$

where l_i means the length, A_i the cross section (design variable) of bar i, σ_{ij} is the corresponding stress caused by loading j and σ_{izul} is the prescribed admissible stress.

The following techniques are examined

1. Penalty techniques
   1.1 Gradient free search procedure of Davies and Swann
   1.2 Method of Fletcher and Powell
2. Sequential Linear Programming
   2.1 Direct Linearisation; a slightly modified version of the dual simplex method published in [22] is used
   2.2 Direct Linearisation, with lower and upper bounding technique for (9c), (9d)
   2.3 'Inverse' Linearisation(12); because no better algorithm was at hand, the reduced gradient method of Wolfe [23] is used with PARTAN acceleration [24] and slack variables.
3. Optimality criteria; for trusses stress constraints can be transformed to stiffness constraints [25] and a procedure similar (17), (18) is used

As the performance criteria for the algorithms the total number of iterations and the CPU-time is used. The termination criterion is

$$\left| (f^{(k+1)} - f^{(k)})/f^{(k)} \right| \leq 10^{-4}$$

The number n of bars is increased from 3 to 6 and then in steps of 5 up to 31, so that n design variables, 2n stress constraints because of two loading conditions and n constraints (19c) exist. Details of the examples and the results are out of interest here, they can be found in [15].

Algorithm / n	1.1	1.2	2.1	2.2	2.3	3
3	210/3.8	12/2.9	4/1.1	4/1.1	4/1.1	6/0.6
6	290/32	34/19	8/4.4	8/4.0	2/1.0	13/2.8
11	580/60	62/52	17/41	16/30	16/22	15/ 8
16	970/660		16/157	16/80	11/85	17/22
21			15/359	13/142		13/26
26			14/524	14/282		17/57
31			15/703	15/487		15/68

Inthe columns of this table the first number means the number of iterations, i. e. the number of system-analyzing and gradient determination, the second means the CPU-time in seconds. As the performance is influenced by the parameters of the algorithms, results are presented which ly somewhat above the mean behaviour and can be reached by an experienced user. The computation was performed on the DEC 1040 (PDP 10) computer of the Fachbereich Maschinenbau, TH Darmstadt. The algorithms of type 1 become inefficient for increasing number of variables. For those of type 2 and 3 the number of iterations seems to be rather independent on n. The outstanding performance of type 3 can be seen and still a better performance can be expected for 2.2 and 2.3 if more efficient algorithms for the solution of (9) or (12), respectively, are used.

Conclusion

Structural Optimization serves as a link from a discipline of mathematics to real world application. So still a lot of problems are still unsatisfactorily solved, such as discrete optimization, parametric optimization for instance if the external loading is time dependent, reliability concepts with optimization etc. This is the reason why engineers are looking for practically usable (which nearly always means computerizable) theorems and algorithms.

References

[1] Fox, R. L., and L. A. Schmit: Advances in the Integrated Approach to Structural Synthesis, Journ. of Spacecraft and Rockets, V 3, No6, June 1966, pp 858 - 866

[2] Whiteman, J. R., ed.: The Mathematics of Finite Elements and Applications, Academic Press, London, New York, 1973

[3] Zienkiewicz, O. C.: The Finite Element Method in Engineering Science, McGraw-Hill, London, 1971

[4] Sheu, C. Y., Prager, W.: Recent Developments in Optimal Structural Design, Applied Mechanics Reviews, Vol. 21, No. 10, Oct. 1968

[5] Caroll, C. W.: The Created Response Surface Technique for Optimizing Nonlinear Restrained Systems, Operations Research, Vol. 9, 169, 1961

[6] Fiacco, A., G. P. McCormick: Nonlinear Programming - Sequential Unconstrained Minimitation Techniques, John Wiley a. Sons, New York 1968

[7] Baier, H., Helwig, G.: Nichtlineare Optimierung für den Anwender, Report No. L0174 of the Fachgebiet Leichtbau, TH Darmstadt, BRD, 1974

[8] Kavlie, D.: Optimum design of statically indeterminate structures, Ph. D. Thesis, University of California, Berkeley, 1971

[9] Hooke, R., T. A. Jeeves: Direct Search Solution of Numerical and Statistical Problems, Journ. of the ACM, Vol. 8, 212, 1961

[10] Powell, M. J. D.: An Efficient Method for Finding the Minimum of a Function of Several Variables without Calculating Derivatives, Computer Journal, Vol. 7, 155, 1964

[11] Davies, D., W. H. Swann: Nonlinear Optimization Techniques, ICI Ltd., Monograph No. 5, Oliver and Boyd Ltd., Edinburgh, 1969

[12] Fletcher, R., M. J. D. Powell: A Rapidly Convergent Descent Method for Minimization, Computer Journal, Vol. 6, 163, 1963

[13] Lasdon, L. S., R. L. Fox, M. W. Ratner: An Efficient One-Dimensional Search Procedure for Barrier Functions, Mathematical Programming, 4, 1973

[14] Griffith, R. E., R. A. Stewart: Nonlinear Programming Technique for the Optimization of Continuous Systems, Management Science, Vol. 7, 379, 1961

[15] Deicke, E.: Zur Lösung nichtlinearer Strukturoptimierungsprobleme durch bereichsweise Linearisierung am Beispiel von Stabwerken, Studienarbeit am Fachgebiet Leichtbau, TH Darmstadt, 1974

[16] Beale, E. M. L.: "Numerical Methods", in: Nonlinear programming, Ed. J. Abadie, North Holland, Amsterdam, 1967

[17] Kuhn, H. W., A. W. Tucker: 'Nonlinear Programming', Proceedings of the 2nd Berkeley Symposium on Math. Statistics and Probability, pp. 481 - 492, University of California Press, Berkely, 1951

[18] Prager, W.: Necessary and Sufficient Conditions for Global Structural Optimality, Proceedings of the 2nd Symposium on Structural Optimization, AGARD Proceeding No. 123, 1973

[19] Berke, L., N. S. Khot: Use of Optimality Criteria Methods for Large Scale Systems, AGARD Lecture Series No. 70 on Structural Optimization,1974

[20] Gellatly, R. A., L. Berke: OPTIM II, A Magic Compatible Large Scale Automated Minimum Weight Design Program, Wright-Patterson Air Force Base, Ohio, 1974

[21] Baier, H.: 'Zur Optimierung elastischer Strukturen durch Lösung eines dualen Optimierungsproblems', Proceedings of the GAMM-Tagung, 1975, Göttingen, to be published in ZAMM.

[22] Künzi, H. P., H. G. Tzschach, C. A. Zehnder: Numerische Methoden der mathematischen Optimierung, Teubner, Stuttgart, 1966

[23] Wolfe, P.: Methods of Nonlinear Programming, in J. Abadie (ed.): Nonlinear Programming, North Holland Publ. Co., Amsterdam, Chap. VI, 1967

[24] Shah, B. V., R. J. Baehler, O. Kempthorne: Some Algorithms for Minimizing a Function of Several Variables, SIAM Journal, Vol. 12, 74, 1964

[25] Nagtegaal, J. C.: A New Approach to Optimal Design of Elastic Structures, Computer Methods in Applied Mechanics and Engineering, Vol. 2, 1973

On Closed Sets Having A Least Element

Peter Bod (Budapest)

This paper deals with some recent results related to the s.c. "indifferency" in mathematical programming, and can be considered as a sequel of my survey paper submitted to the V. Oberwolfach Meeting on Operations Research in 1972 [1].

We say that the set $S \subset \mathbb{R}^n$ has a so called least element iff $\exists x_0 \varepsilon S$ such that $x_0 \leq x : \forall x \varepsilon S$.

The characterization of sets having a least element finds its relevance in optimization theory. Namely, let us consider the following general mathematical programming problem:

$$(P): \quad \begin{aligned} f(x_0) &= \min f(x) \\ & x \varepsilon S \subset \mathbb{R}^n \end{aligned} ,$$

where $f(x)$ is an arbitrary isotonic function.

The function $f(x)$ defined on S is said to be isotonic if : $x^1; x^2 \varepsilon S;\ \ x^1 \leq x^2 \Rightarrow f(x^1) \leq f(x^2)$.

If the set S has a least element $x_0 \varepsilon S$, then this point is a minimum solution of (P) whatever specific isotonic function appears in the problem as objective. We say, concerning the notion introduced by G. Wintgen [2], that the problem (P) is indifferent in connection with the family of the isotonic functions.

The linear functions with non negative coefficients form an important subfamily of the family of the isotonic functions. Indifferency in connection with the family of the non negative linear functions is equivalent with the fact that the set of the feasible solutions has a least element.

G. Wintgen has shown in 1964 that the set

$$L = \{x \mid x \varepsilon \mathbb{R}^n ;\ Bx \geq b\}$$

has a least element if every row of B contains one non negative element while the other ones are non positive.

I gave necessary and sufficient conditions in 1966 [3] characterising those polyhedral sets to have a least element which are the intersections of a finite number of hyperplanes with the non-negative orthant.

In 1971 R.W. Cottle and A.V. Veinott Jr. have completely characterized the convex polyhedral sets having a least element [4].

I am investigating now sets which are the intersections of lower sets of a finite number of differentiable convex functions.

Let be given an open convex set: $S \subset \mathbb{R}^n$ and a finite number of convex functions: $g_i(x) \varepsilon C^1$ $(i=1,2,\ldots,m)$ defined on it. Without loss of generality let be $K_i = \{x | x \varepsilon S;\ g_i(x) \leq 0\}$ and

$$K = \bigcap_{i=1}^{m} K_i .$$

As lower sets of convex functions are closed and convex so do their intersections. Therefore K is closed and convex. Let us suppose

$$K \neq \emptyset .$$

The point $x_o \varepsilon K$ is said to be regular if the gradients of the active constraints in x_o are linearly independent. If we are denoting the set of the indices of the constraints by $J = \{1,2,\ldots,m\}$ let us define two disjoint subsets of this set for each point from K:

$$J_1 = \{i | i\varepsilon J\ ;\ g_i(x_o) = 0\}$$

$$J_2 = \{i | i\varepsilon J\ ;\ g_i(x_o) < 0\} .$$

In order to see what are necessary assumptions to characterize a set with a least element I shall show a lemma with a negative statement in advance.

Lemma: The regular point $x_o \varepsilon K$ cannot be a least element of the set K if

$$k = |J_1| < n .$$

Proof. Let be $J_1 = \{1,2,\ldots,k\}$. We are considering the non-linear equations

$$g_i(x_o) = 0\ ;\ i\varepsilon J_1$$

consisting of k equations in n unknowns. As x_o is a regular point, the Jacobian of the system has full row-rank in x_o. Let us say the first k columns are linearly independent. Then the conditions of the implicit function theorem are fulfilled for

$$g_i(x_o) = 0\ ;\ i\varepsilon J_1$$

and the functions are continuously differentiable. So there exist k single-valued functions

$$\varphi_1(\xi_{k+1}, \dots, \xi_n)$$

$$\varphi_2(\xi_{k+1}, \dots, \xi_n)$$

$$\dots\dots\dots\dots$$

$$\varphi_k(\xi_{k+1}, \dots, \xi_n)$$

which are continuous in a neighbourhood of

$$\hat{x}_o = \begin{bmatrix} \xi^o_{k+1} \\ \cdot \\ \cdot \\ \cdot \\ \cdot \\ \xi^o_n \end{bmatrix} \varepsilon \mathbb{R}^{n-k}$$

and have the following properties:

i. $\xi^o_i = \varphi_i(\xi^o_{k+1}, \dots, \xi^o_n)$; $i \varepsilon J_1$.

ii. for all $\hat{x}$ in the neighbourhood of $\hat{x}_o$ the vector composed as

$$\overline{x} = \begin{bmatrix} \varphi_1(\hat{x}) \\ \varphi_2(\hat{x}) \\ \cdot \\ \cdot \\ \cdot \\ \varphi_k(\hat{x}) \\ \hat{x} \end{bmatrix}$$

satisfies

$$g_i(\overline{x}) = 0 \ ; \ i \varepsilon J_1 \ .$$

Let be $\hat{x} = \hat{x}_o - \lambda\, 1_{n-k}$ where 1_{n-k} is the vector in $\mathbb{R}^{n-k}$ each element of which equals 1. Let us choose $\lambda > 0$ so small that $\hat{x}$ be in the corresponding neighbourhood where the functions φ_i ; $i \varepsilon J_1$ have the above properties. As $g_i(x_o) < 0$; $i \varepsilon J_2$ it is possible to diminish λ if necessary in order to have

$$g_i(\overline{x}) \leq 0 \ ; \ i \varepsilon J_2 \ .$$

As $g_i(\bar{x}) = 0$ for $i \varepsilon J_1$ we have

$$g_i(\bar{x}) \leq 0 \; ; \; i \varepsilon J \text{, and consequently } \bar{x} \varepsilon K .$$

But the last $n-k$ components of $\bar{x}$ are apparently less than the corresponding ones of x_o. Thus x_o cannot be the least element in K.

The following theorem gives sufficient conditions for K to have a least element.

Theorem 1: The set K has a least element if there exists $x_o \varepsilon K$ yielding equality in at least n constraints and if the Jacobian matrix formed by the binding constraints in x_o has a basis of order n with non-positive inverse.

Proof. Let be $J_1 = \{1,2,\ldots,k\}$. According to the assumptions:

$$\text{rank } [\nabla g_{J_1}(x_o)] = n .$$

Supposing the first n rows to be linearly independent:

$$B^{-1} = \begin{bmatrix} \frac{\partial g_1(x_o)}{\partial \xi_1} & \cdots & \frac{\partial g_1(x_o)}{\partial \xi_n} \\ \cdots & \cdots & \cdots \\ \frac{\partial g_n(x_o)}{\partial \xi_1} & \cdots & \frac{\partial g_n(x_o)}{\partial \xi_n} \end{bmatrix}^{-1} \leq 0 .$$

Let be $c \geq 0$ an arbitrary non-negative vector in $\mathbb{R}^n$. We define the vector

$$u_o^T = [u_1^T \; ; \; u_2^T] \; ; \; u_1^T = (-c^T)B^{-1} \; ; \; u_2^T = 0^T \varepsilon \mathbb{R}^{m-n} .$$

It is easy to see that $[x_o, u_o]$ satisfies the sufficient optimality criteria of the following non-linear programming problem:

$$(P'): \qquad f(x_o) = \min(c^T x)$$
$$x \; \varepsilon \; K .$$

This means: x_o is a minimum solution of (P') whatever $c \geq 0$ be in the objective function. Therefore x_o is the least element in K.

It is possible to weaken the conditions of Theorem 1 in connection with convexity assumptions, and one is able to prove it without using the Kuhn-Tucker conditions.

Theorem 1-bis : Let $x_o \varepsilon K' = \{x | x\varepsilon S\ ;\ g(x) \leq 0\}$ have at least n active constraints. If the functions $g_i(x)\ ;\ i\varepsilon J_1$ are locally quasi-convex and continuously differentiable (with respect to K') and if the Jacobian of the binding constraints in x_o has a basis of order n with nonpositive inverse, then x_o is a least element in K'.

Proof. Let be x an arbitrary point in K' for which $g(x) \leq 0$. Let us assume that the first $k \geq n$ constraints are active in $x_o : g_{J_1}(x) \leq 0 = g_{J_1}(x_o)$. From quasiconvexity:

$$g_{J_1}(x) \leq g_{J_1}(x_o)\ ,\quad \nabla g_{J_1}(x_o)(x-x_o) \leq 0\ ;\quad x\varepsilon K'\ .$$

If the first n rows in $\nabla g_{J_1}(x_o)$ are independent let be

$$\nabla g_{J_1}(x_o) = \begin{bmatrix} B \\ N \end{bmatrix}$$

and let us form the $n\times k$ matrix: $[B^{-1}\ ;0]$. As $B^{-1} \leq 0$,

$$[B^{-1}\ ;0] \begin{bmatrix} B \\ N \end{bmatrix} (x-x_o) = x-x_o \geq 0\ ,$$

$$x_o \leq x\ :\ \forall x\varepsilon K'\ .$$

Let me show an illustrative example:

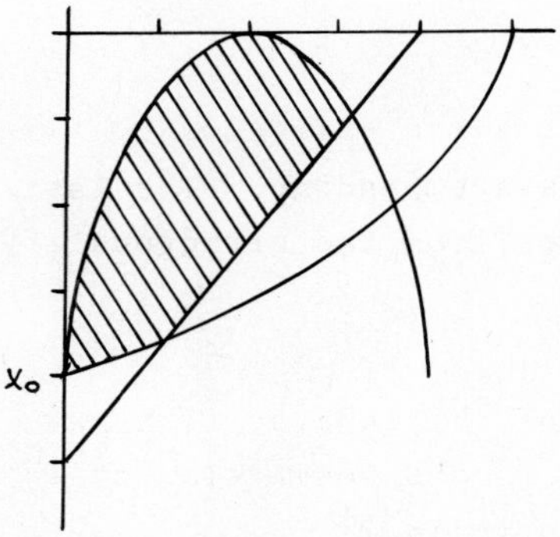

$$K = \{x | x\varepsilon R^2;\ g_i(x) \leq 0;\ i=1,2,3\}$$

$$g_1(x) = 16\xi_1^2 + 25\xi_2^2 - 400$$

$$g_2(x) = \xi_1^2 - 4\xi_1 + \xi_2 + 4$$

$$g_3(x) = 5\xi_1 - 4\xi_2 - 20$$

$$x_o = \begin{bmatrix} 0 \\ -4 \end{bmatrix}\ ;\quad J_1 = \{1,2\};\ J_2 = \{3\}$$

$$\left[\nabla g_{J_1}(x_o)\right]^{-1} = \begin{bmatrix} 0 & -200 \\ -4 & 1 \end{bmatrix}^{-1} = -\frac{1}{800}\begin{bmatrix} 1 & 200 \\ 4 & 0 \end{bmatrix}$$

As it is usual in non-linear programming, the necessary conditions do not coincide with the sufficient ones in this problem. The following example shows a set having a least element, though neither the convexity of the set nor the requirements concerning the Jacobian are satisfied.

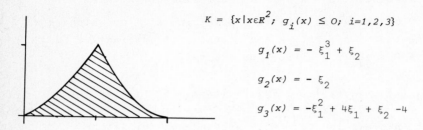

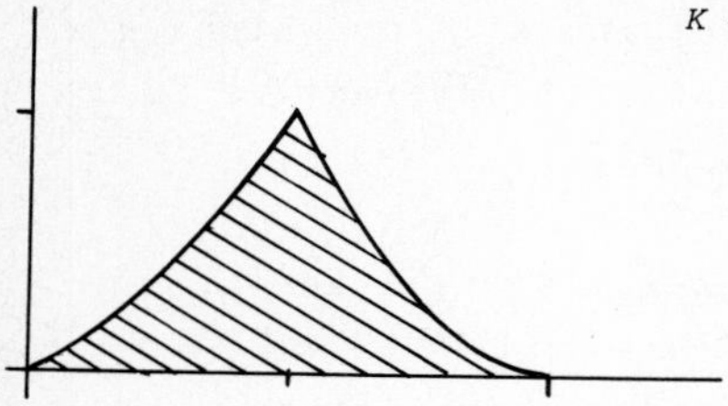

$$K = \{x \mid x \in R^2;\ g_i(x) \le 0;\ i=1,2,3\}$$

$$g_1(x) = -\xi_1^3 + \xi_2$$

$$g_2(x) = -\xi_2$$

$$g_3(x) = -\xi_1^2 + 4\xi_1 + \xi_2 - 4$$

One has to introduce some constraint qualification in order to make the conditions of Theorem 1 necessary.

Theorem 2: If x_o is a least element of the set K so that equality holds in exactly n constraints, and if K satisfies some constraint qualification, then $\nabla g_{J_1}(x_o)$ is non-singular and has a non-positive inverse.

Proof. x_o is a minimum solution of the problem (P') with arbitrary non-negative c. Since a constraint qualification holds there exists to every $c \ge 0$ such an $u_o(c)$ that $[x_o; u_o(c)]$ satisfies the Kuhn-Tucker conditions:

$$c^T + \nabla g(x_o) u_o(c) = 0^T$$

$$g(x_o) \le 0$$

$$u_o^T(c) g(x_o) = 0$$

$$u_o(c) \ge 0 .$$

Let us suppose the first n constraints are binding. The last $m-n$ components of $u_o(c)$ are therefore zeros from the orthogonality to $g(x_o)$,

$$u_o^T(c) = [u_1^T(c);\ 0^T] .$$

One has

$$c^T + \nabla g_{J_1}(x_o)\ u_1(c) = 0^T$$

$$\nabla g_{J_1}(x_o)\ u_1(c) = -c^T .$$

This equality has for every $c \geq 0$ a solution, consequently

$$\nabla g_{J_1}(x_o)$$

is non singular. As $u_1(c)$ has to be non-negative for arbitrary $c \geq 0$ we must have:

$$[\nabla g_{J_1}(x_o)]^{-1} \leq 0$$

— . —

Let us apply our theorems for the special case where K is polyhedral. Let us consider: $L = \{x \mid x \in \mathbb{R}^n ;\ b - Ax \leq 0\}$ where A is an $m \times n$ matrix and $b \in \mathbb{R}^m$.

We obtain

Corollary: x_o is a least element of L if and - degeneracy excluded - only if $x_o \in L$ and x_o is determined by a basis with non-negative inverse.

We say: $\hat{x}$ is determined by the basis A_B in $Ax \geq b$ if it is possible by a suitable interchanging of the rows to partition the problem in the form

$$A_B \hat{x} \overset{\cdot}{=} b_B$$

$$A_N \hat{x} \geq b_N$$

so that

$$\hat{x} = A_B^{-1}\ b_B \ .$$

The proof is immediate for in the linear case the corresponding Jacobian is $[-A_B]$. The Corollary expresses the main statement of the Theorem due to Cottle and Veinott.

What can be said about non-convex sets having a least element? Theorem 2 is of course valid as necessary condition even if K is not convex. Furthermore there exists Wintgen's theorem giving sufficient conditions for a general "indifferency" (see [1] or [2]) which can be specialized in the following from:

Theorem 3: The non-empty, closed set $L \subset \mathbb{R}^n$ which is bounded from below has a least element if it is closed for the vector operation "$\cap$". This means

$$x; y \in L \Rightarrow x \cap y \in L$$

where

$$z = x \cap y; \quad (z)_i = \min [(x)_i; (y)_i] \quad (i=1,2,\ldots,n) .$$

Proof. If L is compact, then the minimization problem

$$\min_{x \in L} (x)_i$$

has a finite solution on L. Let the solution be $\hat{x}_i$. As L is closed for $\cap$ we have

$$\hat{x} = \bigcap_{i=1}^{n} \hat{x}_i \in L$$

and $\hat{x}$ is the least element of L.

If L is not bounded, let be $\bar{x} \in L$. As $L \neq \emptyset$, $\bar{x}$ always exists. We consider the set

$$L' = \{x \mid x \in L; \ x \leq \bar{x}\} \subset L .$$

L' is compact, closed for $\cap$, and possesses a least element which is also the least element of L.

Unfortunately it is difficult to check in most cases whether the conditions of the above theorem hold or not. But if it happens then one can obtain good results by a very easy way.

I should like to show simple proofs for the minimality and complementarity properties of the z-functions discovered by A. Tamir [5].

Definition 1: A mapping

$$f(x) = \begin{bmatrix} f_1(x) \\ \cdot \\ \cdot \\ \cdot \\ f_n(x) \end{bmatrix} : \mathbb{R}_+^n \to \mathbb{R}^n$$

is said to be a z-function if $f(x)$ is extradiagonal antiton, this means that for $\forall x \in \mathbb{R}_+^n$ and for $i \neq j$ the functions $F_{ij}(\tau) = f_i(x+\tau e_j)$ do not increase with $\tau \geq 0$.

Definition 2: The following problem is said to be a complementarity problem: given a mapping $f : \mathbb{R}_+^n \to \mathbb{R}^n$ and a vector $b \in \mathbb{R}^n$. Find the vector x such that

i. $f(x) + b \geq 0 \; ; \; x \geq 0$

ii. $x^T[f(x) + b] = 0$.

x is a feasible solution if it satisfies i.; a feasible solution is a complementary solution if ii. holds, too.

<u>Theorem 4:</u> Let $S = \{x | f(x) + b \geq 0; \; x \geq 0\}$. If f is a z-function then S is closed for the vector operation $\cap$.

Proof. From $x, y \varepsilon S$ follows that

$$\left.\begin{array}{l} f_i(x)+\beta_i \geq 0 \\ \\ f_i(y)+\beta_i \geq 0 \end{array}\right\} \quad \forall \; i \; .$$

Let us denote $z = x \cap y$. As $x \geq 0$, $y \geq 0$ it also holds that $z \geq 0$. We shall show that

$$f_i(z) + \beta_i \geq 0 \qquad (i=1,2,\ldots,n) \; .$$

Let us suppose (without loss of generality) that $(z)_i = (x)_i$. We consider the following sequence of points:

$$z = v_0, \; v_1, \; v_2, \; \ldots \; , \; v_n = x$$

$$v_k = v_{k-1} + \tau e_k \; .$$

Let be

$$\tau_k = 0 \qquad \text{if} \quad (z)_k = (x)_k \; ,$$

$$\tau_k = (x)_k - (y)_k \qquad \text{if} \quad (z)_k = (y)_k \; .$$

One has

$$f_i(v_k) = f_i(v_{k-1} + \tau e_k) \leq f_i(v_{k-1})$$

and therefore

$$f_i(x) \leq f_i(z) \; .$$

Then

$$f_i(x)+\beta_i \geq 0 \, , \qquad f_i(z)+\beta_i \geq 0 \quad \text{for } \forall \mathbf{i},$$

$$f(z) + b \geq 0 \, , \quad z \varepsilon S \; .$$

Theorem 5: S has a least element if f is a continuous z-function and $S \neq \emptyset$.

If f is continuous then S is closed. As $x \geq 0$, S is bounded from below. Therefore $x_0 \varepsilon S$, a least element, exists from Theorem 3.

Theorem 6: If x_0 is the least element of the set S and f is a continuous z-function, then x_0 is a complementary solution.

Proof. Let us suppose that the complementary relation does not hold. Then $\exists i$ such that $\xi_i^0[f_i(x_0) + \beta_i] > 0$, and $\xi_i^0 > 0$; $f_i(x_0) + \beta_i > 0$ from the non-negativity of both expressions. Let $\delta > 0$ be small enough that

$$\bar{x} = x_0 - \delta e_i \geq 0 .$$

Let us consider the functions $F_{ji}(\tau)$ from the point $\bar{x} \varepsilon \mathbb{R}_+^n$,

$$F_{ji}(\tau) = f_j(x_0 - \delta e_i + \tau e_i) \leq f_j(x_0 - \delta e_i) .$$

Let be:

$$\tau = \delta$$

$$F_{ji}(\delta) = f_j(x_0) \leq f_j(x_0 - \delta e_i) = f_j(\bar{x}) .$$

From

$$f_j(x_0) + \beta_i \geq 0 \Rightarrow f_j(\bar{x}) + \beta_j \geq 0 \quad \text{for } \forall j \neq i .$$

As $f_i(x_0) + \beta_i > 0$ and $f_i(x)$ is continuous, we have

$$f_i(\bar{x}) + \beta_i = f_i(x_0 - \delta e_i) + \beta_i \geq 0$$

if δ is small enough. This means however that $\bar{x} \varepsilon S$. But $(x)_i < (x_0)_i$ which contradicts the assumption that x_0 is the least element in S.

– . –

In order to see a practical application of the notions and theorems from above let us consider a non-linear generalization of the open, static, input-output model due to W. Leontieff.

An economy is given consisting of n homogeneous production sectors each operating with a single technology. Every sector is using products of the others in realizing its own activity. The demand of the sector j for the product of the sector i is given by the following function:

$$f_{ij}(\xi_j)$$

where ξ_j is the level of the gross activity performed in the sector j. We make the following assumptions on the input functions:

i. $f_{ij}(\xi_j)$ $(i=1,2,\ldots,n;\ j=1,2,\ldots,n)$

are continuously differentiable.

ii. $f_{ij}(0) = 0$.

iii. $0 \leq \xi_j^1 \leq \xi_j^2 \Rightarrow f_{ij}(\xi_j^1) \leq f_{ij}(\xi_j^2)$.

The balances between total activities and final demands are given by

$$\xi_i = \sum_{j=1}^{n} f_{ij}(\xi_j) + \eta_i \qquad (i=1,2,\ldots,n)$$

where η_i denotes final demand.

Let us denote:

$$f_j(\xi_j) = \begin{bmatrix} f_{1j}(\xi_j) \\ f_{2j}(\xi_j) \\ \cdot \\ \cdot \\ f_{nj}(\xi_j) \end{bmatrix} \qquad (j=1,2,\ldots,n) ,$$

then

$$x = \sum_{j=1}^{n} f_j(\xi_j) + y .$$

Let be

$$F(x) = x - \sum_{j=1}^{n} f_j(\xi_j) .$$

It is easy to show that $F(x)$ is a z-function, and the following statements hold:

1. To each attainable final demand $y_o > 0$ there belongs a gross production such that $F(x_o) = y_o$ and x_o is realizing y_o with minimum social cost.
2. This vector x_o is the least element of the set
$$L_{y_o} = \{x \mid F(x) \geq y_o\} .$$

3. If some constraint qualification holds then the Jacobian of $F(x)$ is in x_o non-singular and has a non-negative inverse.

4. If non-decreasing return to scale holds in general, then the solution of the equation

$$F(x) = y_o$$

is unique.

References

[1] Bod P.: Über "indifferente" Optimierungsaufgaben. Methods of Operations Research XVI, Teil 1, 40-50. Verlag Anton Hain. V. Oberwolfach-Tagung über Operations Research.

[2] Wintgen G.: Indifferente Optimierungsprobleme. KMÖ. Tagung-Konferenzprotokoll, Teil II, 3-6. Akademie Verlag, Berlin. 1964.

[3] Bod P.: A remark to a theorem due to G. Wintgen (written in Hungarian) MTA III. Osztályának Közleményei. 16 (1966) 275-279.

[4] Cottle R.W., and Veinott A.F.Jr.: Polyhedral sets having a least element. Mathematical Programming. Vol. 3. No. 2. 238-249.

[5] Tamir A.: Minimality and complementary properties associated with z-function and m-function. Mathematical Programming. Vol. 7. No.1. 17-31.

Mathematical Institute of the
Hungarian Academy of Sciences
Budapest V
Reáltanoda u. 13-15

On the Convergence of the Variable Metric Method with Numerical Derivatives and the Effect of Noise in the Function Evaluation

L.C.W. Dixon

Numerical Optimisation Centre

The Hatfield Polytechnic

Abstract

The effect of numerical estimates of the gradient on Wolfe's convergency proof for descent algorithms is considered. It is shown that there is a value ε_o such that if the termination criteria is set greater than this value then convergence is assured, but that below this value the behaviour is uncertain. This theoretical result is in agreement with previously published experimental results. A modified scheme for the search along a line is described which both enables a better estimate of the initial slope to be made, and then implies that convergence to a more accurate point can be achieved. It is known that variable metric methods are very unreliable when the function evaluation is subject to noise. It is shown that this is due to two reasons. The numerical estimates of the gradient become very unreliable and the standard line search strategy fails. The new linear search overcomes the second of these difficulties.

Introduction

The variable metric algorithm is widely recognised as one of the most efficient ways of solving the following problem:-

$$\text{Locate } x^* \text{ a local minimum point of } f(x) \quad \underline{x} \in R^n . \tag{1}$$

Considerable attention has been given to the study of the convergence properties of this algorithm especially for the case where analytic expressions are available for the derivatives

$$g_i = \partial f / \partial x_i \quad i = 1 \dots n . \tag{2}$$

In particular we shall mention the results of Wolfe (1969) and Powell (1972), (1975). Wolfe established general conditions under which a descent algorithm will converge to a stationary point and Powell showed that two particular very efficient algorithms that cannot be shown to satisfy Wolfe's conditions do in fact converge to the minimum of convex functions under certain conditions. These results will be stated more completely in Section 2.

In most practical problems analytic expressions for the gradient vector g (Equ. 2) are not available and numerical derivatives are subject to truncation error. In Section 3 we shall consider the effects of these errors on Wolfe's convergent properties and will discuss possible modifications of the algorithms to make them reliable in these circumstances. The effects of rounding error are considered in Section 4, whilst in Section 5 these thoughts are extended to include the case of on-line function minimisation where each function evaluation is subject to random noise. Variable metric algorithms are known to be very unreliable in these conditions, Crombie (1972), and it is shown that this is predictable for existing implementations and the, fairly drastic, modifications needed to produce a reliable algorithm in these circumstances are discussed.

2. Convergence with accurate analytic derivatives

In this section we will first repeat Wolfe's Theorem for the convergence of descent algorithms, when accurate analytic derivatives are available. In fairness it should be stated that the conditions stated are only one of many possible sets of conditions stated by Wolfe and the proof of convergence for this particular combination is repeated as it will form the basis of the discussion of this paper.

The theorem applies to the behaviour of descent algorithms in a wellbehaved function, Dixon (1973).

Definition: A well behaved function

A function is said to be well behaved for minimisation if within a certain region of interest J

(i) a lower bound exists, i.e. $f(x) \geq L$ all $x \in J$ (2.1)

(ii) the Hessian matrix exists at all points $x \in J$ and is uniformly bounded

$$\text{i.e. for all } \|z\| = 1, \quad -M \leq z^T G(x) z \leq M \tag{2.2}$$

where $G(x)$ is the Hessian matrix at x

$$\text{i.e. } G_{ij} = \partial^2 f / \partial x_i \partial x_j$$

(iii) the third derivatives of $f(x)$ exists at all points $x \in J$

(iv) at least one minimum point x^* exists in the interior of J and the Hessian at all such minima x_1 are well conditioned

$$\text{i.e. for all } \|z\| = 1, \quad 0 < m \leq z^T G(x_1) z \leq M \tag{2.3}$$

Definition: A descent method

A descent algorithm is defined as one that proceeds iteratively and in which each iteration consists of two stages:

In the first a direction $p^{(k)}$ is selected and in the second a step size in that direction is determined. The iteration is therefore

$$x^{(k+1)} = x^{(k)} + \alpha p^{(k)} \tag{2.4}$$

and we can introduce d for the step taken during the iteration and then

$$d = x^{(k+1)} - x^{(k)} = \alpha p^{(k)} . \tag{2.5}$$

It is normal to assume that the region of interest J is bounded by the contour

$$f(x) = K f(x^{(o)}) \qquad K \geq 1$$

and that the set $\{x : f(x) \leq K f(x^{(o)})\}$ has a finite maximum diameter. It is also normal to assume that the iteration terminates if an $x^{(k)}$ is found for which

$$\|g(x^{(k)})\| < \varepsilon_o \tag{2.6}$$

for some preselected $\varepsilon_o > 0$.

Theorem 1 (Wolfe (1969, 1971))

If in the application of a descent method to a well behaved function, the step taken at a regular subsequence of iterations is chosen to satisfy conditions I, II, III below, and if on the other iterations the function value does not increase then the iteration will terminate with $\| g(x) \| < \varepsilon_o$.

Condition I

$$- p^{(k)^T} g^{(k)} \geq \varepsilon_1 \| p^{(k)} \| \| g^{(k)} \| \qquad \varepsilon_1 > 0 \tag{2.7}$$

This implies that the direction of search has a significant component in the negative gradient direction.

Condition II

$$\text{Either} \qquad | g^{(k+1)^T} p^{(k)} | \leq \varepsilon_2 | g^{(k)^T} p^{(k)} | \ , \quad 0 < \varepsilon_2 < 1 \tag{2.8}$$

$$\text{or} \qquad | f(x^{(k)}) + g^T d - f(x^{(k+1)}) | > \varepsilon_3 | g^T d |, \quad 0 < \varepsilon_3 < 1 \ . \tag{2.9}$$

Either of these implies that the step taken in this direction is bounded away from zero.

Condition III

$$f(x^{(k)}) - f(x^{(k+1)}) \geq - \varepsilon_4 \, g^T d \tag{2.10}$$

This condition implies that the reduction in function value is bounded away from zero.

Proof of Wolfe's Theorem

Let us commence with condition II (Equ. 2.9)

If we expand $f(x^{(k+1)})$ about $x^{(k)}$ by Rolle's theorem we obtain

$$f(x^{(k+1)}) = f(x^{(k)}) + g^T d + \frac{1}{2} d^T G(\xi) d$$

$$\text{where} \qquad \xi = x^{(k)} + \theta (x^{(k+1)} - x^{(k)}), \quad 0 \leq \theta \leq 1 \ .$$

Now apply the bound (2.2) to obtain

$$f(x^{(k)}) + g^T d - \tfrac{1}{2} M \| d \|^2 \leq f(x^{(k+1)}) \leq f(x^{(k)}) + g^T d + \tfrac{1}{2} M \| d \|^2 \tag{2.11}$$

$$\text{and} \qquad | f(x^{(k)}) + g^T d - f(x^{(k+1)}) | < \tfrac{1}{2} M \| d \|^2$$

and so condition (2.9) implies

$$\| d \|^2 > 2 \, \varepsilon_3 | g^T d | / M \tag{2.12}$$

$$\alpha > 2 \, \varepsilon_3 | g^T p | / M \| p \|^2 \ .$$

If this is substituted into (2.10) we obtain

$$f(x^{(k)}) - f(x^{(k+1)}) \geq 2\,\varepsilon_3\,\varepsilon_4\,|g^T p|^2 / M\,\|p\|^2 \tag{2.13}$$

and from (2.7)

$$f(x^{(k)}) - f(x^{(k+1)}) \geq 2\,\varepsilon_3\,\varepsilon_4\,\varepsilon_1^2\,\|g\|^2 / M\,. \tag{2.14}$$

Now if the sequence were to continue indefinitely with $\|g\| > \varepsilon_o$, this would contradict the existence of the lower bound (2.1). Hence the algorithm must terminate with $\|g\| < \varepsilon_o$.

It is generally accepted that the variable metric algorithms form the most efficient class of deterministic unconstrained optimisation algorithms.

In these algorithms

$$p^{(k)} = -H^{(k)}\,g^{(k)} \tag{2.15}$$

where $H^{(k)}$ is updated by one of a number of formulae, of which that suggested independently by Broyden/Fletcher/Shanno and Goldfarb in 1970

$$\text{i.e.}\quad H^{(k+1)} = H^{(k)} - (H^{(k)}y\,y^T H^{(k)})\,/(y^T H^{(k)}y) + \rho(dd^T)/(d^T y) + y^T H y v^{(k)} v^{(k)T}\,,$$

$$\text{where}\quad v^{(k)} = d/(d^T y) - H^{(k)}y/(y^T H^{(k)}y)\,, \tag{2.16}$$

$$y = g^{(k+1)} - g^{(k)}\,,$$

$$\text{and}\quad \rho = 1$$

is normally one of the most efficient.

<u>Theorem 2 : Powell (1975)</u>

If a descent algorithm is constructed by choosing the direction by (2.15), updating the matrix by (2.16) and then determining the step size to satisfy (2.8) and (2.10) with

$$0 < \varepsilon_4 < \varepsilon_2 < 1\,,\quad \varepsilon_4 < \tfrac{1}{2}$$

then on a convex function the iteration will converge (2.6) in the neighbourhood of the minimum.

This result strengthens his earlier result Powell (1971), (1972) which applied to perfect line searches (i.e. (2.8) with $\varepsilon_2 = 0$).

It will be noted that in both of Powell's proofs condition I is neither proven nor necessary.

As the above result does not apply to nonconvex functions, in our implementations at the N.O.C., we test Condition I at each iteration and when it is not satisfied modify the matrix $H^{(k)}$ to ensure that it is.

Theorem 3 (Goldfarb (1969), Murtagh & Sargent (1969))

If $p^{(k)} = - H^{(k)} g^{(k)}$

and $H^{(k)}$ is a positive definite matrix such that

$\frac{\lambda \min}{\lambda \max} > \varepsilon_1$ where λ are the eigenvalues of $H^{(k)}$

then condition I is automatically satisfied.

Therefore as $H^{(k)}$ is a positive definite sequence and trace $(H^{(k)}) > \lambda_{max}$ then

$$H^{(k)'} = H^{(k)} + \varepsilon_1 \text{ trace } (H^{(k)}) I \qquad (2.17)$$

satisfies the condition of theorem 3.

Theorem 4

Given an algorithm in which (i) the direction $p^{(k)}$ is chosen initially by (2.15) but whenever condition I of Theorem 1 does not hold, it is recalculated with $H^{(k)}$ replaced by $H^{(k)'}$ (2.17)

(ii) the step size is calculated to satisfy (2.9) and (2.10) for instance by the method described in Biggs (1975)

and (iii) $H^{(k)}$ is updated by (2.16) either with $\rho = 1$ or with ρ chosen by the method described in Biggs (1973) then the sequence of points $x^{(k)}$ will, by Wolfe's theorem, converge to a point with $\| g(x) \| < \varepsilon_0$.

Comment: Given the slope $f' = -p^T g$ we require a point satisfying (2.9) and (2.10). Assume $\varepsilon_3 = \varepsilon_4 = .1$ then any point such that $.1 \alpha f' < \Delta f < .9 \alpha f'$ will suffice. If $\Delta f < .1 \alpha f'$ then we need to decrease α, whilst if $\Delta f > .9 \alpha f'$ then we need to increase α. One method based on interpolation to predict α such that $\Delta f = .5 \alpha f'$ is described in Biggs (1975). The principle behind this proceedure is illustrated in Fig. 1.

3. The effect of Numerical Derivative Estimates

In many practical problems the formulae for g are not available analytically and it is then normal to approximate g by some difference formula such as

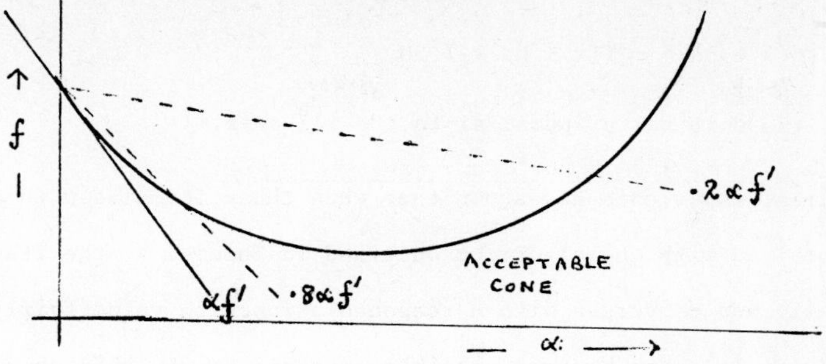

FIG. 1

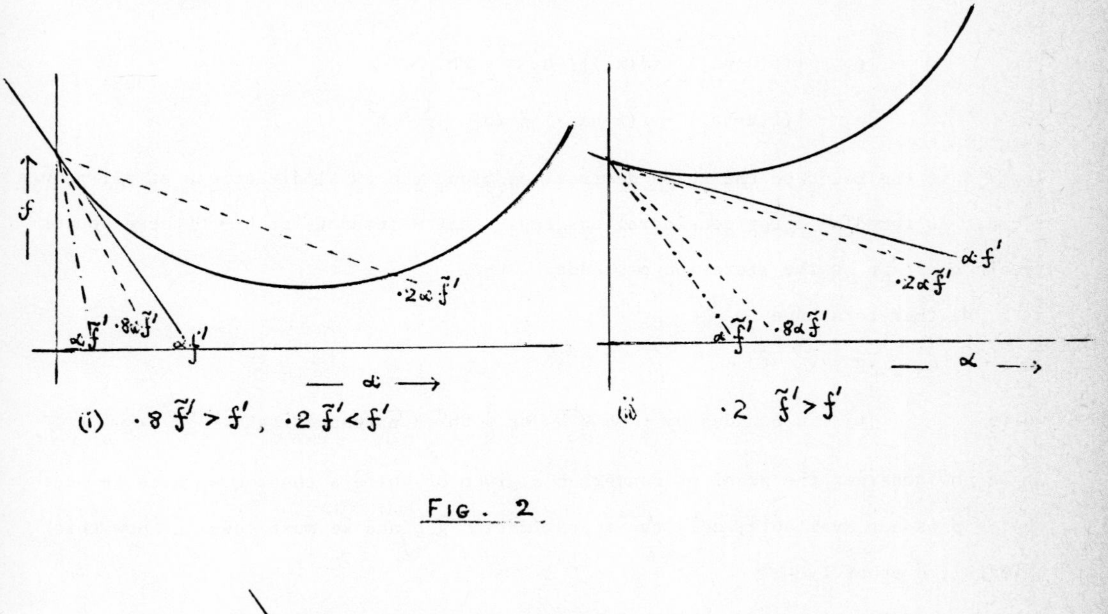

(i) $\cdot 8\,\tilde{f}' > f'$, $\cdot 2\,\tilde{f}' < f'$ (ii) $\cdot 2\,\tilde{f}' > f'$

FIG. 2

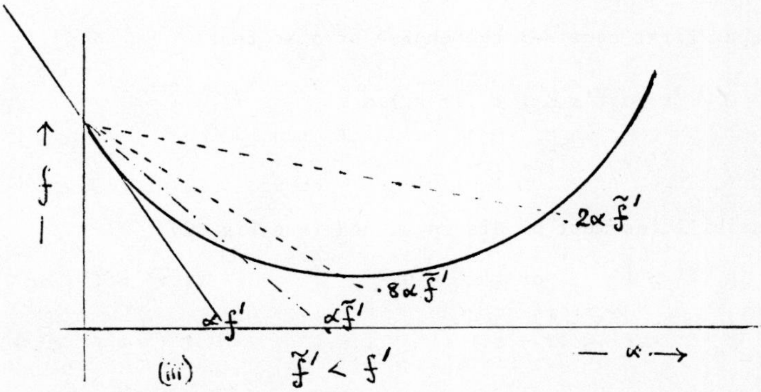

(iii) $\tilde{f}' < f'$

$$\tilde{g}_i = \{f(x + h\hat{a}_i) - f(x)\}/\ h \qquad (3.1)$$

or $$\tilde{g}_i = \{f(x + h\hat{a}_i) - f(x - h\hat{a}_i)\}/\ 2h \qquad (3.2)$$

where $\hat{a}_i$ is the unit vector parallel to the ith axis.

Practical experience has shown that when these formulae with appropriate values of h are combined with the algorithm outlined in Theorem 4, the iteration usually proceeds normally and converges with a reasonable function value fairly efficiently (see Dixon (1973)). It seems however sensible to consider why this should be so and what adverse affects we can expect.

It is fairly easy to show that the truncation errors in (3.1) and (3.2) imply that

$$|g_i - \{(f(x+ha_i) - f(x))\}/\ h| < \frac{1}{2}\,Mh$$

and $$|g_i - \{(f(x+ha_i) - f(x-ha_i))\}/\ 2h| < \frac{1}{6}\,S\,h^2 \qquad (3.3)$$

where S is the bound on the third derivatives along the axial directions at all points $x \in J$. As rounding error considerations imply that we cannot let $h \to 0$ these bounds remain constant as the iteration proceeds.

We therefore have

$$g - \tilde{g} = t \qquad (3.4)$$

where $\|t\|$ is bounded by $\frac{1}{2}\,Mh\,\sqrt{n}$ or $\frac{1}{6}\,Sh^2\,\sqrt{n}$ respectively.

If we now consider the proof of convergence given of Wolfe's theorem we note at once that g^Tp is not available, only the approximation $\tilde{g}^Tp$ and we must consider how this affects the proof.

Let us first consider the choise of α so that

$$\varepsilon_4\,\alpha\,\tilde{f}' < \Delta\,f < (1-\,\varepsilon_3)\,\alpha\,\tilde{f}'$$

where $$\tilde{f}' = -\,p^T\tilde{g}\ . \qquad (3.5)$$

Three possibilities must be distinguished (see Fig. 2)

a) $\varepsilon_4\,\tilde{f}' > f'$ or $\tilde{f}'f' < 0$ $\qquad$ $.1\,\tilde{f}' > f'$ or $\tilde{f}'\,f' < 0$

b) $\varepsilon_4\,\tilde{f}' < f' < (1 - \varepsilon_3)\,\tilde{f}'$ i.e. $\qquad$ $.1\,\tilde{f}' < f' < .9\,\tilde{f}'$ $\qquad$ (3.6)

c) $(1-\,\varepsilon_3)\,\tilde{f}' < f'$ $\qquad$ $.9\,\tilde{f}' < f'$

In case a) no value of α can satisfy (3.5) for a convex function and the line search routine will steadily reduce α to zero.

In case b) values of α satisfying (3.5) can be located but these no longer imply a bound on $\|d\|$ of the type given at (2.12) as (3.5) will now be satisfied as $\alpha \to 0$.

In case c) (3.5) can be satisfied and a bound equivalent to (2.12) is valid, so at least there is no difficulty over the choice of α, but the difficulty now lies in proving that the decrease in f is significant. In contrast to the above pessimistic statements we may state the following theorem which refers to a subset of case (c).

<u>Theorem 5</u>

If in the application of a descent method with approximate derivatives to a well behaved function, the step taken at a regular subsequence of iterations is chosen to satisfy conditions I, II, III below and if on the other iterations the function value does not increase then the iteration will terminate with $\|\tilde{g}(x)\| < \varepsilon_o$ providing ε_o is set sufficiently large to ensure that

(i) $$f' - (1 - \varepsilon_3)\,\tilde{f}' > \sigma \tilde{f}' \quad \text{for some} \quad 0 < \sigma < 1 \,. \tag{3.7}$$

and (ii) $$f'\,\tilde{f}' > 0 \,.$$

<u>Condition I</u>

$$\tilde{f}' \geq \varepsilon_1 \,\|p^{(k)}\|\,\|\tilde{g}^{(k)}\| \tag{3.8}$$

<u>Condition II</u>

$$f(x^{(k+1)}) - (f(x^{(k)}) - \alpha \tilde{f}') \geq \varepsilon_3\,\alpha\,\tilde{f}' \tag{3.9}$$

and

<u>Condition III</u>

$$f(x^{(k)}) - f(x^{(k+1)}) \geq \varepsilon_4\,\alpha\,\tilde{f}' \tag{3.10}$$

<u>Proof</u>

Proceeding as in the proof of theorem 1

$$f\,(x^{(k+1)}) - (f(x^{(k)}) - \alpha\, f') \leq \tfrac{1}{2}M\,\|d\|^2$$

therefore

$$\tfrac{1}{2}M\|d\|^2 + \alpha(\tilde{f}' - f') \geq \varepsilon_3\,\alpha\,\tilde{f}'$$

$$\tfrac{1}{2}M\|d\|^2 \geq \alpha\, f' - (1-\varepsilon_3)\alpha\,\tilde{f}'$$

$$\geq \alpha\,\sigma\,\tilde{f}'$$

$$\alpha \geq 2\,\sigma\,\tilde{f}' \,/\, M\,\|p\|^2 \tag{3.11}$$

then from condition III $\Delta f \geq 2\sigma \varepsilon_4 (\tilde{f}')^2 / M \|p\|^2$ (3.12)

and from condition I

$$\Delta f \geq 2\sigma \varepsilon_4 \varepsilon_1^2 \|\tilde{g}^{(k)}\|^2 / M . \quad (3.13)$$

Condition (3.7) can be rearranged to give

$$(1 - \varepsilon_3) \tilde{f}' < f' - \sigma \tilde{f}'$$

which implies

$$(1 - \varepsilon_3) \tilde{f}' < f'$$

and hence possibility (c) of (3.6). Thence values of α can be found satisfying conditions II and III.

We can estimate a bound on ε_0 above which both this condition and $\tilde{f}' f' > 0$ must be true.

Theorem 6

If $\tilde{f}'$ is calculated by $\tilde{f}' = -\tilde{g}^T p$ and $\tilde{g}$ is a numerical estimate of the derivative and t is defined by (3.4) then condition (3.7) is satisfied at all iterations provided

$$\varepsilon_0 > \|t\| / (\varepsilon_3 - \sigma)\varepsilon_1 . \quad (3.14)$$

is satisfied.

Proof

a) Let us consider $f' \tilde{f}' > 0$ first.

Now $g^T p - \tilde{g}^T p = t^T p$

and $f' = -g^T p$ and $\tilde{f}' = -\tilde{g}^T p$

so $f' = \tilde{f}' - t^T p$

and hence $f' \tilde{f}'$ is positive unless $|t^T p| > \tilde{f}' = |\tilde{g}^T p|$

which is definitely true if

$$\varepsilon_0 > \|t\|$$

the error in the estimation of $\tilde{g}$.

b) Now consider $(1 + \sigma - \varepsilon_3) \tilde{f}' < f'$.

This implies

$$(1 + \sigma - \varepsilon_3) \tilde{f}' < \tilde{f}' - t^T p$$

which is true providing

$$t^T p < (\varepsilon_3 - \sigma) \tilde{f}' = (\varepsilon_3 - \sigma) \tilde{g}^T p$$

which is definitely satisfied if

$$\varepsilon_o > \|t\| / \varepsilon_1(\varepsilon_3 - \sigma) > \|t\| .$$

Comment 1: The N.O.C. Optima package algorithm OPVMX Biggs (1975) satisfies these conditions and is therefore convergent to a point where

$$\| \tilde{g}^{(k)} \| < \varepsilon_o$$

providing ε_o satisfies (3.14).

Comment 2: When numerical differences are in use it is advisable to increase the value of ε_3 and ε_1 to say 0.25 as this decreases the bound (3.14).

Comment 3: If we wish to proceed to a more accurate solution then we must incorporate safeguards in the program as condition (3.7) will not be satisfied, and all three cases (a), (b) and (c) may occur. In case (a) the line search routine Biggs (1975) will reduce α steadily to zero, whilst in case (b) an unacceptably small value of α may occur. Both may be recognised by a test based on (3.11),

$$\text{i.e. if} \quad \alpha\|p\| < \varepsilon_5 \tilde{f}' / \|p\| \qquad \varepsilon_5 < 2\sigma / M \tag{3.15}$$

then a more accurate estimate of f' can be made. The central difference estimate

$$\tilde{\tilde{f}}' / \|p\| = f(x + \frac{hp}{\|p\|}) - f(x - \frac{hp}{\|p\|}) / 2h$$

is itself likely to be more accurate than the slope calculated from (3.1) and (3.2) and can be improved either by using the set of values α_i, $f(\alpha_i)$ accumulated during the interpolation and fitting a model through this data, or by using a higher order central difference scheme to estimate f' more accurately.

Now let us consider two alternatives, first when $|f'| > \varepsilon_6 \|p\|$ secondly when $|f'| < \varepsilon_6 \|p\|$. In the first case as $\tilde{\tilde{f}}'$ is improved the estimate will eventually be such that (3.7) is satisfied and hence the test based on (3.15) will be satisfactory, then (3.11) and (3.12) follow with $\tilde{\tilde{f}}'$ replacing $\tilde{f}'$. We note that (3.12) is then sufficient to give a lower bound on Δf and that testing whether Condition I is still satisfied (hence (3.12)) is unnecessary. In contrast in the second case we have now shown that the information leading to the direction $p^{(k)}$ was based on an inaccurate $\tilde{g}^{(k)}$ and we have information that could be used to improve the estimate to

$$\tilde{\tilde{g}}^{(k)} = (I - p^{(k)}p^{(k)^T} / (p^{(k)^T}p^{(k)}))\, \tilde{g}^{(k)} + \tilde{\tilde{f}}' p^{(k)} / (p^{(k)^T}p^{(k)}) \tag{3.16}$$

The new value $\| \approx{g}^{(k)} \|$ could then be tested for convergence, and if this convergence test is not satisfied, then $\approx{g}$ can be substituted into (2.15) to obtain an improved direction $\tilde{p}^{(k)}$. The improved gradient $\approx{g}^{(k)}$ should obviously also be used in calculating y in the updating formula.

We note that the sequence of matrices $H^{(k+1)}$ only remains positive definite if

$$d^T y > 0$$

and whilst this is implied by condition (2.8) and (2.10), it is not even automatically true of (2.9) and (2.10) let alone (3.9), (3.10).

In these circumstances Biggs (1975), to ensure the series $H^{(k)}$ remains positive definite, chooses not to update $H^{(k)}$. As an alternative to this procedure we might choose to estimate

$$\approx{f}^{+\prime} = g^{(k+1)^T} p^{(k)}$$

in the same way as $\tilde{f}'$ using interpolation model along the line $p^{(k)}$, and to adopt a similar strategy to (3.16) to correct $\tilde{g}^{(k+1)}$

$$\approx{g}^{(k+1)} = (I - p^{(k)} p^{(k)^T} / p^{(k)^T} p^{(k)})\, \tilde{g}^{(k+1)} + \approx{f}^{+\prime} p^{(k)} / p^{(k)^T} p^{(k)} \tag{3.17}$$

This would imply that the value obtained in calculating $d^T y$ is compatible with the information obtained during the linear search and is hence presumably positive. The updating formula for M the approximation to the Hessian which corresponds to (2.16) is

$$M^{(k+1)} = M^{(k)} - M^{(k)} d\, d^T M^{(k)} / d^T M^{(k)} d + y\, y^T / \rho y^T d \tag{3.18}$$

so along the direction d we have the information that

$$M^{(k+1)} d = y/\rho \tag{3.19}$$

and

$$d^T M^{(k+1)} d = d^T y/\rho = (\approx{f}^{+\prime} - \approx{f}')\, \alpha/\rho \tag{3.20}$$

The scalar ρ is either set to unity or calculated according to the philosophy described by Biggs (1973).

Whilst the information contained in (3.20) is justified by the linear search the accuracy of the information in the orthogonal directions implied by (3.19) is highly doubtful. The justification for continuing to use (2.16) i.e. (3.18) when the errors in calculating y are significant is very doubtful but no more reliable procedure has yet been suggested, except, briefly, in Powell (1974).

We may also note that if we increase ε_1 and modify H by (2.17) whenever necessary then $f' < \varepsilon_6 \|p\|$ is less likely to occur and this should not slow down the rate of convergence on well conditioned problems.

4. The effect of Rounding Error in the Function Evaluations

In many practical problems the assumption that f(x) can be calculated reasonably accurately at any point x is not completely justified. As the computer holds a finite length approximation of any number, most arithmetical operations involve round off errors. In major calculations such as that of the function described by Brown (1975), we may expect that these round off errors may accumulate and become significant compared to the word length of the computer. This may even introduce local minima in the calculated function $F_R(x) = f(x) + r(x)$, where r(x) is the round off error.

If an analytic derivative method were started within the basin of one of these local minima, (Dixon, Gomulka & Hersom (1975)) then the analytic derivatives would frequently be incompatible with function value information and the algorithm would terminate with an error message. Experience indicates that numerical derivative formula basing their estimates of the derivatives on a diameter of 2h tend to smooth out basins less than h in diameter. No theoretical basis for this was available before Archetti & Betro (1975) introduced their convolution integral, showed that numerical differences could be considered an approximation to it, and showed that it smoothed out minor discontinuities.

In Fig. 3 a one dimensional example is shown, the concept of the basin of x_1 is illustrated and the estimate of the slope at x_2. If we now consider the line search started at x_2 (Fig. 4) we see that it is easy to escape from such small ripples. With the standard linear search, provided the first step is further from x_2 than α_3, then the line search would eventually find an acceptable point in region A. However, if the first step were less than α_3 then the standard search would be trapped in the basin. If however we follow section 3 then a more accurate estimate of the slope at x_2 would eventually be required to be calculated as $\alpha \to 0$. Figure 4 further emphasises that in the region where roundoff error can cause significant effects a least squares interpolation model fitted over a large range of α is more

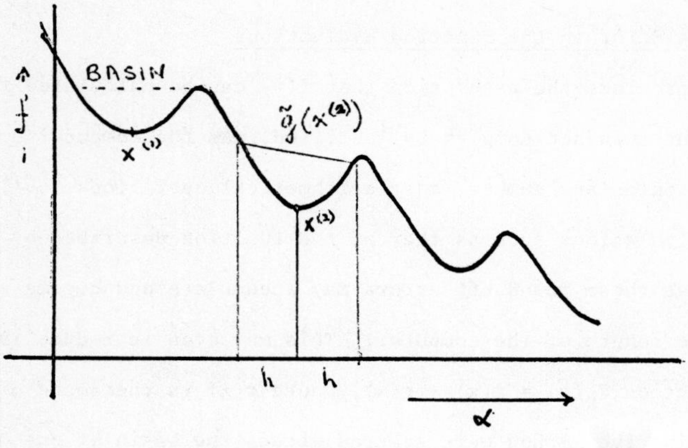

FIG. 3

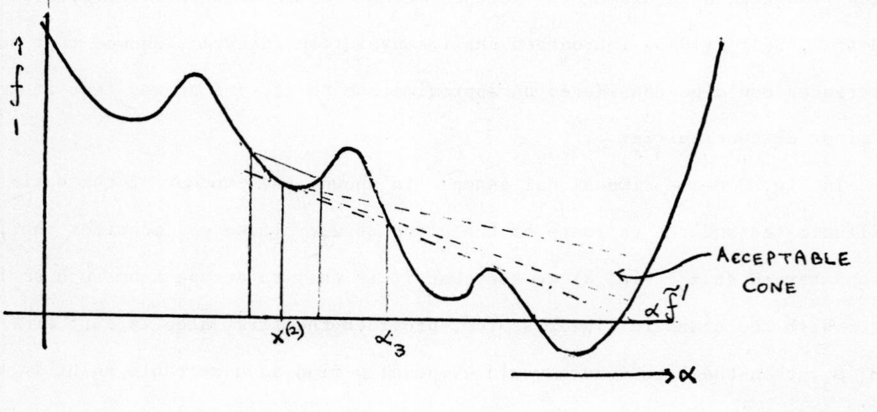

FIG 4

likely to represent the true function than a deterministic polynomial fitted to the minimal amount of data. This therefore influences the choice between an interpolation model or high order central differences formula which was left open in the previous section.

The figure should also serve as a reminder that the interpolation model may be very different dependent on the range of α used. In particular the estimate of the slope based on a model in the range $x_2 \pm \alpha_3$ would be very different and less helpful than those based on $x_2 \pm \alpha_5$. The result of using such an interpolation polynomial is discussed in detail in the next section after the effects of noise have been discussed.

Rounding errors in the function evaluation do also influence the accuracy of g and f', but essentially such errors only increase the size of $\|t\|$ and do not otherwise effect the analysis of the previous section.

5. The effect of noisy function evaluation

When optimisation is attempted on line i.e. real time control of an industrial plant, then the function evaluation is subject to noise and the standard variable metric algorithms are very unreliable, Crombie (1972). Typically he found that they rapidly reduced the function initially but then floundered about hopelessly. The effect of noise is similar to that of round off error. Typically if we calculate the function subject to round off then we obtain

$$F_R(x) = f(x) + r(x) \tag{5.1}$$

where $F_R(x)$ is the value obtained

$f(x)$ the true function value

and $r(x)$ the round off error which will be a given function of x. Similarly if the function evaluation is subject to noise then we obtain

$$F_N(x) = f(x) + w \tag{5.2}$$

where $F_N(x)$ is the value obtained

$f(x)$ the true value

and w a random variable probably drawn from a normal distribution $N(0,\sigma)$.

This implies that at a given value of x, $F_R(x)$ is always repeatable but that $F_N(x)$ is not. In the initial stages of the optimisation when the gradient magnitude is high we may expect the effect of the noise to be negligible. Indeed if we define t by (3.4) then its bound is increased due to the noise but we can still obtain an initial estimate f', and from theorem 6 we could not expect continued consistent progress below the value of $\|\tilde{g}(x)\|$ indicated by (3.14). However, the situation in the line search might prevent progress still earlier as the random nature of the variations w could completely defeat the logic of any completely deterministic linear search.

5.1 A least squares linear search

The solution to the linear search problem appears to be to introduce a small order model, which might be a cubic

$$\phi(z,\alpha_i) = z_1 + z_2\alpha + z_3\alpha^2 + z_4\alpha^3 \tag{5.3}$$

evaluate $F_N(x^{(k)} + \alpha_i p^{(k)})$ for a number of values, m, of α_i , $F_N(\alpha_i)$ and perform a least squares fit to minimise

$$E(z) = \sum_i^m (\phi(z,\alpha_i) - F_N(\alpha_i))^2 \tag{5.4}$$

with respect to y. If we let the minimum be z*, then the function $\phi(z^*,\alpha)$ would be treated as representing $f(x^{(k)} + \alpha p^{(k)})$ accurately in the range of α covered by the least squares fit. The slope along the line at $x^{(k)}$ would be represented by $\phi'(z^*,0)$ and a value of α determined, α^* say such that

$$\varepsilon_4\alpha^*\phi'(z^*,0) < \Delta\phi < (1-\varepsilon_3)\alpha^*\phi'(z^*,0) \tag{5.5}$$

Note 5.1 One difficulty in the initial conception of this search lay in determining what to do at $x^{(k)}$, $\alpha = 0$. There seems no justification in setting $z_1 = 0$ i.e. insisting

$$\phi\ (z^*,0) = F_N(x^{(k)}) \tag{5.6}$$

and indeed there seems little justification in doing this even if we are concerned with rounding error rather than noise. If we are assuming that f(x) is calculated correctly as in section 3 then (5.6) is a natural constraint. It is more logical to replace $F_N(x^{(k)} + \alpha^* p^{(k)})$ by $\Phi = \phi\ (z^*,\alpha^*)$ and to include a term $(\phi(z,0) - \Phi)^2$ in

E(z) at (5.4) suitably weighted to take into account the additional confidence in that value

$$\text{i.e.}\quad E(z) = m(\phi(z,0) - \Phi)^2 + \sum_i^m (\phi(z,\alpha_i) - F_N(\alpha_i))^2 \qquad (5.7)$$

But although (5.7) seems theoretically more justified it became apparent that we can state more about the ultimate convergence of the iteration if we impose

$$\phi(z^*,0) = \Phi \qquad (5.8)$$

the value determined at the end of the previous search, as a constraint.

Note 5.2 The cubic model is very simple and quite practical, however, if we wish to be able to ensure convergence of ϕ' to f' as m, the number of observations, increases then we might need a much more sophisticated model. (Mikhal-skii (1974).)

Note 5.3 Assuming that a uniformly convergent model is being used then we can ensure that (3.7) is satisfied with ϕ' replacing f', and hence if

$$\phi'(z^*,0) \geq \varepsilon_6 \, \| p^{(k)} \| \qquad (5.9)$$

we can guarantee a significant reduction in ϕ during that iteration. If the line search results in a value of ϕ' such that

$$\phi'(z^*,0) \leq \varepsilon_6 \, \| p^{(k)} \|$$

then the value of $\tilde{g}^{(k)}$ could be improved by (3.16) and a new direction of search calculated.

If the new value of $\|\approx{g}^{(k)}\| < \varepsilon_o$ then convergence could be verified by undertaking a linear search of type (5.1) along each axial direction in turn hence confirming that the point is in some sense a statistical minimum.

Note 5.4 If $\tilde{g}^{(k)}(x)$ is calculated by (3.1) or (3.2) then the point x(h) at which $\|\tilde{g}^{(k)}(x)\| = 0$ is biased and does not correspond with the minimum of f(x). There is therefore little point in finding such a point. However, if a uniformly convergent linear search is used along each axis, then this difficulty is not experienced.

6. Conclusions

(1) We have examined the behaviour of the standard deterministic variable metric algorithm when applied with numerical derivatives, round off error on the function evaluation and noise on the function. It has been shown, theorem 5, that

$\exists\ \varepsilon_o^*$ such that the methods will enter a region with $\|\tilde{g}(x)\| < \varepsilon_o$ for all values of $\varepsilon_o > \varepsilon_o^*$. It has further been shown that if they are asked to locate a point with $\|\tilde{g}(x)\| < \varepsilon_7$, $\varepsilon_7 < \varepsilon_o^*$ they are likely to fail. This is in agreement with published experimental data.

(2) It has been shown that by modifying the line search strategy it is possible to improve the estimation of the slope along the line and hence enable a more accurate result to be obtained.

(3) It has been shown that the standard updating formulae are likely to be inappropriate in the extended stage of the search and that a new technique is required. This will be pursued in a later report.

References

1. F. Archetti & B. Betro (1975)
   Convex Programming via Stochastic Regularization
   Presented at 2nd Workshop on Global Optimisation, Varenna (1975)
2. M.C. Biggs (1973) J.I.M.A. Vol 12 No. 3 pp 337-9
3. M.C. Biggs (1975)
   The Numerical Optimisation Centre, The Hatfield Polytechnic,
   Technical Report No. 69.
4. A.H.O. Brown (1975) The Development of Computer Optimisation Procedures for use in Aero Engine Design.
   To appear in L.C.W. Dixon Ed., "Optimisation in Action", Academic Press.
5. D.B. Crombie (1972) Review of the Performance of Hill Climbing Strategies when Applied in the Hybrid Computing Environment.
   M.Sc. Thesis, The Hatfield Polytechnic.
6. L.C.W. Dixon (1973) Nonlinear Optimisation: A survey.
   In D.J. Evans Ed., "Software for Numerical Mathematics", Academic Press.
7. L.C.W. Dixon, J. Gomulka & S.E. Hersom (1975) Reflections on the Global Optimisation Problem.
   To appear in L.C.W. Dixon Ed., "Optimisation in Action", Academic Press.
8. D. Goldfarb (1969) Sufficient conditions for convergence of a variable metric algorithm.
   In R. Fletcher Ed., "Optimisation", Academic Press.
9. A.I. Mikhal-skii (1974) The method of averaged splines in the problem of approximating dependences on the basis of empirial data.
   Avtomatika i Telemekhanika 3 pp 45-50.

10. B.A. Murtagh & R.W.H. Sargent (1969) A constrained minimisation method with quadratic convergence.
In R. Fletcher Ed., "Optimisation", Academic Press.

11. M.J.D. Powell (1971) J. Inst. Maths. Applics 7 pp 21-36.

12. M.J.D. Powell (1974) C.S.S.9, A.E.R.E. Harwell.
Presented at "Mathematical Software II", Purdue University.

13. M.J.D. Powell (1975) C.S.S.15, A.E.R.E. Harwell.
Presented at "AMS/SIAM Symposium on Nonlinear Programming".

14. P. Wolfe (1969) Siam Review 11 pp 226-235.

15. P. Wolfe (1971) Siam Review 13 pp 185-188.

PARALLEL PATH STRATEGY
THEORY AND NUMERICAL ILLUSTRATIONS

Hilmar Drygas [1)]
Institut für Angewandte Mathematik
D 6 Frankfurt am Main, Robert-Mayer-Str. 10

Introduction

This paper deals with parallel-planning strategies for the development of a new item. Let us assume that several proposals for develepping the new item are available. Usually one is interested to choose the "best" of these proposals with respect to some criterion. But it may happen that the data available at this point of the development process are very inaccurate and uncertain and therefore unreliable. For example a proposal promising low costs and low time to bring the project to the end may turn out to be very expensive after its completion. In such situations a wrong decision may be avoided by pursuing a parallel path approach. Several, say m, of the proposals are pursued to a review-point at which point the best project is selected and brought to completion while the other still remaining approaches are stopped. This is the simplest model of a parallel path-approach; generalization to several review points, however, are possible and done in the literature (see Marschak [3]).

It is the purpose of this paper to answer the following question: How much must the uncertainty at the review point be reduced in order that a parallel-approach is at all worthwhile to be pursued? A lower bound on the variation of the cost-estimate is obtained in order that a parallel-approach with m proposals to start at the beginning of the project may be worthwhile at all. It is shown that this lower bound is attained only if the cost-estimates are discrete, allowing only two values.

To obtain the indicated results a stochastic model is built up and analysed in the next section. The problem of computation of the optimal strategy is discussed in the following section, while numerical and graphical illustrations are given in the last paragraph.

[1)] This work was done, when the author was a research worker at "Studiengruppe für Systemforschung", D 69 Heidelberg 1, Werderstr. 35

1. A General Model

Let us have a set {1,2,...} of Research and Development (R&D) approaches which can start at point t, a set $\Gamma = \{i_1,...,i_m\}$ of m approaches of them are pursued to a review-point $t+\theta$, at which point the best project, i.e., that project having the smallest total money and time cost-estimate is selected and brought to the end while the other remaining projects are stopped.

In order that such a procedure can work the cost-estimate must in some sense be consistent with the actual costs, i.e., there must be some (stochastic) relationship between cost-estimators and actual costs. This relationship is unbiasedness and is formulated as follows: Let $K_{i,t+\theta}$ be the total time and money-cost that finally is obtained by bringing proposal i from review-point $t+\theta$ to completion and let moreover $K_{t+\theta}^{(i)}$ be the total time and money cost-estimate obtained at the point $t+\theta$. Following Th. Marschak ([2],[3]) we again assume that $K_{t+\theta}^{(i)}$ is an unbiased estimator of $K_{i,t+\theta}$, i.e., that

(1) $$E(K_{i,t+\theta} | K_{t+\theta}^{(i)}) = K_{t+\theta}^{(i)}.$$

This assumption may seem strange since often in practice cost-estimates are much lower than actual costs. To make assumption (1) more realistic it may be assumed that

(1') $$E(K_{i,t+\theta} | K_{t+\theta}^{(i)}) = g(K_{t+\theta}^{(i)}),$$

where g is a monotonic non-decreasing function, e.g., $g(x) = 3x$. Nothing in our analysis will be changed if we replace assumption (1) by assumption (1') if $K_{t+\theta}^{(i)}$ is replaced by $g(K_{t+\theta}^{(i)})$ whenever it occurs.

Let Γ be a set of approaches and let w_γ be the money-cost which is necessary to bring approach $\gamma \in \Gamma$ to review-point $t+\theta$. If C_Γ is the total time and money-cost which is necessary to follow a parallel-approach with approach-set Γ, then under assumption (1):

(2) $$E(C_\Gamma) = \Sigma_{\gamma\in\Gamma} E(w_\gamma) + E(\min_{\gamma\in\Gamma} K_{t+\theta}^{(\gamma)})$$

Indeed, let $\gamma_o \epsilon \Gamma$ be such that $K_{t+\theta}^{(\gamma_o)} = \min_{\gamma \epsilon \Gamma} K_{t+\theta}^{(\gamma)}$, then by (1):

(3) $C_\Gamma = \Sigma_{\gamma \epsilon \Gamma} w_\gamma + K_{\gamma_o, t+\theta}$

(4) $E(C_\Gamma) = \Sigma_{\gamma \epsilon \Gamma} E(w_\gamma) + E(E(K_{\gamma_o, t+\theta} | K_{t+\theta}^{(\gamma_o)}))$

$= \Sigma_{\gamma \epsilon \Gamma} E(w_\gamma) + E(K_{t+\theta}^{(\gamma_o)}) = \Sigma_{\gamma \epsilon \Gamma} E(w_\gamma) + E(\min_{\gamma \epsilon \Gamma} K_{t+\theta}^{(\gamma)}$

If $\Gamma = \{i_1, \ldots, i_m\}$ and

(5) $F_\Gamma(x) = F_{i_1 \ldots i_m}(x) = P(\bigcap_{\gamma \epsilon \Gamma} \{K_{t+\theta}^{(\gamma)} > x\})$,

then by a well-known fundamental lemma (Th. Marschak [3], p. 217), (4) can be rewritten as

(6) $E(C_\Gamma) = \Sigma_{\gamma \epsilon \Gamma} E(w_\gamma) + \int_o^\infty F_\Gamma(x)\, dx$.

Suppose that the projects are ordered according to increasing expected looking costs, i.e., $E(w_1) \leq E(w_2) \leq \ldots$. We now make the following monotonity-assumption, already introduced by Marschak in a similar context, namely that the distributions of the $K_{t+\theta}^{(i)}$ are consistent with the values of $E(w_i)$. More precisely: If $E(w_i) \geq E(w_j)$, we require that there is a higher probability that $K_{t+\theta}^{(i)}$ exceeds a given value x than that $K_{t+\theta}^{(j)}$ exceeds the value x and this should hold for any x, given an arbitrary set of alternative approaches.

<u>Definition:</u> The given set of approaches is said to have a monotonous class of distribution functions if for any subset of approaches Γ and any pair (i,j); $i,j \notin \Gamma$ and any real number x

(7) $P(K_{t+\theta}^{(i)} > x | \bigcap_{\gamma \epsilon \Gamma} K_{t+\theta}^{(\gamma)} > x) \geq P(K_{t+\theta}^{(j)} > x | \bigcap_{\gamma \epsilon \Gamma} K_{t+}^{(\gamma)} > x)$, if $i \geq j$.

<u>Theorem:</u> Under the assumptions (1) and (7) we have

a) $E(C_{\Gamma \cup i}) \geq E(C_{\Gamma \cup j})$ for any subset of approaches Γ if $i \geq j$.

b) $\Gamma = \{1\}$ is the optimal set of approaches to be pursued if

(8) $\int_o^\infty P(K_{t+\theta}^{(2)} \leq x, K_{t+\theta}^{(1)} > x)\, dx < E(w_2)$.

c) If there is a last integer $m \geq 2$ such that

(9) $$E(w_m) \leq \int_0^\infty P(K_{t+\theta}^{(m)} \leq x, \bigcap_{i=1}^{m-1} K_{t+\theta}^{(i)} > x)\, dx,$$

then $\Gamma = \{1,2,\ldots,m\}$ is the optimal set of approaches to be pursued.

Interpretation of a): It says that under the monotonity-assumption it is always worthwhile to substitute a given approach by an approach possessing a lower cost-estimate; by such a substitution the total expected time and money-cost is not increased.

Proof of a):

$$E(C_{\Gamma\cup i}) - E(C_{\Gamma\cup j}) = E(w_i) - E(w_j) + \int_0^\infty [P(K_{t+\theta}^{(i)} > x \mid \bigcap_{\gamma\in\Gamma} K_{t+}^{(\gamma)} > x)$$

$$- P(K_{t+\theta}^{(j)} > x \mid \bigcap_{\gamma\in\Gamma} K_{t+\theta}^{(\gamma)} > x]\, P(\bigcap_{\gamma\in\Gamma} K_{t+\theta}^{(\gamma)} > x)\, dx \geq 0$$

if $i \geq j$ by assumption (7).

Proof of b) and c): By a) it is sufficient to consider

(10) $$R_m = E(C_{\{1,2,\ldots,m\}}) = \Sigma_{i=1}^m E(w_i) + \int_0^\infty P(\bigcap_{i=1}^m K_{t+\theta}^{(i)} > x)\, dx.$$

We have

(11) $$V_m = \Delta R_m = R_m - R_{m-1}$$

$$= E(w_m) - \int_0^\infty [P \bigcap_{i=1}^{m-1} K_{t+\theta}^{(i)} > x) - P(\bigcap_{i=1}^m K_{t+\theta}^{(i)} > x)]\, dx$$

$$= E(w_m) - \int_0^\infty P(K_{t+\theta}^{(m)} \leq x, \bigcap_{i=1}^{m-1} K_{t+\theta}^{(i)} > x)\, dx$$

(12) $$\Delta^2 R_m = \Delta V_m = V_m - V_{m-1}$$

$$= E(w_m) - E(w_{m-1}) + \int_0^\infty [P(K_{t+\theta}^{(m-1)} \leq x, \bigcap_{r=1}^{m-2} K_{t+\theta}^{(r)} > x)$$

$$- P(K_{t+\theta}^{(m)} > x, \bigcap_{r=1}^{m-1} K_{t+\theta}^{(r)} > x)]\, dx$$

Now by assumption (7)

(13) $$P(K_{t+\theta}^{(m-1)} \leq x, \bigcap_{r=1}^{m-2} K_{t+\theta}^{(r)} > x)$$

$$= P(K_{t+\theta}^{(m-1)} \leq x \mid \bigcap_{r=1}^{m-2} K_{t+\theta}^{(r)} > x)\, P(\bigcap_{r=1}^{m-2} K_{t+\theta}^{(r)} > x)$$

$$\geq P(K_{t+\theta}^{(m)} \leq x \mid \bigcap_{r=1}^{m-2} K_{t+\theta}^{(r)} > x)\ P(\bigcap_{r=1}^{m-2} K_{t+\theta}^{(r)} > x)$$

$$= P(K_{t+\theta}^{(m)} \leq x,\ \bigcap_{r=1}^{m-2} K_{t+\theta}^{(r)} > x)$$

and so by inserting this relation into (12):

$$(14)\qquad \Delta^2 R_m \geq E(w_m) - E(w_{m-1}) + \int_o^\infty P(K_{t+\theta}^{(m)} \leq x,\ \bigcap_{r=1}^{m-2} K_{t+\theta}^{(r)} > x)$$

$$- P(K_{t+\theta}^{(m)} \leq x,\ \bigcap_{r=1,\dots,m-1} K_{t+\theta}^{(r)} > x)\ dx$$

$$= E(w_m) - E(w_{m-1}) + \int_o^\infty P(\bigcap_{i=m-1}^{m} K_{t+\theta}^{(i)} \leq x,\ \bigcap_{r=1}^{m-2} K_{t+\theta}^{(r)} > x)\ dx \geq 0.$$

So R_m has monotonous increasing marginal costs and so m and hence $\Gamma = \{1,2,\dots,m\}$ is optimal if $\Delta R_m \leq 0$, i.e., (9) holds, provided such an $m \geq 2$ exists. The assertion of b) is now also evident.

2. Upper and Lower Bounds for the Optimal Number of Approaches to be Pursued

In this section we simplify the general model of the previous section by assuming that the cost-estimates $K_{t+\theta}^{(\gamma)}$ are stochastically independent random variables. Under this assumption the optimality-criterion of theorem 1.2 simplifies to

$$(2.1)\qquad \int_o^\infty P(K_{t+\theta}^{(m)} \leq x) \prod_{i=1}^{m-1} (1-P(K_{t+\theta}^{(i)} \leq x))\ dx \geq E(w_m),$$

or we define

$$(2.2)\qquad F_i(x) = P(K_{t+\theta}^{(i)} \leq x)$$

to

$$(2.3)\qquad \int_o^\infty F_m(x) \prod_{i=1}^{m-1} (1-F_i(x))\ dx \geq E(w_m).$$

The monotonity-assumption implies that

$$(2.4)\qquad 1-F_i(x) \leq 1-F_m(x),\ i = 1,2,\dots,m-1.$$

We now make the rather realistic assumption that all considered random variables (cost-estimates) $K_{t+\theta}^{(i)}$ are restricted to a finite interval $[M_1, M_2]$, i.e., that

(2.5) $P(M_1 \leq K_{t+\theta}^{(i)} \leq M_2) = 1; \; i = 1,2,\ldots$

This implies that $F_m(x) = 0$ if $x \leq M_1$, and $F_m(x) = 1$ if $x \geq M_2$. Using this and (2.4) we get from (2.3)

(2.6) $E(w_m) \leq \int_0^\infty F_m(x) \prod_{i=1}^{m-1} (1-F_i(x)) \, dx \leq \int_{M_1}^{M_2} F_m(x)(1-F_m(x))^{m-1} \, dx$

The key for understanding the following considerations is

<u>Lemma 2.1.</u> The function

(2.7) $f(P) = P^m(1-P), \; 0 \leq P \leq 1$

is monotonously increasing for $P \in [0, m(m+1)^{-1}]$ and monotonously decreasing for $P \in [m(m+1)^{-1}, 1]$

(2.8) $\max_{0 \leq P \leq 1} f(P) = f(\frac{m}{m+1}) = A_m = (1+\frac{1}{m})^{-m} \frac{1}{m+1} = (1+\frac{1}{m})^{-(m+1)} \frac{1}{m}$

(2.9) $((m+1)e)^{-1} \leq A_m \leq (me)^{-1}, \; \lim_{m\to\infty} m \cdot A_m = e^{-1}$

where $e = \lim_{m\to\infty} (1+\frac{1}{m})^m = \Sigma_{m=0}^{\infty} \frac{1}{m!} = 2{,}71818459\ldots$

Proof: The first assertion concerning the maximum follows from

(2.10) $f'(P) = (m+1)P^{m-1} (m(m+1)^{-1}-P) \gtrless 0 \quad \text{iff } P \lessgtr m(m+1)^{-1}$.

Evidently

(2.11) $f(m(m+1)^{-1}) = A_m = (1+m^{-1})^{-m}(m+1)^{-1} = (1+m^{-1})^{-(m+1)} \; m^{-1}$.

From this and well-known facts from a first course in calculus the remaining statements of the lamma follow immediately.

<u>Corollary 2.2.</u> If the $K_{t+\theta}^{(\gamma)}$, $\gamma = 1,2,\ldots$ are stochastically independent random variables and $P(M_1 \leq K_{t+\theta}^{(\gamma)} \leq M_2) = 1$, then a necessary condition that there exist distribution functions $F_1(x),\ldots,F_m(x)$ such that

(2.12) $F_i(x) = P(K_{t+\theta}^{(i)} \leq x) \leq F_{i-1}(x) = P(K_{t+\theta}^{(i-1)} \leq x)$

for $i = 2,3,\ldots,m$ and $\Gamma = \{1,2,\ldots,m\}$ is the optimal set of approaches to be pursued is that

(2.13) $M_2-M_1 \geq E(w_m)(A_{m-1})^{-1}$.

Remark 2.3. Since $me \le (A_m)^{-1} \le (m+1)e$, M_2-M_1 is to be larger than $E(w_m) \cdot K_m$, where $(m-1)e \le K_m \le me$.

Proof of the Corollary: By (2.6) we must have

$$(2.14) \qquad E(w_m) \le \int_{M_1}^{M_2} (1-F_m(x))^{m-1} F_m(x)\, dx .$$

By lemma 2.1 $(1-F_m(x))^{m-1} F_m(x) \le A_{m-1}$ (by letting $P = 1-F_m(x)$).
So we get

$$(2.15) \qquad E(w_m) \le A_{m-1}(M_2-M_1)$$

and (2.13) is obtained.

Remark 2.4. Note that if $(M_2-M_1) = (A_{m-1})^{-1} E(w_m)$ the considered integral can be equal to $A_{m-1}(M_2-M_1)$ if and only if $1-F_m(x) = m(m+1)^{-1}$ for all $x \in [M_1,M_2]$ or $F_m(x) = (m+1)^{-1}$ for all $x \in [M_1,M_2]$. The only distribution meeting this condition is one for which $P(X=M_1) = (m+1)^{-1}$, $P(X=M_2) = m(m+1)^{-1}$, the case of a two-class-parallel planning model such as considered by Nelson [4]. If a two class-situation can be excluded then it is to expect that much higher lower bounds for M_2-M_1 can be obtained. This will be studied in the sequel.

It has just been proved, that $(E(w_m))^{-1}(M_2-M_1) \ge (A_{m-1})^{-1} = a_m = m(1+(m-1)^{-1})^m$ and $a_m \in [(m-1)e, me]$. This approximation is still too crude and will be improved by the following lemma:

Lemma 2.5. $\lim_{m\to\infty} (m+1)\,[(1+\frac{1}{m})^m - e] = -e/2$

Proof: $\lim_{m\to\infty} (m+1)\,[(1+m^{-1})^m - e] = \lim_{m\to\infty} (1+m^{-1}) \cdot \lim_{m\to\infty} m[(1+m^{-1})^m - e]$

$= \lim_{m\to\infty} m[(1+m^{-1})^m - e] = \lim_{x\to 0} x^{-1}[(1+x)^{1/x} - e]$

by letting $m^{-1} = x$.

So we get

$$(2.16) \qquad \lim_{m\to\infty} (m+1)[(1+m^{-1})^m - e] = \lim_{x\to 0} x^{-1}\,[(1+x)^{1/x} - e] = \frac{d}{dx}(1+x)^{1/x}\Big|_{x=0}.$$

But $(1+x)^{1/x} = \exp(1/x \log(1+x))$, implying

(2.17) $\frac{d}{dx}(1+x)^{1/x}|_{x=o} = \exp(1/x \log(1+x))[-\frac{1}{x^2}\log(1+x)$

$+ (x(1+x))^{-1}]|_{x=o} = e\frac{d}{dx}[1/x \log(x+1)]|_{x=o}$

$= e\frac{d}{dx}[-\frac{1}{x}\log(1+x) + (1+x)^{-1}]|_{x=o}$

Comparing the last two expressions in the chain of equalities we get

(2.18) $\frac{d}{dx} 1/x \log(1+x)|_{x=o} = 1/2 \frac{d}{dx}(1+x)^{-1}|_{x=o} = -1/2.$

This proves the lemma.
This lemma yields the approximation

(2.19) $(M_2-M_1)(E(w_m))^{-1} \geq a_{m-1} \cong (m-\frac{1}{2})e,$

which will be used in a subsequent paragraph.

To study the problem indicated in the previous remark, we want to make the more stringent assumption that all $K_{t+\theta}^{(i)}$ are not only independent, but have also all the same distribution $F(x)$. Under this assumption, which is usually made in literature, the optimal number of approaches to be pursued is equal to the last number (which in this case must exist) m such that

(2.20) $\int_{M_1}^{M_2} F(x)(1-F(x))^{m-1}\,dx \geq E(w_m).$

We discretize the sample-space and consider k+1-class-parallel-planning-models, i.e., we assume that $F(x)$ belongs to a random variable X such that $N_1 < N_2 < \ldots < N_{k+1}$ and $P(X=N_i) = p_i$, $i = 1,\ldots,k+1$, where of course $p_i \geq 0$, $1 \leq i \leq k+1$ and $\Sigma_{i=1}^{k+1} p_i = 1$. Let

(2.21) $P_i = 1 - \Sigma_{j=1}^{i} p_j$, $i = 1,2,\ldots,k.$

Then $F(x)$ is equal to $1-P_i$ in the interval $[N_i, N_{i+1}[$. Evidently $N_1 = M_1$ and $M_2 = N_{k+1}$. So we get from (2.18) if we let $M_i^* = N_{i+1} - N_i$, $i = 1,2,\ldots,k$:

(2.22) $\Sigma_{i=1}^{k} M_i^* (1-P_i) P_i^{m-1} \geq E(w_m).$

Let us assume that $M_1^*, M_2^*, \ldots, M_{k-1}^*$ as well as $p_1,\ldots,p_{k-1}$ (and hence also $P_1,\ldots,P_{k-1}$) are given. What is the lowest value of M_k^* such that we can find a real number $p_k \geq 0$ (or equivalently a P_k such that $P_k \leq P_{k-1}$) with (2.20) to hold? Evidently

$$(2.23) \qquad h_m(P_{k-1}) = \max_{P_k \leq P_{k-1}} (1-P_k)P_k^{m-1}$$

$$= \begin{cases} A_{m-1}, \text{ if } P_{k-1} \geq (m-1)m^{-1} \\ (1-P_{k-1})\, P_{k-1}^{m-1}, \text{ if } P_{k-1} \leq (m-1)m^{-1} \end{cases}$$

So the stated problem has a solution if and only if

$$(2.24) \qquad \Sigma_{i=1}^{k-1} M_i^*(1-P_i)\, P_i^{m-1} + M_k^*\, h_m(P_{k-1}) \geq E(w_m)$$

or

$$(2.25) \qquad M_k^*\, h_m(P_{k-1}) \geq E(w_m) - \Sigma_{i=1}^{k-1} M_i^*(1-P_i)\, P_i^{m-1} ,$$

i.e.,

$$(2.26) \qquad M_k^* \geq [E(w_m) - \Sigma_{i=1}^{k-1} M_i^*\, (1-P_i)\, P_i^{m-1}](h_m(P_{k-1}))^{-1} = M_{k,min}^*$$

$$(2.27) \qquad M_2-M_1 = \Sigma_{i=1}^{k} M_i^* \geq (h_m(P_{k-1}))^{-1}[E(w_m) + \Sigma_{i=1}^{k-1} M_i^*(h_m(P_{k-1}) - (1-P_i)\, P_i^{m-1})]$$

Especially if $M_1^* = M_2^* = \ldots = M_{k,min}^* = M^*$ we get

$$M^* = E(w_m)(h_m(P_{k-1}) + \Sigma_{i=1}^{k-1} (1-P_i)\, P_i^{m-1})^{-1}$$

and

$$(2.28) \qquad M_2-M_1 \geq kM_{k,min}^* = kE(w_m)[h_m(P_{k-1}) + \Sigma_{i=1}^{k-1} (1-P_i)\, P_i^{m-1}]^{-1} .$$

Especially if $k = 2$ we get

$$(2.29) \qquad M_2-M_1 \geq 2M_{2,min}^* = \begin{cases} 2E(w_m)(A_{m-1} + p_1(1-p_1)^{m-1})^{-1}, \text{ if } p_1 \leq m^{-1} \\ E(w_m)\, [p_1(1-p_1)^{m-1}]^{-1}, \text{ if } p_1 \geq m^{-1} \end{cases}$$

This curve will be plotted in section 4 as a function of p_1 for $m = 1$ and $E(w_m) = 1$.

It may also happen that a certain strategy m_o will never apply for given values of $M_1^*, M_2^*, \ldots, M_{k-1}^*$; $P_1, P_2, \ldots, P_{k-1}$ because the strategy $m = m_o + 1$ is always superior to the strategy $m = m_o$, whatever may be the value of p_k (P_k). This can happen if and only if

$$(2.30) \qquad M_k^* \; (1-P_k) \; P_k^{m_o} \geq E(w_{m_o+1}) - \Sigma_{i=1}^{k-1} \; M_i^* \; (1-P_i) \; P_i^{m_o}$$

for all $P_k \leq P_{k-1}$. Especially $P_k = 0$ implies

$$(2.31) \qquad E(w_{m_o+1}) - \Sigma_{i=1}^{k-1} \; M_i^* \; (1-P_i) \; P_i^{m_o} \leq 0$$

If especially $M_1^* = M_2^* = \ldots = M_{k-1}^* = M^*$ (equidistance of cost-estimates), then (2.29) means

$$(2.32) \qquad M^* \geq [\Sigma_{i=1}^{k-1} \; (1-P_i) \; P_i^{m_o}]^{-1} \; E(w_{m_o+1})$$

and thus

$$(2.33) \qquad M_2 - M_1 = (k-1)M^* + M_k^* \geq (k-1) \; E(w_{m_o+1})[\Sigma_{i=1}^{k-1} \; (1-P_i) \; P_i^{m_o}]^{-1} + M_k^* \; .$$

If moreover $M^* = M_k^*$, too and $k = 2$, we get

$$(2.34) \qquad (M_2 - M_1) \;\; (E(w_{m_o+1})) \geq 2 \cdot M^* \geq 2[p_1 (1-p_1)^{m_o}]^{-1}.$$

This curve will also be plotted in section 4 as a function of p_1 for $m_o = 1$. For large m (and if $p_1 \geq (m+1)^{-1}$ in general) it is twice the minimum-value of $M_2 - M_1$ for which the strategy $m_o + 1$ will apply at all, while (2.34) gives the minimum values of $M_2 - M_1$ for which no strategy $m \leq m_o$ will apply. The tables that we will construct will also reveal this situation.

3. The Computation of Probability Intervals Corresponding to Given Optimal Strategies

Given a certain set of values $M_1^*, \ldots, M_k^*$; $P_1, P_2, \ldots, P_{k-1}$ the question may arise how the probability intervals for p_k can be computed for which a given strategy $m \geq 2$ is optimal. (The interval in which $m = 1$ is the optimal strategy is then evidently the interval $(0, P_{k-1}) - \bigcup_{m=2}^{\infty} I_m$, if I_m denotes the interval in which strategy m is optimal). We have evidently

$$(3.1) \qquad P_k = P_{k-1} - p_k,\ p_k = P_{k-1} - P_k ,$$

implying $P_k \leq P_{k-1}$. By (2.24), for the given values, P_k and hence p_k is optimal if m is the last integer such that

$$(3.2) \qquad (1-P_k)\ P_k^{m-1} \geq M_k^{*-1}[E(w_m) - \Sigma_{i=1}^{k-1} M_i^*\ (1-P_i)\ P_i^{m-1}] = N_m^* .$$

If N_m^* happens to be smaller or equal to zero, then any $P_k \leq P_{k-1}$ satisfies the inequality (3.2). Let us denote by J_m the set of values $P_k \leq P_{k-1}$ such that inequality (3.2) is met. If $J_m = [a,b]$, then evidently $J_m^{(o)} = [P_{k-1}-b,\ P_{k-1}-a]$ is the p_k-interval satisfying (3.2). Then evidently $J_m^{(o)} \supseteq J_{m+1}^{(o)}$ and

$$(3.3) \qquad I_m = J_m^{(o)} - J_{m+1}^{(o)}$$

We know from the last paragraph that there is a lower bound for N_m^* in order that $J_m^{(o)} \neq \emptyset$, namely that

$$(3.4) \qquad N_m^* \leq h_m\ (P_{k-1}) = \begin{cases} A_{m-1}, & \text{if } P_{k-1} \geq (m-1)m^{-1} \\ P_{k-1}^{m-1}\ (1-P_k), & \text{if } P_{k-1} \leq (m-1)m^{-1} \end{cases}$$

If $N_m^* \leq 0$, then evidently $J_m = [0,\ P_{k-1}]$, $J_m^{(o)} = [0,\ P_{k-1}] = [0,\ 1 - \Sigma_{i=1}^{k-1} p_i]$. Now let us assume that $0 < N_m^* < h_m\ (P_{k-1})$. (If $N_m^* = h_m\ (P_{k-1})$, then $P_k = P_{k-1}$ is the only element of $J_m^{(o)}$ in which case the k+1-class-parallel-planning-model degenerates to a k-class-parallel-planning-model). We now make use of the results of the last paragraph, namely that $g_{m-1}(P) = (1-P)P^{m-1}$ is monotonously increasing in $[0,\ m^{-1}(m-1)]$ and monotonously decreasing in $[(m-1)m^{-1},\ 1]$; the maximum is at $P = (m-1)m^{-1}$, a saddle-point at $(m-2)m^{-1}$.

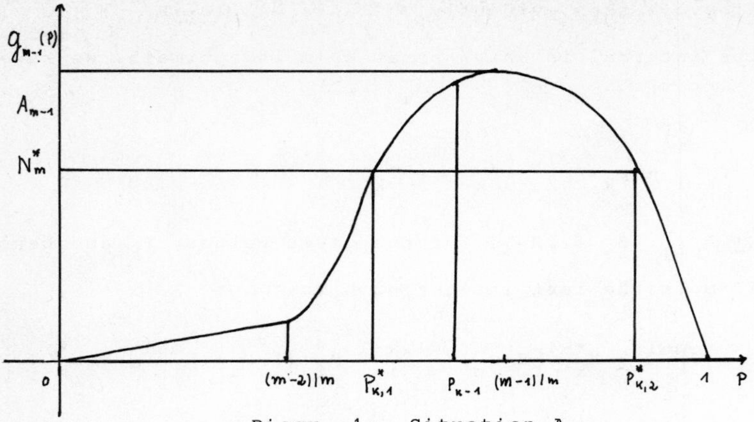

Diagr. 1: Situation A

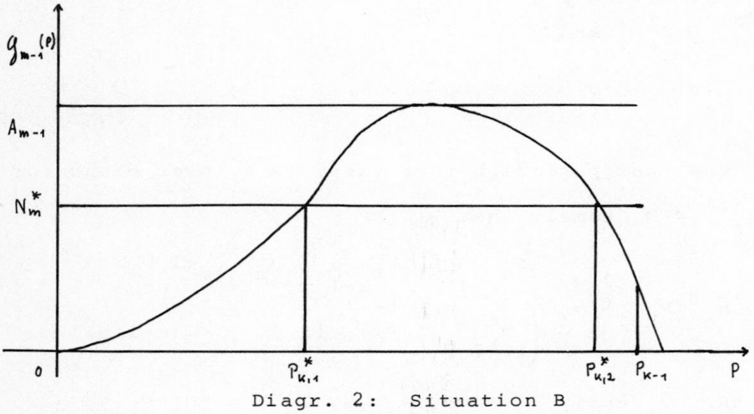

Diagr. 2: Situation B

If $0 < N_m^* < h_m(P_{k-1})$, there will exist - diagram and analysis of $g_{m-1}(P)$ show it - two values $P_{k,1}^*$, $P_{k,2}^*$ such that $P_{k,1}^* < P_{k,2}^*$ and therefore

$$(1-P_{k,2}^*)(P_{k,2}^*)^{m-1} = (1-P_{k,1}^*)(P_{k,1}^*)^{m-1} = N_m^*.$$

Thus $(1-P_k)\ P_k^{m-1} \geq N_m^*$ iff $P_k \in [P_{k,1}^*, P_{k,2}^*]$.

$N_m^* < h_m(P_{k-1})$ implies $P_{k,1}^* < P_{k-1}$. What $P_{k,2}^*$ is concerned two situations may arise. In situation A (Figure 1), $P_{k,2}^* > P_{k-1}$; hence

$$(3.5) \qquad J_m = [P^*_{k,1}, P_{k-1}], \; J_m^{(o)} = [0, P_{k-1} - P^*_{k,1}].$$

In situation B (Figure 2), $P^*_{k,2} > P_{k-1}$ and hence

$$(3.6) \qquad J_m = [P^*_{k,1}, P^*_{k,2}], \quad J_m^{(o)} = [P_{k-1} - P^*_{k,2}, P_{k-1} - P^*_{k,1}]$$

In general we have

$$(3.7) \qquad J_m = [P^*_{k,1}, \min(P_{k-1}, P^*_{k,2})],$$

$$(3.8) \qquad J_m^{(o)} = [P_{k-1} - \min(P^*_{k,2}, P_{k-1}), P_{k-1} - P^*_{k,1}]$$

$$= [\max(0, P_{k-1} - P^*_{k,2}), P_{k-1} - P^*_{k-1}].$$

So our computing procedure will be as follows:

<u>Computing Procedure 3.1.</u> Compute N^*_m, starting with $m = 2$ and $J_m^{(o)}$ according to

$$(3.9) \qquad J_m^{(o)} = \begin{cases} \emptyset & \text{if } N^*_m > h_m(P_{k-1}) \\ \{0\} & \text{if } N^*_m = h_m(P_{k-1}) \\ [\max(0, P_{k-1} - P^*_{k,2}), P_{k-1} - P^*_{k-1}] & \text{if } 0 < N^*_m < h_m(P_{k-1}) \\ [0, P_{k-1}] & \text{if } N^*_m \leq 0 \end{cases}$$

where $P^*_{k,1}$ denotes the smaller and $P^*_{k,2}$ denotes the larger of the two roots of the equation $P^{m-1}(1-P) = N^*_m$, provided $0 < N^*_m < 1$. If $J_m^{(o)} \neq \emptyset$ go to m+1; if $J_m^{(o)} = \emptyset$ stop the computation. If the computation stops at $m = m_o$, then

$$(3.10) \qquad I_{m_o} = \emptyset, \; I_{m_o-1} = J^{(o)}_{m_o-1}, \; I_m = J_m^{(o)} - J^{(o)}_{m+1}, \; m = 2,3,\ldots,m_o-2$$

$$I_1 = [0, P_{k-1}] - \bigcup_{r=2}^{m_o-1} I_r.$$

The only thing that now must be still done is the computation of the two roots of the equation $P^{m-1}(1-P) = N^*_m$ provided $0 < N^*_m < A_{m-1}$. This problem will be solved by the following theorem:

<u>Theorem 3.2</u> a) Let $\hat{A}_{m-1} = g_{m-1}((m-2)m^{-1}) = 2m^{-1}(1-2m^{-1})^m$ and $P_o = (m-2)m^{-1}$ if $0 < N_m^* \leq \hat{A}_{m-1}$, $P_o = (m-1)m^{-1}$ if $A_{m-1} > N_m^* > \hat{A}_{m-1}$ and moreover

$$(3.11) \qquad P_{n+1} = (m-1)^{-1} \left[(m-2)P_n + N_m^* \{P_n^{m-2}(1-P_n)\}^{-1} \right],$$

$n = 0,1,2,\ldots$. Then $P_n \geq P_{n+1}$ and $\lim_{n\to\infty} P_n = P_1^*$, where P_1^* is the smaller of the two roots of the equation $F^{m-1}(1-P) = N_m^*$.

b) Let $\hat{P}_o = (m-1)m^{-1}$ and

$$(3.12) \qquad \hat{P}_{n+1} = 1 - N_m^* \hat{P}_n^{-(m-2)}, \quad n = 0,1,2,\ldots,$$

then $\hat{P}_{n+1} \geq \hat{P}_n$ and $\lim_{n\to\infty} \hat{P}_n = P_2^*$, the larger of the two roots of the equation $P^{m-1}(1-P) = N_m^*$.

<u>Remark 3.3</u> The part a) of the theorem 3.2 could also be correct if $P_o = (m-1)m^{-1}$ also if $0 < N_m^* \leq \hat{A}_{m-1}$. But the modification of a) will later allow to obtain better bounds on the error $P_n - P_1^*$.

The algorithm is a modification of the algorithm for the determination of the m-th root of positive number. Readers who are interested in a detailed proof may consult [6].

Moreover, one might be interested to get a bound for the error $P_n - P_1^*$ and $P_2^* - \hat{P}_n$. The need for such bounds arises when we are wondering when the iterative computing procedure should be stopped in order to get a required degree of accuracy. The detailed computations of [6] are again omitted here. We only quote the result:

<u>Corollary 3.3.</u> We have the following error bounds:

α) For a:

$$P_n - P_1^* \leq (P_{n+1} - P_n)\, h_o(P_n), \text{ if } P_o = (m-1)m^{-1}$$

$$P_n - P_1^* \leq (P_{n+1} - P_n)\, h_1(P_n), \text{ if } P_o = (m-2)m^{-1}$$

where

$$h_o(P_n) = (1-P_n)((m-1) - mP_n)^{-1}$$

$$h_1(P_n) = (1-P_n)^2((m-1) - mP_n)^{-1} P_n^{m-1} (N_m^*)^{-1}$$

β) For b: $P_2^* - \hat{P}_n \leq (\hat{P}_{n+1} - \hat{P}_n)\, h_2(\hat{P}_n)$

where $h_2(\hat{P}_n) = \hat{P}_n (m\hat{P}_n - (m-1))^{-1}$.

4. Graphical and Numerical Illustrations

In this section we want to illustrate the theory of the previous three sections by some numerical and graphical examples. In (2.19) it had been whown that a parallel-approach with m approaches pursued to the review-point can be the optimal strategy only if

$$(2.19) \quad (M_2 - M_1)(E(w_{m_o}))^{-1} \geq a_{m_o-1} \cong (m_o - \tfrac{1}{2})e,$$

where $M_2 - M_1$ is the variation of the cost-estimate and Ew_{m_o} the expected cost of carrying approach m to the review-point. Moreover, we know that $a_{m-1} \in [(m-1)e, me]$.

In the following table (Table 1) a_{m_o} and $(m_o - \frac{1}{2})e$ as well as $(m_o - 1)e$ and $m_o e$ are listed. It turns out that even for $m_o = 2$ the error between a_{m_o} (= 4) and $(m_o - \frac{1}{2})e$ is less than 2 %. It may also be noted that computing a_{m_o} by logarithmus according to the formula

$a_{m_o} = \exp(m_o \log m_o - (m_o - 1)\log(m_o - 1))$ (using a four decimals-table) may be not more accurate than the approximate value $(m_o - \frac{1}{2})e$ if m_o becomes larger.

It has been shown in section 2 that if $a_{m_o} = (M_2 - M_1) \mid E(w_{m_o})$, then there is only one distribution function F(x), concentrated in $[M_1, M_2]$, such that $m = m_o$ can be the optimal number of approaches to be pursued, namely the distribution which takes as values only $X = M_1$ (with probability m_o^{-1}) and $X = M_2$ (with probability $1 - m_o^{-1} = (m_o - 1)m_o^{-1}$). If this very special two-class-problem can be excluded, then there must be a much higher value of $(M_2 - M_1) \mid E(w_{m_o})$ in order that $m = m_o$ can act as optimal strategy.

a_{m_o} / m_o	$(m_o-1)e$	a_{m_o}	a_{m_o}, logarith. comp.	$(m_o-\frac{1}{2})e$	$m_o e$	relative error of $(m_o-\frac{1}{2})e$
2	2,718	4	3,999	4,077	5,436	2 %
3	5,436	6,75	6,735	6,795	8,154	1 %
4	8,154	9,481	9,575	9,513	10,872	3 ‰
5	10,872	12,207	12,095	12,231	13,590	2,0‰
6	13,590	14,929	14,933	14,949	16,308	1,3‰
7	16,308	17,651	17,64	17,667	19,026	0,9‰
8	19,026	20,371	20,375	20,385	21,744	0,7‰
9	21,744	23,092	23,07	23,103	24,462	0,5‰
10	24,462	25,811	25,835	25,821	27,180	0,4‰
11	27,180	28,531	28,539	28,535	29,898	0,3‰

Table 1: Minimal number of $(M_2-M_1)|E(w_{m_o})$, ratio of cost-estimate range and expected inspection cost, that is necessary in order that $m = m_o$ can be the optimal number of approaches to be pursued.

To win a further insight into the nature of this problem, let us assume that we have an equidistant three-class-parallel-planning model, i.e. that $P(X=M_1) = p_1$, $P(X=M_1+(M_2-M_1) \mid 2) = p_2$, $P(X=M_2) = 1-p_1-p_2$, where $p_1, p_2 \geq 0$. If $p_1 > 0$ is given then necessary for the existence of some $p_2 \geq 0$ such that $m = m_o$ may act as optimal strategy is that (see 2.29))

$$(4.1) \qquad (E(w_{m_o}))^{-1}(M_2-M_1) \geq h_{m_o}(p_1) = \begin{cases} 2a_{m_o}(1+a_{m_o}(p_1(1-p_1)^{m_o-1})^{-1} & \text{if } p_1 \leq m_o^{-1} \\ [p_1(1-p_1)^{m_o-1}]^{-1} & \text{if } p_1 \geq m_o^{-1} \end{cases}$$

Note that if $p_1 \geq m_o^{-1}$ and the equality-sign holds in (4.1), then $p_2 = 0$ is the only value meeting the required condition, so that the problem degenerates in this case to a two-class problem. If $p_1 \to 1$, then $h_{m_o}(p_1) \to \infty$. This is quite clear because if $p_1 = 1$, then we have a one-class problem in which all uncertainty is removed and so in this case there will be no need for a parallel-strategy. $h_{m_o}(p_1)$ has its

minimum at $p_1 = m_o^{-1}$ with minimum-value $h_{m_o}(m_o^{-1}) = a_{m_o}$. But the corresponding value of p_2, if $(E(w_m))^{-1}(M_2-M_1) = a_{m_o}$ is $p_2 = 1-m_o^{-1}$ implying $P(X=M_2) = 0$. So the problem degenerates to a two-class problem and the minimum value a_{m_o} is actually not attained in a three-class problem. The same holds if $p_1 = 0$: the value $h_m(0) = 2a_{m_o}$ is not really correct since in this case the distribution is concentrated in $[(M_1+M_2)|2,M_2] = [\hat{M}_1^*, \hat{M}_2]$ and so $(\hat{M}_2-\hat{M}_1) \mid E(w_{m_o}) \geq a_{m_o}$ must hold in order that $m = m_o$ can act as optimal strategy.
So more correctly $h_{m_o}(p_1) = a_{m_o}$ causing a discontinuity at $p_1 = 0$.

The curve $h_{m_o}(p_1)$ has been plotted in Fig. 3 for $m_o = 2$. On the same diagram lying over $h_{m_o}(p_1)$ the curve l_{m_o} is plotted; this curve is equal to

$$(4.2) \qquad l_{m_o}(p_1) = 2[p_1(1-p_1)^{m_o-1}]^{-1}$$

and is the minimal number of $(M_2-M_1) \mid E(w_{m_o})$, given p_1, in order that the strategy $m = m_o-1$ will be never applied as optimal strategy whatever is the value of p_2. If $(M_2-M_1) \mid E(w_{m_o}) \geq l_{m_o}(p_1)$ then either $m = m_o$ or $m = m_o+1$, m_o+2, ... will be applied as optimal strategies. If $p_1 \geq m_o^{-1}$ then $l_{m_o}(p_1) \mid h_{m_o}(p_1) = 2$. If $p_1 < m_o^{-1}$ this ratio will be much larger and approach infinity if p_1 approaches zero. This is quite clear for if $p_1 = 0$ the problem again degenerates to a two-class problem. In Table 2 we will show that in a two-class problem always $m_o = 1$ is an admissible strategy (i.e., an optimal strategy for some appropriate value of the probability p_2).

As an example of a three-class problem we present the table for $p_2 = .30$ (Table 3), the middle-class probability. More material can be found in [6].

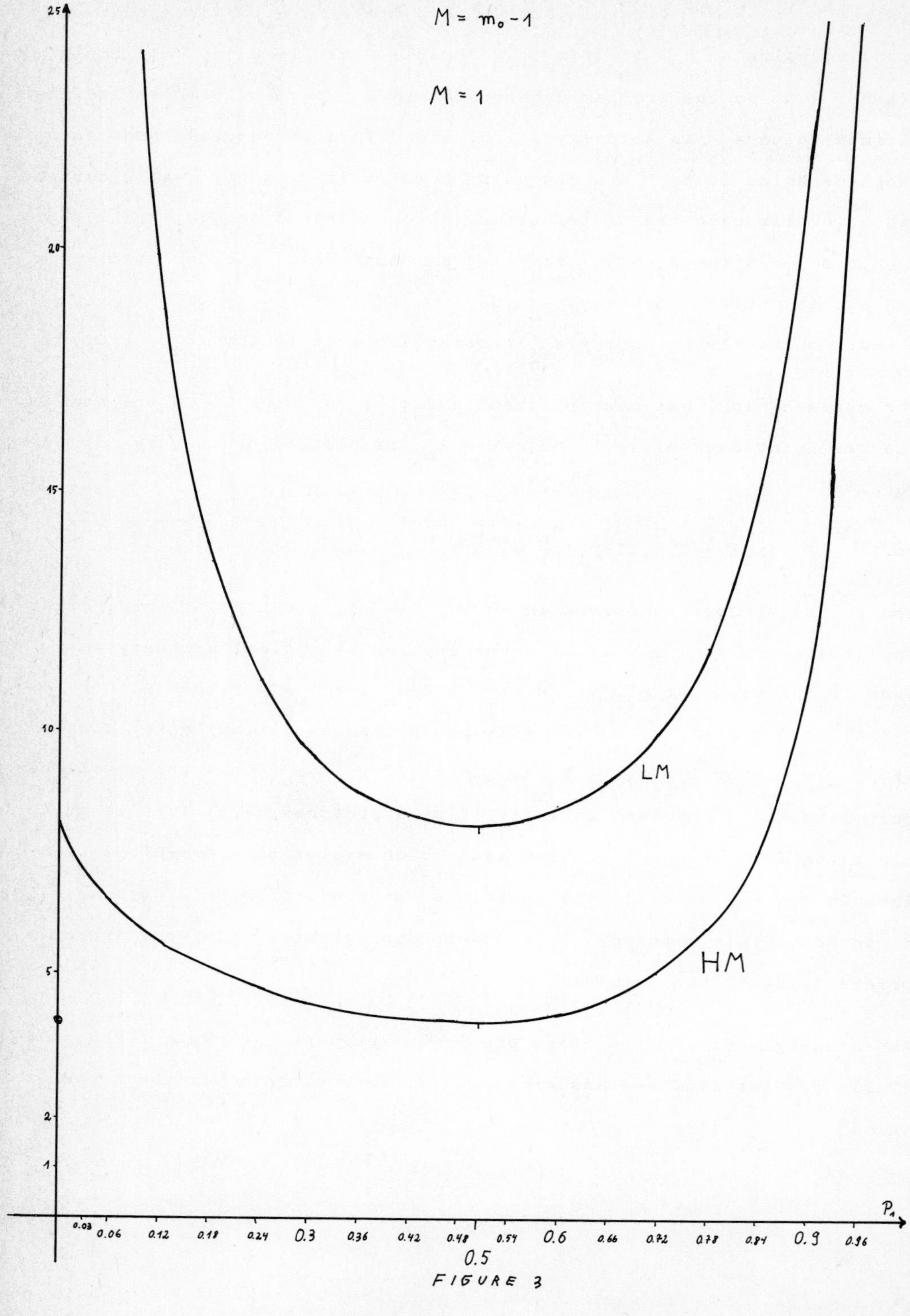
M = m₀ - 1
M = 1
LM
HM
P₁
25
20
15
10
5
2
1
0.03
0.06
0.12
0.18
0.24
0.3
0.36
0.42
0.48
0.54
0.6
0.66
0.72
0.78
0.84
0.9
0.96
0.5

FIGURE 3

Table 2: The two-class problem, $P = P(X=M_1)$, $1-P = P(X=M_2)$, $(M_2-M_1) \mid E(w_{m_0}) = M^*>0$, $E(w_1) \leq E(w_2) \leq \ldots$

M^* \ m_o	P ∈ 1	P ∈ 2	P ∈ 3	P ∈ 4	P ∈ 5	P ∈ 6	P ∈ 7
4,0	P ≠ 1/2	P = 1/2	–	–	–	–	
4,5	(0,667,1,0) (0,0,333	(0,333,0,667)	–	–	–	–	–
5,0	(0,724,1) (0,0,276)	(0,276,0,724)	–	–	–	–	–
6,0	(0,789,1) (0,0,211)	(0,211,0,789)	–	–	–	–	–
7,0	(0,827,1,0) (0,0,173)	(0,409,0,827) (0,173,0,263)	(0,263,0,409)	–	–	–	–
8,0	(0,854,1,0) (0,0,146)	(0,500,0,854) (0,146,0,191)	(0,191,0,500)	–	–	–	–
9,0	(0,873,1,0) (0,0,127)	(0,551,0,873) (0,127,0,156)	(0,156,0,551)	–	–	–	–
9,5	(0,880,1,0) (0,0,120)	(0,570,0,880) (0,120,0,143)	(0,264,0,570) (0,143,0,237)	(0,237,0,264)	–	–	–
10,5	(0,893,1,0) (0,0,107)	(0,602,0,893) (0,107,0,124)	(0,355,0,602) (0,124,0,162)	(0,162,0,355)	–	–	–
11,5	(0,904,1,0) (0,0,096)	(0,628,0,904) (0,096,0,110)	(0,397,0,628) (0,110,0,134)	(0,134,0,397)	–	–	–
12,0	(0,908,1,0) (0,0,092)	(0,639,0,908) (0,092,0,104)	(0,414,0,639) (0,104,0,124)	(0,124,0,414)	–	–	–
12,5	(0,912,1,0) (0,0,088)	(0,649,0,912) (0,088,0,098)	(0,428,0,649) (0,098,0,116)	(0,241,0,428) (0,116,0,163)	(0,163,0,241)	–	–
14,0	(0,923,1,0) (0,0,077)	(0,675,0,923) (0,077,0,085)	(0,464,0,675) (0,085,0,097)	(0,303,0,464) (0,097,0,118)	(0,118,0,303)	–	–
15,0	(0,928,1,0) (0,0,072)	(0,689,0,928) (0,072,0,079)	(0,483,0,689) (0,079,0,088)	(0,329,0,483) (0,088,0,103)	(0,182,0,329) (0,103,0,152)	(0,152,0,182)	–
17,0	(0,927,1,0) (0,0,063)	(0,713,0,937) (0,063,0,068)	(0,515,0,713) (0,068,0,074)	(0,367,0,515) (0,074,0,083)	(0,253,0,367) (0,083,0,099)	(0,099,0,253)	–
18,0	(0,941,1,0) (0,0,059)	(0,723,0,941) (0,059,0,063)	(0,528,0,723) (0,063,0,069)	(0,383,0,528) (0,069,0,076)	(0,272,0,383) (0,076,0,088)	(0,170,0,272) (0,088,0,118)	(0,118,0,170)
20,0	(0,947,1,0) (0,0,053)	(0,740,0,947) (0,053,0,056)	(0,550,0,740) (0,056,0,060)	(0,409,0,550) (0,060,0,066)	(0,302,0,409) (0,066,0,073)	(0,217,0,302) (0,073,0,085)	(0,085,0,217)

Table 2: Probability-intervals for given optimal strategies: The table gives probability-intervals for P, the probability of the cheaper class if m_0 and M^*, ratio between the range cost-estimates and the expected looking-costs of the m_0th cheapest approach are given, in order that $m = m_0$ is the optimal number of approaches to be pursued.

Table 3: Three-class equidistant PPM, p_1 = .30

M^* \ m_o	$p_2 \in$ 1	$p_2 \in$ 2	$p_2 \in$ 3	$p_2 \in$ 4	$p_2 \in$ 5	$p_2 \in$ 6	$p_2 \in$ 7
2,0	(0,.70)	–	–	–	–	–	–
2,1	(0,.70)	–	–	–	–	–	–
2,2	(.27,.70) (0,.13)	(.13,.27)	–	–	–	–	–
2,3	(.36,.70) (0,.041)	(.041,.36)	–	–	–	–	–
2,6	(.47,.70)	(0,.47)	–	–	–	–	–
3,0	(.56,.70)	(0,.56)	–	–	–	–	–
3,5	(.62,.70)	(.14,.62)	(0,.14)	–	–	–	–
4,0	(.66,.70)	(.28,.66)	(0,.28)	–	–	–	–
4,5	(.69,.70)	(.36,.69)	(0,.36)	–	–	–	–
5,0	–	(.43,.70)	(.044,.43)	(0,.044)	–	–	–
6,0	–	(.55,.70)	(.20,.55)	(0,.20)	–	–	–
6,5	–	(.61,.70)	(.25,.61)	(0,.25)	–	–	–
7,0	–	–	(.29,.70)	(.007,.29)	(0,.007)	–	–
8,0	–	–	(.38,.70)	(.095,.38)	(0,.095)	–	–
9,0	–	–	(.48,.70)	(.16,.48)	(0,.16)	–	–
10,0	–	–	–	(.22,.70)	(.004,.22)	(0,.004)	–
10,5	–	–	–	(.25,.70)	(.029,.25)	(0,.029)	–
12,0	–	–	–	(.33,.70)	(.09,.33)	(0,.09)	–
13,0	–	–	–	(.41,.70)	(.13,.41)	(0,.13)	–
14,0	–	–	–	–	(.16,.70)	(0,.16)	–
15,0	–	–	–	–	(.20,.70)	(.021,.20)	(0,.021)

Table 3: Probability intervals for p_2 (middle-class-probability) for m_0 being the optimal strategy in a three-class equidistant PPM if $M^* = (M_2 - M_1) \mid 2E(w_{m_0})$ (= ratio between the variation of cost-estimates and twice the expected looking costs of the m_oth lowest approach) and p_1 = .30 are given.

Acknowledgements

I am indebted to Dr. Herbert Paschen, also from Studiengruppe für Systemforschung, who has encouraged the research presented and whose useful suggestions in several discussions have improved the presentation considerably. I am also grateful to Mr. Horst Schneider, who has done the programming-work on a IBM/360-45-computer and Mr. Peter Schmitt, who has done some calculations on a desk-calculator and has moreover drawn the curves of section 4. Last, but not least, I am indebted to Mr. Losem (Abteilung für Statistik des Instituts für Wirtschafts und Gesellschaftswissenschaften, Bonn) and to Mrs. M. Schmidt (Institut für Angew. Math., Frankfurt/M) for the careful typing of this manuscript.

References

[1] T.A. Marschak and J.A. Yahav: "The sequential selection of approaches to a task", Management Science, Vol. 11, No. 9, May 1966, p. 627 - 647.

[2] T.A. Marschak: "Models, Rules of Thumb, and Development Decisions", Operations Research in Research and Development Decisions, B.V. Dean, editor, John Wiley, New York, 1962, p. 247 - 263.

[3] Th. Marschak: "Towards a normative theory of Development" in Th. Marschak et. al. (Ed.) "Strategy for R & D" Econometrics and Operations Research, Vol. VIII, Springer Verlag, Berlin-Heidelberg-New York (1967).

[4] R.R. Nelson: "The Economics of Parallel R and D efforts: A Sequential-Decision Analysis", Rand Research Memorandum Rm-2482, The Rand Corp., Santa Monica, Calif., Nov. 12, 1959.

[5] H. Drygas und H. Paschen: "Parallelplanung bei Forschungs- und Entwicklungsprojekten", in H. Krauch, H. Paschen (Eds.) "Methoden und Probleme der Forschungs- und Entwicklungsplanung, unter besonderer Berücksichtigung der Prioritätsbestimmung", Oldenbourg-Verlag, München 1972.

[6] H. Drygas: "Parallel Path Strategy - Theory and Numerical Illustrations", Diskussionspapier der Studiengruppe für Systemforschung, Januar 1973.

On minimization under linear equality constraints

Jürgen Fischer, Stuttgart

Abstract

The problem of minimizing a general function subject to linear equalit. constraints is transformed into an equivalent unconstrained problem. In order to solve the first one may apply any unconstrained minimization method to the second problem. By translating this into the language of the first problem one obtains equivalent methods, together with the equivalent convergence properties. An application is given to Goldfarb's extension of the variable metric method to linearly constrained problems.

The problem of minimizing a function subject to linear inequality constraints is often solved by a so-called "active set strategy". During several iterations some of the inequality constraints are regarded as equalities (the "active" constraints); therefore there are subproblems with linear equality constraints to solve, which will be treated in the following sections.

1. The equivalence. We consider the problem

minimize $f(x)$ subject to $Ax=b$

with a function $f:R^n \to R$, a mxn matrix A with rank m ($m<n$) and $b \in R^m$. If we define

$$N:=\{x \in R^n \mid Ax=0\}, \qquad M:=\{x \in R^n \mid Ax=b\}$$

we can write this problem as

(P1) $\qquad \min\{f(x) \mid x \in M\}$.

The linear space N has dimension $n-m$. Let $\{t_1,\ldots,t_{n-m}\}$ be any basis of N and define the $nx(n-m)$ matrix T by $T:=(t_1,\ldots,t_{n-m})$. If x_0 is any point in M, the linear manifold M may be represented as $M = x_0+N$. Since the columns of T form a basis of N, each $y \in N$ may be written as $y = Tz$ with $z \in R^{n-m}$. Thus we have a bijective mapping between R^{n-m} and M, given by $z \to x_0+Tz$. If we define $F(z):=f(x_0+Tz)$, then (P1) is equivalent to the unconstrained problem

(P2) $\qquad \min\{F(z) \mid z \in R^{n-m}\}$.

This equivalence implies: If z^* solves (P2), then $x^* = x_0+Tz^*$ solves (P1) and vice versa; if z_k converges to z^*, then $x_k = x_0+Tz_k$ converges to x^*. Therefore, in order to solve (P1), we solve the unconstrained problem (P2) by any suitable method $\mathfrak{M}$; using the equivalence between

(P1) and (P2) we obtain a method $\mathfrak{M}'$ solving (P1), which has the same convergence properties as $\mathfrak{M}$.

2. Examples. Obviously, F(z) has the same differentiability properties as f(x). If f is twice continuously differentiable with gradient g(x) and Hessian matrix G(x), then we obtain for the first and second derivatives of F (with respect to z)

$\nabla F(z) = T'g(x_0+Tz)$, $\qquad \nabla^2 F(z) = T'G(x_0+Tz)T$

(the gradients are written as column vectors; T' is the transpose of T).

Now we use several methods to solve the unconstrained problem (P2):

(a) *the method of steepest descent:*
the search direction in (P2) is $\bar{s} = -\nabla F(z)$;
the corresponding direction in (P1) is given by $s = T\bar{s} = -TT'g(x)$,
where x corresponds to z by $x = x_0 + Tz$.
TT' is a nxn matrix; for each $y \in R^n$ we have $TT'y \in N$, but in general $TT'|_N \neq I|_N$.

(b) *Newton's method:* the search direction in (P1) is

$$s = -T(T'G(x)T)^{-1}T'g(x)$$

(c) *conjugate gradient methods:* the Fletcher-Reeves algorithm [4], for instance, computes in (P2) the direction

$$\bar{s}_{k+1} = -\nabla F(z_{k+1}) + \frac{\|\nabla F(z_{k+1})\|^2}{\|\nabla F(z_k)\|^2}\,\bar{s}_k;$$

in (P1) this yields

$$s_{k+1} = -TT'g(x_{k+1}) + \frac{g(x_{k+1})'TT'g(x_{k+1})}{g(x_k)'TT'g(x_k)}\,s_k \quad .$$

The reset version is obtained in a similar way.

(d) *variable metric methods:* see the following section.

In all of these methods one-dimensional searches along the search direction are needed. They may be done in the language of (P1) too, since the function values along the line $x_k+\lambda s_k$ correspond to those along $z_k+\lambda\bar{s}_k$.

3. Variable metric methods. In the unconstrained case the variable metric method of Davidon-Fletcher-Powell [2,3] generates a sequence of points (x_k) and a sequence (H_k) of positive definite symmetric matrices such that $s_k = -H_k g(x_k)$, x_{k+1} is chosen to minimize f in the direction s_k, and

(1) $$H_{k+1} = H_k - \frac{H_k\gamma_k\gamma_k' H_k}{\gamma_k' H_k\gamma_k} + \frac{\delta_k\delta_k'}{\delta_k'\gamma_k}$$

with $\delta_k = x_{k+1}-x_k$, $\gamma_k = g(x_{k+1}) - g(x_k)$.

If we apply this method to (P2) and transform the resulting sequences (z_k), $(\overline{H}_k)$ into (P1), we obtain sequences (x_k), (H_k), where the nxn matrices H_k are given by

(2) $$H_k = T\overline{H}_k T' .$$

The $\overline{H}_k$ are symmetric positive definite (n-m)x(n-m) matrices. $\overline{H}_1$ is the initial matrix for (P2); with $\overline{H}_1=I$ we obtain $H_1=TT'$.
It is easy to show that, if we start the iteration with $H_1=T\overline{H}_1T'$, we can compute H_{k+1} from H_k directly by formula (1), without explicitly using T at each step. The search direction is given again by $s_k = -H_k g(x_k)$. The matrices H_k are no longer positive definite, but positive semidefinite. From (2) it follows that $H_k y \in N$ for all $y \in R^n$; therefore $s_k \in N$, and the sequence (x_k) remains "automatically" in M, if we start with $x_1 \in M$.

An analogous result can be obtained for other rank-one or rank-two updating formulas, for instance for the members of Broyden's class of rank-two algorithms [1].

4. The computation of T. T may be obtained by *eliminating* variables from the equations Ax=b. Because of rank A=m, after rearranging the columns of A a decomposition A=(B,C) is possible in such a way that B contains m columns and B^{-1} exists.

If we write $x \in R^n$ as $x = \binom{y}{z}$ with $y \in R^m$, $z \in R^{n-m}$, then for $x \in M$

$$Ax = By+Cz = b,$$

hence $$y = -B^{-1}Cz + B^{-1}b$$

and $$x = \binom{y}{z} = \binom{-B^{-1}C}{I_{n-m}}z + \binom{B^{-1}b}{0};$$

thus we may choose $T = \binom{-B^{-1}C}{I_{n-m}}$.

The actual computation may be done in a tableau similar to that used in the simplex method; to get $B^{-1}C$ from A m pivotal operations are required. If in a problem with linear inequality constraints the set of active constraints is changed, that is, if one constraint is added to or removed from this set, the corresponding new $\hat{T}$ is obtained from the old T by one pivotal operation. From the tableau one may obtain an estimate for the Lagrange multipliers, too, which allow to decide whether to remove an active constraint.

Another possibility is to choose an *orthonormal* basis $t_1,...,t_{n-m}$ of N. Then TT' is the orthogonal projection P onto N. In this special case the method of steepest descent (section 2(a)) turns out to be Rosen's gradient projection method [7]. The Fletcher-Reeves conjugate gradient method (section 2(c)) has now the form

$$s_{k+1} = -Pg(x_{k+1}) + \frac{||Pg(x_{k+1})||^2}{||Pg(x_k)||^2} s_k .$$

If in a problem with linear inequality constraints the set of active constraints is changed, the new projection $\hat{P}=\hat{T}\hat{T}'$ is obtained from P by the well known formulas that may be found in [7] for instance.

5. A remark on Goldfarbs's method [5]. While the active constraints remain unchanged, Goldfarb uses the variable metric method updating formula (1); he starts the iteration with a matrix H_1 which is always "admissible" in the sense that it may be written in the form $H_1=T\overline{H}_1T'$ with a positive definite symmetric (n-m)x(n-m)matrix $\overline{H}_1$ and a basis T of N. If the active constraints remain unchanged, we obtain therefore at once - under assumptions corresponding to those used by Powell [6] - the superlinear convergence of the sequence (x_k). Goldfarb only proofs a "quadratic termination property", that follows here directly from the equivalence to the unconstrained problem.

References

[1] C.G.BROYDEN, "The convergence of a class of double - rank minimization algorithms. 1. General considerations", J.Inst.MathsApplics 6(1970), 76-90.

[2] W.C.DAVIDON, "Variable metric method for minimization", A.E.C.Research and Development Report, ANL-5990,1959.

[3] R. FLETCHER and M.J.D.POWELL, "A rapidly convergent descent method for minimization", Comput.J. 6(1963), 163-168.

[4] R. FLETCHER and C.M.REEVES, "Function minimization by conjugate gradients", Comput.J. 7(1964), 149-154.

[5] D. GOLDFARB, "Extension of Davidon's variable metric method to maximization under linear inequality and equality constraints", SIAM J.Appl.Math. 17(1969), 739-764.

[6] M.J.D.POWELL, "On the convergence of the variable metric algorithm", J.Inst.Maths.Applics 7(1971), 21-36.

[7] J.B. ROSEN, "The gradient projection method for nonlinear programming, Part I. Linear Constraints", SIAM J.Appl.Math. 8(196o), 181-217.

Jürgen Fischer
Mathematisches Institut A
Universität Stuttgart
7 Stuttgart 80
Pfaffenwaldring 57

ON CONSTRAINED SHORTEST-ROUTE PROBLEMS *)

Wolfgang Gaul

Summary: Among the many constrained shortest-route problems are only few for which time-dependent restricitons are taken into consideration. Instead of determining a shortest route through a network with travel times depending on the departure time and with additional time-dependent constraints on movement and parking a dual problem and an algorithm for solving this dual program is considered giving sufficient information for the route problem. A comparison with known time-dependent shortest-route formulations is made.

Introduction

Shortest-route problems (with constraints) and algorithms for determining such routes have successfully been used for the formulation and solution of lots of network problems (see for example [6],[8],[9],[14],[16],[19]). Naturally, attempts have been made to take into consideration the known results and methods of determining shortest routes without additional restrictions (For unconstrained shortest-route (u.s.r.) problems see for example [1],[5],[6] and the references of [19].). Introductory, some of the most important (but not always very efficient) attempts are mentioned.

Let G(N,A,l) be a finite, directed graph where N denotes the nodes, $A \subset N \times N \setminus \cup\{(i,i) \mid i \in N\}$ the (directed) arcs and $l \in R_+^{*A}$ the length of the arcs. If G is connected, it is also called a network. A route $R(p,q) = \{p=n_o,n_1,\dots,n_m = q\}$ from p to q, $p,q \in N$, is described by the sequence of nodes which are visited on the way from p to q in the given order. $A(R(p,q)) = \{(n_o,n_1),\dots,(n_{m-1},n_m)\}$ is the arc set belonging to R(p,q). $l(R(p,q)) = \sum_{(i,j) \in A(R(p,q))} l_{ij}$, $i,j \in N$, is the length of R(p,q).

R(p,p) is called a cycle and positive, negative or zero if its length l(R(p,p)) is positive, negative or equal to zero (Throughout this paper all values assigned to the elements of the graph G are assumed nonnegative to avoid negative cycles.). A route R(p,q) is called simple or elementary if A(R(p,q)) or R(p,q) consists of distinct elements only. An intuitive idea is to sort all routes from p to q with respect to their lengths $(l(R_1(p,q)) \leq l(R_2(p,q)) \leq \dots \leq l(R_k(p,q)) \leq \dots)$ and to search for the minimum value of k for which $R_k(p,q)$ will satisfy the additional

*) This research has been supported by Sonderforschungsbereich 72, University Bonn.

constraints. If $n(i,j,k)$, $i,j\epsilon N$, defines the number of routes, among the k shortest routes from i to q, that begin by going from i to j, the algorithm has to compute

$$l(R_k(i,q)) = \min_{j \neq i} l(i,j,k) \quad , k \geq 2$$

where

$$l(i,j,k) := l_{ij} + l(R_{n(i,j,k-1)+1}(j,q)) \quad , k \geq 2$$

describes the length of the deviation $(i,j) \cup A(R_{n(i,j,k-1)+1}(j,q))$ (which corresponds to $R_h(i,q)$ for some $h \geq k$) from the k-1 shortest routes from i to q (The existence of zero-cycles must be avoided by ϵ-pertubating the l-data.). In most applications one is interested in elementary solutions (without cycles) (see for example [18] where u.s.r.-problems have to be determined in subgraphs of the given graph G). Naturally, the efficiency of all procedures for computing the k shortest (elementary) routes depends on the network structure and the l-data, but can be recommended if k is bounded by a small value such as in planning models for urban traffic.

There are many other possibilities of applying branch and bound procedures and the dynamic programming technique.

A class of problems which can be solved in this way is that of determining a shortest (elementary) route from p to q in a graph $G(N,A,l)$ where all nodes of a set $N_o \subset N$ must be visited exactly once, abbreviated (G,p,N_o,q) (see [15]). Here, too, the possibility of determining such problems is constrained by the magnitude of $\#N$, as it is apparent from the well-known traveling-salesman problem $(G,p,N \setminus \{p\},p)$ (see [2],[13], [20]). A dynamic programming solution for the other special case $(G,p,\emptyset,q)$ is due to Bellman [1] himself.

Determining the longest (elementary) route in a graph is another well-known constrained shortest-route problem which has relations to the traveling-salesman problem (see [12]). Here, the existence of positive cycles prohibits the application of u.s.r.-procedures. Additional restrictions to exclude routes with (positive) cycles from the set of solutions (such as the subtour-elimination constraints for the traveling-salesman problem) have to be taken into consideration. Only for project planning models when the underlying graph $G(N,A,l)$ has no cycles a critical (longest) route can easily be found determining a u.s.r.-problem in $G(N,A,-l)$.

A class of problems where a reduction to u.s.r.-problems was successful is that of determining shortest routes with arc-changing costs (Such problems arise for example when reloading between different transport facilities or, in urban traffic, turning to the left or prohibitions of turning must be taken into consideration.) (see [3],[16]). While in all former mentioned problems only arcs $a \varepsilon A$ have been assigned lengths l_a now additional values

$$p_{a_1 a_2} = \begin{cases} p_{ijk} & (a_1,a_2)\varepsilon A_o \\ \infty & \text{otherwise} \end{cases}$$

with $A_o = \{(a_1,a_2) \mid a_1 = (i,j),\ a_2 = (j,k)\} \subset A\times A$, have to be considered. $p_{ijk} = \infty$ indicates a prohibited arc change from (i,j) to (j,k). Such problems can be solved interpreting the arcs of G(N,A,l) as nodes and the A_o-arc pairs as arcs of a new graph $\tilde{G}(\tilde{N},\tilde{A},\tilde{l})$ with $\tilde{l}_{a_1 a_2} = l_{a_1} + p_{a_1 a_2}$ and applying an u.s.r.-procedure to $\tilde{G}$ where $\tilde{G}$, advantageously, needs not to be constructed explicitly (The elementary solutions of $\tilde{G}$ correspond to the simple ones of G, and it can be shown that the subset of simple routes contains all optimal solutions.).

As the most network problems are formulated from a static viewpoint (as all ones mentioned up to now) attention is now transferred to problems with time-dependent constraints. To the first authors who have formulated network problems from a dynamic viewpoint belong Ford/Fulkerson with their work on maximal dynamic flows (see [7],[8]). They, too, pointed out that, with much more effort, consideration could be restricted to the static case introducing a so-called "time-expanded" network. This is also true for the following discussion on shortest-route problems with time-dependent constraints. There are some authors who have dealt with these topics. Cooke/Halsey [4] use dynamic programming, Dreyfus [6] suggests the Dijkstra [5]-method, Klafszky [17] applies duality principles and gives a procedure, which, too, is closely related to paper [5], and recently, Halpern/Priess [11] have considered additional constraints of parking in the nodes, Gaul [10] has treated the problem of computing optimum routes within prescribed time-periods in case of time-dependent capacities, costs and arc-lengths.

In this paper an algorithm is presented which handles a slightly more general situation than that solved by [11], if parking constraints are excluded from consideration corresponding simplifications are pointed out and a necessary modification of [17] is given. Also, little additional expenditure is included to lighten the backtracking process of an optimal route.

Formulation of the Problem

Let the nonnegative integers $l_{ij}(t)$, $t \varepsilon \mathcal{T} = \{0,1, \ldots ,T\}$, $i,j \varepsilon N$, denote the traversal times from i to j which depend on the departure time t from node i. [1] Now, the corresponding length for R(p,q) can be described by $l(R(p,q)) = \sum_{(n_i,n_{i+1})\in A(R(p,q))} l_{n_i n_{i+1}}(l(R(p,n_i)) + t_i)$ where $t_i \in \mathcal{T}$ is the waiting time in n_i.

Searching for a shortest route within T time-periods means to consider the triples (R(p,q), a(R(p,q)), d(R(p,q))) where

$a(R(p,q)) = (a(n_o), \ldots, a(n_m))$ describes the arrival times,

$d(R(p,q)) = (d(n_o), \ldots, d(n_{m-1}))$ with $d(n_i) \in \mathcal{T}$ the departure times with

$$\begin{aligned} d(n_i) &\geq a(n_i), \quad n_i \in R(p,q) \sim \{n_m\} \\ a(n_o) &= 0 \\ l(R(p,q)) = a(n_m) &\leq T \\ a(n_i) &= d(n_{i-1}) + l_{n_{i-1}n_i}(d(n_{i-1})), \quad i = 1, \ldots, m \end{aligned} \tag{1}$$

The parking (waiting) time in node n_i is $d(n_i) - a(n_i)$. If no parking is allowed in node i within the time intervalls $[t^i_{1k}, t^i_{2k}\rangle$, k=1,...,r(i) where " $\rangle$ " symbolizes the appropriate upper bound, one has the additional constraints

$$\begin{aligned} a(i) \in [t^i_{1k}, t^i_{2k}\rangle &\Rightarrow d(i) = a(i) \\ t^i_{2k} < a(i) < t^i_{1(k+1)} &\Rightarrow d(i) \leq t^i_{1(k+1)} \end{aligned} \tag{2}$$

If no connection from node i to node j is possible at time t let be $l_{ij}(t) = M > T$.

If $\mathfrak{A}_T$ denotes the set of feasible solutions (R(p,q), a(R(p,q)), d(R(p,q))) with respect to the constraints (1), (2), the problem is to find $(R_o(p,q), a^o(R_o(p,q)), d^o(R_o(p,q)))$ (if $\mathfrak{A}_T \neq \emptyset$) so that $a^o(n_m)$ describes the minimum arrival time at node $n_m = q$.

The Solution Procedure

Instead of dealing with $\mathfrak{A}_T$ consider the problem

$$\max\ b(q)$$

under the constraints

$$\begin{aligned} b(p) &= 0 \\ b(j) &\leq \min_{(i,j)\in A} \min_{t\in T_i} \{l_{ij}(t) + t\} \end{aligned} \tag{3}$$

---

[1] For computational convenience a restriction is made to integer values (and a discrete measure of time). Also, observation time is constrained, and there are good reasons that exceeding a given time T is of no interest.

where $T_i = \{t \varepsilon \mathcal{T} \mid \exists\ R(p,i)$ with $l(R(p,i)) = t$ or $l(R(p,i)) = \tau < t$ and $[\tau,t) \cap \bigcup [t^i_{1k}, t^i_{2k}\rangle = \emptyset\}$.

If $\mathcal{L}$ denotes the set of feasible solutions of (3), one has $\mathcal{L} \neq \emptyset$, for $0 \varepsilon \mathcal{L}$ because of $l_{ij}(t) \geq 0$.

<u>Lemma 1:</u> $\min a(q) \geq \max b(q)$

<u>Proof:</u> Either $\mathcal{O}_T = \emptyset$ yields $\min a(q) = +\infty$, or for $(R(p,q), a(R(p,q)), d(R(p,q))) \varepsilon \mathcal{O}_T$, $b \varepsilon \mathcal{L}$

$$a(q) = a(n_m) = d(n_{m-1}) + l_{n_{m-1}q}(d(n_{m-1})) \geq \min_{t \varepsilon T_{n_{m-1}}} \{ l_{n_{m-1}q}(t) + t\}$$

$$\geq \min_{(i,q)\varepsilon A} \min_{t \varepsilon T_i} \{l_{iq}(t) + t\} \geq b(q).$$

The following procedure for determining an optimum solution $b^o \varepsilon \mathcal{L}$ will indicate $\mathcal{O}_T = \emptyset$ (if $b^o(q) > T$) or construct an optimum route $R_o(p,q)$. If $N_o \subset N$ is the set of nodes with known earliest arrival times, $\beta(N_o) \varepsilon \{0,1\}^{*N}$ with

$$[\beta(N_o)]_i = \begin{cases} 0 & i \varepsilon N_o \\ 1 & \text{otherwise} \end{cases}$$

and α_i or γ_i denotes the set of feasible arrival or departure times for node i one gets

<u>Algorithm:</u>

<u>Step 1:</u> $N_o = \{p\}$, $(b(i) = 0,\ \alpha_i = \emptyset,\ \gamma_i = \emptyset,\ \forall\, i \varepsilon N)$

$\delta_i = \emptyset,\ i \varepsilon N \setminus \{p\}$, $\delta_p = \{[\,0, t^p_{11}\,]\} \cap \mathcal{T}$

$i(\tau) = \emptyset,\ i \varepsilon N,\ \tau \varepsilon \mathcal{T}$ $\qquad s = p$

<u>Step 2:</u>

2.1 : $\alpha^s_i = \{\tau \mid \tau = l_{si}(t) + t,\ t \varepsilon \delta_s\}$

2.2 : $i(\tau) = i(\tau) \cup \{s\},\ \tau \varepsilon \alpha^s_i$ $\qquad$ (2.1–2.3: $i \varepsilon N$)

2.3.: $\alpha_i = \alpha_i \cup \alpha^s_i \setminus \gamma_i$

2.4 : $\varepsilon = \min_{i \varepsilon N} \min\{\tau \varepsilon \alpha_i\} - \max_{j \varepsilon N} b(j) \longrightarrow$ yields $i_o \varepsilon N$

2.5 : $b(i) = b(i) + [\beta(N_o)]_i \varepsilon,\ \ i \varepsilon N$

2.6 : $\max_{j \varepsilon N} b(j) > T\ ? \longrightarrow$ stop

2.7 : $N_o = N_o \cup \{i_o\}$

2.8 : $i_o = q? \longrightarrow$ stop

2.9 : $\delta_{i_o} = \{d\varepsilon \mathcal{T} | d\varepsilon \bigcup[\tau,\theta],\ \tau\varepsilon\alpha_{i_o},\ \bigcup[\tau,\theta) \cap \bigcup[t_{1k}^{i_o}, t_{2k}^{i_o}\rangle = \emptyset\} - \gamma_{i_o}$

2.10: $i_o(t) = \bigcup_{\varkappa \leq t} i_o(\varkappa),\ \varkappa, t\varepsilon\ [\tau,\theta] \cap \delta_{i_o}$

2.11: $\gamma_{i_o} = \gamma_{i_o} \cup \delta_{i_o}$

2.12: $\alpha_{i_o} = \emptyset$

2.13: $s = i_o$

2.14: repeat step 2

In order to see how the algorithm works one has to check the single operations. Step 1 gives the starting conditions. In step 2 in 2.1, 2.3 feasible arrival times are computed where consideration is restricted to such times which have not served for determining departure times in 2.9, 2.11.

If n is the iteration index (for repeating step 2 of the algorithm) it is shown

<u>Lemma 2:</u> Computing 2.4 yields $\varepsilon \geq 0$.

<u>Proof:</u> Because of $b^1 \equiv 0$ and $l_{ij}(t) \geq 0$ one has

$$\varepsilon^1 = \min_{i\varepsilon N} \min \{\tau\varepsilon\alpha_i^1\} = \min \{\tau = l_{si_o(1)}(t) + t \mid t\varepsilon\delta_s^1\} \geq 0$$

From $\varepsilon^{n-1} = \min \{\tau\varepsilon\alpha_{i_o(n-1)}^{n-1}\} - \max_{j\varepsilon N} b^{n-1}(j)$ it follows

$\min \{\tau\varepsilon\alpha_i^{n-1}\} \geq \min \{\tau\varepsilon\alpha_{i_o(n-1)}^{n-1}\}$ and from

$\min \{\tau\varepsilon\delta_s^n\} = \min \{\tau\varepsilon\alpha_{i_o(n-1)}^{n-1}\}$ and 2.1

$\min \{\tau\varepsilon\alpha_i^n\} \geq \min \{\tau\varepsilon\alpha_{i_o(n-1)}^{n-1}\}$ (remember $\alpha_{i_o(n-1)}^n = \alpha_s^n = \emptyset$ because of 2.12). Thus

$$\begin{aligned} \varepsilon^n &= \min_{i\varepsilon N} \min \{\tau\varepsilon\alpha_i^n\} - \max_{j\varepsilon N} b^n(j) \\ &\geq \min\{\tau\varepsilon\alpha_{i_o(n-1)}^{n-1}\} - \max_{j\varepsilon N} b^{n-1}(j) - \varepsilon^{n-1} = 0 \end{aligned}$$

as $\max_{j\varepsilon N} b^n(j)$ is always taken on $\bigcap_{\rho=1}^{n-1} \overline{N_o^\rho}$ because of

$b^n = \sum_{\rho=1}^{n-1} \beta(N_o^\rho)\varepsilon^\rho$ from 2.5.

<u>Lemma 3:</u> If $b\varepsilon \mathcal{L}$ then $b + \varepsilon \cdot \beta(N_o)\ \varepsilon \mathcal{L}$.

<u>Proof:</u> For $\varepsilon^n = 0$, nothing is to be shown. For $\varepsilon^n > 0$, only $i\varepsilon \overline{N_o^n}$ are

of interest. One has

$$\min_{i\varepsilon \overline{N_o^n}} \min_{(\alpha,i)\varepsilon A} \min_{t\varepsilon T_\alpha} \{l_{\alpha i}(t) + t\} \geq \min_{i\varepsilon N} \min\{\tau\varepsilon\alpha_i^n\} \qquad (4)$$

for otherwise there exists $k_o \varepsilon \overline{N_o^n}$ (which takes the minimum of (4)) and $R(p,k_o)$ with $l(R(p,k_o)) = t_o < \min_{i\varepsilon N} \min \{\tau\varepsilon\alpha_i^n\}$. But then there exists a node j_o with $A(R(p,k_o)) = \{A(R(p,j_o)),(j_o,k_o)\}$. $j_o \varepsilon N_o^n$ forces $t_o \geq \min_{i\varepsilon N} \min \{\tau\varepsilon\alpha_i^n\}$, a contradiction, but $j_o \varepsilon \overline{N_o^n}$ forces $t_o = l(R(p,j_o))$ because of $l_{ij}(t) \geq 0$ and the minimality of k_o and further backtracking on $R(p,j_o)$ must give a node $\rho_o \varepsilon N_o^n$ (because of $p \varepsilon N_o^n$), a contradiction. Thus, (4) is valid and

$$\begin{aligned} \varepsilon^n &= \min_{i\varepsilon N} \min \{\tau\varepsilon\alpha_i^n\} - \max_{j\varepsilon N} b^n(j) \\ &\leq \min_{i\varepsilon \overline{N_o^n}} \min_{(\alpha,i)\varepsilon A} \min_{t\varepsilon T_\alpha} \{l_{\alpha i}(t) + t\} - \max_{j\varepsilon N} b^n(j) \\ &\leq \min_{(\alpha,i)\varepsilon A} \min_{t\varepsilon T_\alpha} \{l_{\alpha i}(t) + t\} - b^n(i) \quad , \forall\, i\varepsilon \overline{N_o^n} \end{aligned}$$

When performing the described algorithm one can get the following possibilities (for the n-th iteration) :

$$\varepsilon^n > 0, \; \# N_o^{n+1} = \# N_o^n + 1 \qquad (5)$$

$$\varepsilon^n > 0, \; \# N_o^{n+1} = \# N_o^n \qquad (6)$$

$$\varepsilon^n = 0, \; \# N_o^{n+1} = \# N_o^n + 1 \qquad (7)$$

$$\varepsilon^n = 0, \; \# N_o^{n+1} = \# N_o^n \qquad (8)$$

In view to the criteria 2.6 and 2.8 the best which can happen is (5), the worst (8).

<u>Lemma 4:</u> If (8) occurs at iteration n there is $k(n) \leq \# N_o^n - 1$ so that for iteration $n + k(n)$ (5), (6) or (7) must hold.

<u>Proof:</u> If $\varepsilon^n = 0$ and $i_o(n) \varepsilon N_o^n$ (which means $N_o^{n+1} = N_o^n$) there exists an integer ν_n with $1 \leq \nu_n \leq \# N_o^n - 1$ which is maximal with $i_o(n-\mu) \varepsilon N_o^n$ and $\max_{j\varepsilon N} b^n(j) \varepsilon \gamma_{i_o(n-\mu)}$, $\mu = 1,\ldots,\nu_n$.

$$i_o(n-\mu_1) \neq i_o(n-\mu_2), \; 0 \leq \mu_1 < \mu_2 \leq \nu_n \qquad (9)$$

because of 2.3 and there must exist an integer $k_n \geq 1$ that if $\varepsilon^{n+\varkappa} = 0$ and $i_o(n+\varkappa) \; \varepsilon \; N_o^{n+\varkappa} = N_o^n$, $\varkappa = 0,1, \ldots ,k_n-1$

$$N_o^{n+k_n-1} \setminus \bigcup_{\rho=-\nu_n}^{k_n-1} \{i_o(n+\rho)\} = \emptyset$$

(as $i_o(n+\rho_1) \neq i_o(n+\rho_2)$, $-\nu_n \leq \rho_1 < \rho_2 \leq k_n-1$, which follows from (9) by induction on n) and

$$\min_{i\epsilon N_o^{n+k_n-1}} \min\{\tau\epsilon\alpha_i^{n+k_n}\} > \max_{j\epsilon N} b^{n+k_n}(j) = \max_{j\epsilon N} b^n(j)$$

must hold. Thus, $k(n) \leq k_n \leq \# N_o^{n+k_n-1} - 1 = \# N_o^n - 1$.

Because of (6), a restriction to integer values [2] is necessary to ensure the finitness of the algorithm (of course, taking into account additional difficulties (when determining inf/sup in 2.4) and restrictions on the family of functions $l_{ij}(t)$ (to ensure convergency of the algorithm) a continuous version of the problem is possible (see [11])). In [11] the lengths of the arcs are of the special form

$$l_{ij}(t) = \begin{cases} \infty & t\epsilon V \\ l_{ij} & \text{otherwise} \end{cases}$$

where V describes times of forbidden movement in arc (i,j) which does not take into consideration (as it is done here) that a later departure from i may yield an earlier arrival at j via arc (i,j).
If no parking constraints (in the nodes) are considered a simplification, which yields elementary optimum solutions by waiting in the nodes as long as the most convenient departure time is at hand, is possible. This simplified situation is considered in [17], but to maintain feasibility (for the dual problem), the correct formula for ε would be (in notations of [17])

$$\varepsilon = \min_{x\epsilon S, y\epsilon T} \min_{\theta} \{\gamma(x,y,\mu(x) + \theta) + \theta + \mu(x) - \mu(y)\}$$

In 2.2 and 2.10 the nodes are assigned predecessor nodes to lighten the backtracking process of an optimum route. An example how the algorithm works is available in [10], a computer program is under preparation.

---

[2] This is,for computational convenience, assumed in the formulation of the problem.

References:

[1] Bellman, R.: On a Routing Problem. Quart. Appl. Math. 16 (1958) 87-90.

[2] Bellmore, M. and G.L. Nemhauser: The Traveling Salesman Problem: A Survey. J. ORSA 16 (1968) 538-558.

[3] Caldwell, T.: On Finding Minimum Routes in a Network with Turn Penalties. Comm. ACM 4 (1961) 107-108.

[4] Cooke, K.L. and E. Halsey: The Shortest Route through a Network with Time-Dependent Internodal Transit Times. J. Math. Anal. and Appl. 14 (1966) 493-498.

[5] Dijkstra, E.W.: A Note on Two Problems in Connexion with Graphs. Num. Math. 1 (1959) 269-271.

[6] Dreyfus, S.E.: An Appraisal of Some Shortest-Path Algorithms. J. ORSA 17 (1969) 395-412.

[7] Ford, L.R. and D.R. Fulkerson: Constructing Maximal Dynamic Flows from Static Flows. J. ORSA 6 (1958) 419-433.

[8] Ford, L.R. and D.R. Fulkerson: Flows in Networks. Princeton University Press, Princeton, N.Y. (1962).

[9] Gaul, W.: Über Flußprobleme in Netzwerken. to appear in ZAMM (1975).

[10] Gaul, W.: Optimale Wege bei zeitabhängigen Nebenbedingungen. working paper (1975).

[11] Halpern, J. and I. Priess: Shortest Path with Time Constraints on Movement and Parking. Networks 4 (1974) 241-253.

[12] Hardgrave, W.W. and G.L. Nemhauser: On the Relation between the Traveling Salesman and the Longest-Path Problems. J. ORSA 10 (1962) 647-657.

[13] Helbig Hansen, K. and J. Krarup: Improvement of the Held-Karp Algorithm for the Symmetric Traveling Salesman Problem. Math. Prog. 7 (1974) 87-96.

[14] Hu, T.C.: Integer Programming and Network Flows. Addison-Wesley, Reading, Mass. (1969).

[15] Ibaraki,T.: Algorithms for Obtaining Shortest Paths Visiting Specified Nodes. Siam Review 15 (1973) 309-317.

[16] Kirby, R.F. and R.B. Potts: The Minimum Route Problem for Networks with Turn Penalties and Prohibitions. Transp. Research 3 (1969) 397-408.

[17] Klafszky, E.: Determination of Shortest Path in a Network with Time-Dependent Edge-Lengths. Math. Operationsforsch. u. Statist. 3 (1972) 255-257.

[18] Lawler, E.L.: A Procedure for Computing the k Best Solutions to Discrete Optimization Problems and its Application to the Shortest Path Problem. Manag. Sc. 18 (1972) 401-405.

[19] Pierce, A.R.: Bibliography on Algorithms for Shortest Path, Shortest Spanning Tree, and Related Circuit Routing Problems (1956-1974). Networks 5 (1975) 129-149.

[20] Thompson, G.L.: Algorithmic and Computational Methods for Solving Symmetric and Asymmetric Travelling Salesman Problems. presented at'Workshop on Integer Programming', Bonn, Sept. (1975).

Wolfgang Gaul

Inst. Angew. Mathematik
u. Informatik
der Universität Bonn

53 BONN
Wegelerstr. 6

BANG-BANG SOLUTION OF A CONTROL PROBLEM FOR THE HEAT EQUATION

Klaus Glashoff
TH Darmstadt
Fachbereich Mathematik
D-61 Darmstadt
Schloßgartenstr. 7

1. Introduction. In this note we consider a boundary control problem for the heat equation in arbitrary dimensions. For simplicity we treat the L_2-minimum norm case only and discuss the results which we proved in [1],[4]. We show how controllability- and normality-properties of a certain linear operator enter into the proof of a bang-bang theorem.

2. The initial-boundary value problem. Let Ω be a bounded domain in $\mathbb{R}^N$ $(N \geqq 1)$ with C_∞-boundary $\partial\Omega$. We consider the parabolic initial-boundary value problem

$$\frac{\partial}{\partial t} y(x,t) - \Delta_x y(x,t) = 0, \qquad (x,t) \in G = \Omega \times (0,T), \tag{1}$$

$$y(x,0) = 0, \qquad x \in \Omega, \tag{2}$$

$$\alpha \frac{\partial}{\partial n_\xi} y(\xi,t) + y(\xi,t) = g(\xi) u(t), \quad (\xi,t) \in \Gamma = \partial\Omega \times (0,T). \tag{3}$$

Here Δ denotes the Laplacian in $\mathbb{R}^N$. $n = n(\xi)$ is the outer normal in $\xi \in \partial\Omega$. The subscripts in Δ_x resp. $\frac{\partial}{\partial n_\xi}$ indicate that these differential operators are acting with respect to the space variable only. $g \in L_\infty(\partial\Omega)$ and $\alpha > 0$ are fixed. We are going to discuss the question of existence and regularity of solutions of (1)-(3).

3. Solutions of (1)-(3). In [4] we showed that it is possible to define a generalized solution of (1)-(3) for any $u \in L_\infty(0,T)$ by

$$y(u;x,t)=\int_0^t G(t,\tau;x)u(\tau)d\tau \qquad (0\le t\le T,\ x\in\bar{\Omega}) \tag{4}$$

where

$$G(t,\tau;x)=\alpha^{-1}\sum_{k=1}^{\infty} g_k \exp(-\lambda_k(t-\tau))v_k(x). \tag{5}$$

Here $\{\lambda_k\}_{k\geqq 1}$ and $\{v_k\}_{k\geqq 1}$ are the eigenvalues resp. the normed eigenfunctions of

$$\Delta v(x)+\lambda v(x)=0, \qquad x\in\Omega,$$

$$\alpha\frac{\partial}{\partial n}v(\xi)+v(\xi)=0, \qquad \xi\in\partial\Omega,$$

and

$$g_k=\int_{\partial\Omega} g(\xi)v_k(\xi)d\xi.$$

We also proved that (4),(5) defines the classical solution of (1)-(3) in case g and u are sufficiently smooth. A further result from [4] is

$$y(u;\cdot,T)\in C(\bar{\Omega}) \qquad \forall\ u\in L_\infty(0,T).$$

4. <u>The control operator.</u> We define a linear operator (control operator)

$$S:L_\infty(0,T)\longrightarrow L_2(\Omega)$$

by $(Su)(x)=y(u;x,T)$, $x\in\Omega$. One can show that S is continuous w.r.t. the weak topologies $\sigma(L_\infty,L_1)$ in $L_\infty(0,T)$ resp. $\sigma(L_2,L_2)$ in $L_2(\Omega)$. This property (which is stronger than continuity of S w.r.t. the norms $\|\cdot\|_\infty$ in $L_\infty(0,T)$ resp. $\|\cdot\|_2$ in $L_2(\Omega)$) is equivalent to

$$S'(L_2(\Omega))\subset L_1(0,T) \tag{6}$$

where S' is the adjoint of S.

5. <u>Control problem (P).</u> Given a fixed function $z\in L_2(\Omega)$, minimize

$$\|Su-z\|_2$$

under the constraint $\|u\|_\infty\le 1$. (This is the task to approximate a desired end-temperature distribution z by choosing a boundary temperature u(t) under the constraint $|u(t)|\leqq 1, t\in[0,T]$.)

6. <u>Existence Theorem.</u> There is a solution $\hat{u}$ of (P). (This is directly implied by the weak continuity of S and the weak compactness of the unit sphere in $L_\infty(0,T)$, c.f. [4]).

7. <u>Characterization Theorem</u>. It is well known that $\hat{u}$ is a solution of (P) if and only if $\hat{u}$ satisfies the following variational inequality:

$$(S\hat{u}-z, Su-S\hat{u}) \geq 0 \quad \forall\, u \in L_\infty(0,T),\ \|u\|_\infty \leqq 1.$$

Here $(\cdot,\cdot)$ is the scalar product in $L_2(\Omega)$.

As $\|\cdot\|_2$ is uniformly convex, the vector

$$\hat{l}=z-S\hat{u} \tag{7}$$

is uniquely determined. We define

$$\hat{\lambda}=S'\hat{l} \tag{8}$$

which is an $L_1(0,T)$-function by (6). The characterization theorem gives us

8. <u>The maximum principle.</u>

$$\max_{\|u\|_\infty \leqq 1} \int_0^T \hat{\lambda}(t)u(t)dt = \int_0^T \hat{\lambda}(t)\hat{u}(t)dt$$

for any solution $\hat{u}$ of (P). This implies

$$\hat{u}(t)=\operatorname{sgn} \hat{\lambda}(t) \quad \text{a.e. on } [0,T]\setminus N_{\hat{\lambda}} \tag{9}$$

where $N_{\hat{\lambda}}$ is the set of zeroes of $\hat{\lambda}$.

9. <u>Normality of S.</u> It is not difficult to show that $\hat{\lambda}$ is an analytic function on $[0,T)$. Therefore either $\hat{\lambda}(t)\equiv 0$ or measure $(N_{\hat{\lambda}})=0$ (Normality of S; [2],[4]). In the latter case $\hat{u}$ is uniquely determined and we have the bang-bang principle

$$|u(t)| = 1 \quad \text{a.e. on } [0,T]. \tag{10}$$

Thus we are interested in conditions which imply

$$\hat{\lambda} \neq \theta_{L_1(0,T)} \tag{11}$$

where $\theta_{L_1(0,T)}$ denotes the zero vector of $L_1(0,T)$.

10. <u>A necessary condition for (11)</u>. We define $K=\{k\in\mathbb{N}/g_k\neq 0\}$. Let W be the closure in $L_2(\Omega)$ of the linear subspace spanned by $\{v_k\}_{k\in K}$ and $P:L_2(\Omega)\to W$ the projector belonging to W. Completeness of the system $\{v_k\}_{k\in\mathbb{N}}$ implies the representation

$$f=Pf+(I-P)f=f_1+f_2$$

for any $f \in L_2(\Omega)$, where $f_1 \in W, f_2 \in W^{\perp}$. Now

$$S\hat{u} \neq Pz$$

is a necessary condition for (9). This follows immediately by (7),(8) and

$$W^{\perp} \subset (\text{range } S)^{\perp} = \text{kernel } S'$$

(which is implied by (4),(5)) because from $S\hat{u}=Pz$ we get

$$\hat{\lambda} = S'(z-S\hat{u}) = S'z_2 = \theta .$$

In order to obtain sufficient conditions for (11) we introduce the concept of controllability of a linear operator. We remark that by (4),(5) and 10. range S is contained in W.

11. <u>Controllability.</u> S is called controllable (w.r.t.W) if range S is dense in W. In [4] we proved the following <u>Theorem:</u> If

$$\lambda_i \neq \lambda_k \qquad \text{for } i,k \in K (i \neq k)$$

then S is controllable (w.r.t.W). Now it is very easy to see that controllability of S (w.r.t.W) implies

$$P\hat{1} \neq \theta_{L_2(\Omega)} \quad \Longrightarrow \quad S'\hat{1} \neq \theta .$$

Observing that $S\hat{u} \neq Pz$ (for any solution $\hat{u}$ of (P)) is equivalent to $\inf\{\|Su-Pz\|_2, \|u\|_\infty \leqq 1\} > 0$ we obtain by 9.-11. our main theorem:

12. <u>Bang-bang principle.</u> Assume (a) S is controllable and (b) $\inf\{\|Su-Pz\|_2, \|u\|_\infty \leqq 1\} > 0$. Then (P) has a unique solution $\hat{u}$ and

$$|u(t)| = 1 \qquad \text{a.e. on } [0,T].$$

We remark that the analyticity properties of $\hat{\lambda}$ mentioned in 9. imply that $\hat{u}$ is piecewise continuous and assumes the values +1 and -1 with finitely many 'jumps' on $[0,T-\varepsilon]$ for any $\varepsilon > 0$.

We repeat the bang-bang theorem for the important special case where $z \in W$ using the controllability result which we stated in 11., and the definition of W.

13. <u>Theorem.</u> (cf.[4]) Assume

(i) $\inf\{\| Su-z\|_2, \|u\|_\infty \le 1\} > 0$ (i.e. z is 'not reachable')

(ii) $(z,v_k)=0$ for $k \in \mathbb{N} \setminus K$ (i.e. $z \in W$)

(iii) $\lambda_i \neq \lambda_k$ for $i,k \in K$ $(i \neq k)$ (i.e. S is controllable). Then there is a unique solution $\hat{u}$ of (P) and

$$|u(t)|=1 \qquad \text{a.e. on } [0,T].$$

14. <u>Example.</u> (cf.[1]) We show that for an important special case it is easy to obtain (ii) of the Theorem given above <u>without</u> any information about the eigenfunctions v_k. We take

$$g(\xi)=1 \qquad \forall\ \xi \in \partial\Omega$$

and

$$z(x)=\eta \neq 0 \qquad \forall\ x \in \Omega$$

Using Green's second formula we obtain by definition of v_k, g and z

$$(z,v_k)=-\lambda_k^{-1}(z,\Delta v_k)=-\lambda_k^{-1}\int_{\partial\Omega} z(\xi)\frac{\partial}{\partial n} v_k(\xi)d\xi$$

$$=-\lambda_k^{-1}\eta\int_{\partial\Omega}\frac{\partial}{\partial n}v_k(\xi)d\xi=\lambda_k^{-1}\eta\alpha^{-1}\int_{\partial\Omega}v_k(\xi)d\xi$$

$$=\lambda_k^{-1}\eta\alpha^{-1}\int_{\partial\Omega}g(\xi)v_k(\xi)d\xi=\lambda_k^{-1}\eta\alpha^{-1}g_k\ .$$

Thus

$$(z,v_k)=0 \iff g_k=0$$

which implies (ii).

An example where condition (iii) of the Theorem is met is discussed in [4] where we proved the result stated in 13. for the more difficult supremum norm problem too (where $L_2(\Omega)$ is replaced by $C(\overline{\Omega})$).

The controllability- normality- approach to the bang-bang principle may be applied to a much more general class of control problems (cf.[2]).

A number of further references can be found in [1].

References

[1] K.GLASHOFF Über Kontrollprobleme bei parabolischen Anfangs-Randwertaufgaben. Habilitationsschrift, Darmstadt 1975.

[2] K.GLASHOFF Controllability, normality, and the Bang-bang principle. To appear.

[3] K.GLASHOFF and S.Å. GUSTAFSON Numerical treatment of a parabolic boundary-value control problem. J.Optim.Th.Appl. (to appear).

[4] K.GLASHOFF u. N. WECK Boundary control of parabolic differential equations in arbitrary dimensions: supremum-norm problems. SIAM J.Control (to appear).

Generalized Stirling-Newton Methods

J. Gwinner
Universität Mannheim
D-68 Mannheim/Germany

In this contribution we generalize Newton's method and related methods for the solution of fixed point problems. We replace the linear approximations to a given nonlinear operator which result from differentiability assumptions by arbitrary "tangent" mappings. Therefore our iteration algorithms apply to nondifferentiable operators. As these tangent mappings may also be chosen affine, we include the differentiable case, thereby extending some recent results in [2].

Before starting our convergence analysis we state a slight generalization of the mean value theorem in infinite dimensional spaces in the first section. Afterwards we deal with a general class of iteration methods, then as particular examples Stirling's, and Newton's method are in detail examined in the main part of this paper (sections 3-5).

In our study of these methods Banach's fixed point theorem turns out to be the basic tool no longer obscured by linearity arguments e.g. norm calculations of linear operators in virtue of differentiability assumptions. These hypotheses simplify the proof of th. 3.2, only, where Stirling's method is concerned.

1. Generalized Mean Value Statements

Proposition 1.1. Let $I = [\alpha,\beta]$ denote a compact interval in $\mathbb{R}$. Let f be a continuous operator from I to a normed space Y, further let μ be a differentiable real-valued function, defined on I. Suppose there is a "tangent" mapping $\varphi: I \times \mathbb{R} \to Y$, i.e.

$$f(s) - \varphi(t,s) = o_t(s-t) \quad ; \quad \forall t\in I,\ s\in I$$

such that for all $t \in \mathit{int}\ I$, all $u\in J_t$ (an interval which may depend on t and which contains 0 in its interior)

$$\|\varphi(t,t+u) - \varphi(t,t)\| \le \mu'(t) \cdot |u|$$

is valid. Then we have

$$\|f(\beta) - f(\alpha)\| \le \mu(\beta) - \mu(\alpha) .$$

This proposition can be derived in the same manner as the usual mean value theorem, see e.g. prop. 8.5.1 in [1]; namely we define

$$A = \{t\in I \mid \|f(s) - f(\alpha)\| \le \mu(s) - \mu(\alpha) + \varepsilon(s-\alpha),\ \forall s\in[\alpha,t)\}$$

for a fixed $\varepsilon > 0$ and show by a contradiction argument that the non-void set A posesses β for its least upper bound.

Remark 1.2. If $\gamma_t(s) = \|\varphi(t,t+s) - \varphi(t,t)\|$ is convex in s $(t\in I)$, there exists a subdifferential $\theta\gamma_t(s)$, and we can set

$$\mu'(t) = \sup_{\substack{l\in\theta\gamma_t(s) \\ s\in J_t}} \|l\| ,$$

provided the resulting function is integrable.

Proposition 1.3. Let X,Y be normed spaces, F a continuous mapping into Y from a neighborhood W of a segment S joining two points x_0, x_0+h of X. If there exists a mapping $\Phi: S\times W \to Y$ which satisfies for any $x\in S$

$$F(\xi) - \Phi(x,\xi) = o_x(\xi-x),\ \forall \xi\in W$$

$$\|\Phi(x,x+u) - \Phi(x,x)\| \le m(x)\cdot\|u\| ,\ \forall u\in U_x(0) ,$$

then

$$\|F(x_0+h) - F(x_0)\| \leq \|h\| \cdot \int_0^1 m(x_0+th)dt ,$$

provided the function $m(x_0+th)$ is integrable in t.

Proof. Consider

$$g(t) = F(x_0+th)$$
$$\psi(t,s) = \Phi(x_0+th,\ x_0+sh) .$$

At once we have the estimate

$$\|\psi(t,t+s) - \psi(t,t)\| \leq m(x_0+th) \cdot |s| \cdot \|h\| .$$

Now apply prop. 1.1.

Remark 1.4. From prop. 1.3 we obtain immediately that Lipschitz continuity of the tangent operator $\Phi(x,\cdot)$ which is uniform with respect to x implies Lipschitz continuity of F with the same constant.

Corollary 1.5. Let X,Y, and the operators F, and Φ be as in prop. 1.3. If furthermore for all $x\in S$, $x'\in S$; $\xi\in W$, $\xi'\in W$

$$(*) \quad \|[\Phi(x,\xi) - \Phi(x,\xi')] - [\Phi(x',\xi) - \Phi(x',\xi')]\| \leq K \cdot \|\xi-\xi'\| \cdot \|x-x'\|$$

holds with a constant $K > 0$, then

$$\|F(x') - \Phi(x,x')\| \leq \frac{1}{2} K \|x-x'\|^2 \ ; \forall x\in S,\ x'\in W .$$

Proof. Define $G: \xi'\in W \to F(\xi') - \Phi(x,\xi')$, and $\psi: (\xi,\xi')\in S\times W \to \Phi(\xi,\xi') - \Phi(x,\xi')$. Then ψ is tangent to G, and because of (*) we obtain

$$\|\psi(\xi,\xi+\eta) - \psi(\xi,\xi)\| \leq K \cdot \|\xi-x\| \cdot \|\eta\| ,$$

therefore we may set

$$m(\xi) = K \cdot \|\xi-x\|$$

and apply prop. 1.3.

Remark 1.6. If F is differentiable, and the tangent mapping Φ is affine with respect to its second argument, (*) means the Lipschitz continuity of the derivative F'. Thus cor. 1.5 extends prop. 4.4 in [3].

2. A Class of Iteration Methods

Let us assume that the continuous operator F, resp. the operator Φ is defined in the Banach space X, resp. in $X \times X$ and that the relations

(T) $$F(y) - \Phi(x,y) = o_x(y-x)$$

(H1) $$\|\Phi(x,\xi) - \Phi(x,\xi')\| \leq L \cdot \|\xi-\xi'\|$$

hold for all $x \in X$, $y \in X$, $\xi \in X$, $\xi' \in X$. The Lipschitz condition (H1) replaces the hypothesis of uniform boundedness of the derivative F' in [2].

Now we consider the following fundamental *algorithm:* Let the n-th iterate x_n be given. Choose a $y_n \in X$. Set

$$\tilde{\Phi}(x_n,\xi) = \Phi(y_n, \xi+y_n - x_n) + F(x_n) - F(y_n) ,$$

and compute x_{n+1} as a solution of the auxiliary fixed point problem

(2.1) $$\tilde{\Phi}(x_n,x) = x .$$

Remark 2.1. By Banach's fixed point theorem the algorithm above is defined, if $L < 1$.

Remark 2.2. Special cases of this algorithm, namely Newton's method ($y_n = x_n$), modified Newton's method ($y_n = x_o$), Stirling's method ($y_n = F(x_n)$) turn out if $\Phi(x_n,\cdot)$ is affine for each iteration n. Moreover, if a y^* exists which satisfies $\Phi(y^*,\cdot) \equiv F(y^*)$, then the choice $y_n = y^*$ reduces the algorithm to the method of successive approximations $x_{n+1} = F(x_n)$.

All the subsequent theorems in this, and in the following two sections generalize the corresponding theorems in [2]. Notice, that our results are arranged in an order which is different from that in [2].

Theorem 2.3. Let F, and Φ be as above with $L < \frac{1}{3}$. Then the sequence $\{x_n\}$ generated by the algorithm above converges to the unique fixed point x^* of F for arbitrary choice of the starting point x_o and the sequence $\{y_n\}$. Furthermore

(2.2) $$\|x_n - x_{n+1}\| \leq \frac{1}{1-L} \|x_n - F(x_n)\|$$

(2.3) $$\|x^* - x_n\| \le \left(\frac{2L}{1-L}\right)^n \frac{\|x_0 - F(x_0)\|}{1 - 3L} .$$

Proof: We have

$$\begin{aligned} \|x_n - x_{n+1}\| &= \|x_n - \tilde{\Phi}(x_n, x_{n+1})\| \\ &\le \|x_n - F(x_n)\| + \|\Phi(y_n, y_n) - \Phi(y_n, x_{n+1} + y_n - x_n)\| \\ &\le \|x_n - F(x_n)\| + L \cdot \|x_{n+1} - x_n\| . \end{aligned}$$

Therefore (2.2) holds. Furthermore by prop. 1.3 and remark 1.4 F is a contraction with contraction constant L, and we can calculate

$$\|x_n - F(x_n)\| = \|\tilde{\Phi}(x_{n-1}, x_n) - F(x_n)\|$$

(2.4) $$= \|\Phi(y_{n-1}, y_{n-1} + x_n - x_{n-1}) - F(y_{n-1}) + F(x_{n-1}) - F(x_n)\|$$

$$\le 2L \cdot \|x_n - x_{n-1}\| .$$

We combine (2.2) and (2.4) into

$$\|x_n - x_{n+1}\| \le \frac{2L}{1-L} \|x_{n-1} - x_n\| .$$

As the ratio $q \equiv \frac{2L}{1-L}$ is less than 1, standard arguments show the convergence of the sequence $\{x_n\}$ to the unique fixed point x^* of F, and provide

$$\|x^* - x_n\| \le q^n \cdot \frac{\|x_1 - x_0\|}{1 - q} .$$

From this and from the initial distance

$$\|x_1 - x_0\| \le \frac{1}{1-L} \|x_0 - F(x_0)\|$$

we infer that (2.3) is valid.

Remark 2.4. Inequality (2.2) gives an estimate of the next iterate x_{n+1} by means of the error $\|x_n - F(x_n)\|$.

An example in [2], p. 14 shows that, even in the differentiable case, the condition $L < \frac{1}{3}$ is best possible for this general convergence result.

If the continuous operator F is defined on an open set D of X, only, and likewise Φ is defined on $D \times X$, then the convergence result above can be reformulated as follows:

Theorem 2.5. Suppose that the relations (T) and (H1) with $L < \frac{1}{3}$ hold for all $\xi \in X$, $\xi' \in X$, and for all x,y in the ball

$$B_L = \{x: \|x - x_o\| \le \frac{1}{1-3L} \|F(x_o) - x_o\|\} \subset D .$$

Then for each sequence $\{y_n\} \subset B_L$, the sequence $\{x_n\}$ defined by (2.1) converges to the unique fixed point x^* of F in B_L at the rate given by (2.3).

Proof: From (2.2) we gather

$$\|x_1 - x_o\| \le \frac{1}{1-L} \|x_o - F(x_o)\| ,$$

thus $x_1 \in B_L$. Assume that $x_k \in B_L$, $y_k \in B_L$ is chosen for $k = 0,1, \ldots,n$. Then x_{k+1} is defined, and because

$$\|x_{k+1} - x_o\| \le L \cdot \|x_{k+1} - x_k\| + L \cdot \|x_k - x_o\| + \|F(x_o) - x_o\|$$

we obtain for all $k = 0,1,\ldots,n$

$$\|x_{k+1} - x_o\| \le \frac{2L}{1-L} \|x_k - x_o\| + \frac{1}{1-L} \|F(x_o) - x_o\| .$$

Hence

$$(2.5) \qquad \|x_{n+1} - x_o\| \le \frac{\|F(x_o) - x_o\|}{1-L} \cdot \sum_{k=o}^{n} \left(\frac{2L}{1-L}\right)^k$$

$$\le \frac{\|F(x_o) - x_o\|}{1-3L} .$$

Thus $x_{n+1} \in B_L$, from which $\{x_n\} \subset B_L$ follows by mathematical induction. Now the same arguments as in the proof of th. 2.3 show the claimed linear convergence to the unique fixed point x^* of F in B_L.

Remark 2.6. Stirling's method (and of course, Newton's method) is included in the theorem above, as the sequence $\{y_n\} = \{F(x_n)\} \subset B_L$; for from (2.5) we derive

$$\|F(x_n) - x_o\| \le \|F(x_n) - F(x_o)\| + \|y_o - x_o\|$$

$$\le \frac{L}{1-L} \cdot \sum_{k=o}^{n-1} \left(\frac{2L}{1-L}\right)^k \|y_o - x_o\| + \|y_o - x_o\|.$$

This inequality yields at once

$$\|y_n - x_o\| \le \frac{1}{1-3L} \cdot \|F(x_o) - x_o\| .$$

3. Stirling's Method

Let D be an open, convex set in the Banach space X. Here we take for granted that the domain of the continuous operator F contains D, and Φ is defined on $D \times X$ such that for all $x \in D$, $x' \in D$, $\xi \in X$, $\xi' \in X$

(H1) $\|\Phi(x,\xi) - \Phi(x,\xi')\| \leq L \cdot \|\xi - \xi'\|$,

(H2) $\|F(x') - \Phi(x,x')\| \leq \frac{1}{2} K \cdot \|x - x'\|^2$,

(H3) $\|\Phi(x,x + \xi - \xi') - F(x) + \Phi(x,\xi') - \Phi(x,\xi)\|$

$$\leq M \|\xi' - x\| \{\|\xi - x\| + \|\xi' - x\|\}$$

hold with constants $L \in [0,1)$, $K > 0$, and $M \geq 0$.

By the corollary 1.5 (H2) can also be derived from the relation (*) in section 1, provided Φ is tangent to F, i.e. (T) is satisfied. But since only (H2) is needed in the subsequent convergence analysis, we postulate this weeker condition.

<u>Remark 3.1.</u> The hypothesis (H3), which is trivially satisfied for affine approximations with $M = 0$, is motivated by locally quadratic approximations. If we define for any element x of some neighborhood U of x_0

$$\Phi(x_0,x) = F(x_0) + F'(x_0)\,(x-x_0) + \frac{1}{2}\,\psi(x_0)\,(x-x_0,\; x-x_0) ,$$

where $\psi(x_0)$ denotes a symmetric bilinear continuous mapping (e.g. $\psi(x_0) = F''(x_0)$, if F is twice differentiable, see 8.12.2 in [1]), then for $x_0+v-u \in U$, $u \in U$, $v \in U$ we have

$$\Phi(x_0,x_0+v-u) - F(x_0) - \Phi(x_0,v) + \Phi(x_0,u)$$

$$= \psi(x_0)\cdot(v-x_0,x_0-u) + \psi(x_0)\cdot(u-x_0,u-x_0) .$$

Thus (H3) can be satisfied, if $\|\psi(x)\| \leq M$ uniformly in D.

<u>Theorem 3.2.</u> Suppose, $D = X$. Then the Stirling iterates x_n, generated by the algorithm with the choice $y_n = F(x_n)$, converge to the unique fixed point x^* of F from any $x_0 \in X$ such that

(3.1) $$q_S \equiv \frac{1}{1-L}\,[K+M(1+L)(1+2L)]\cdot\frac{\|x_0 - F(x_0)\|}{1-L} < 1 .$$

Furthermore, the convergence is quadratic, with

$$(3.2) \qquad \|x_n - x^*\| \le q_S^{2^n-1} \cdot \frac{\|x_o - F(x_o)\|}{1-L} .$$

Proof: In view of the remark 1.4 the assumptions (H1) and (H2) imply that F is a contraction. Banach's fixed point theorem guarantees the existence and uniqueness of the fixed point x^*, with

$$(3.3) \qquad \|x^* - x_o\| \le \frac{\|x_o - F(x_o)\|}{1-L} .$$

From our assumptions (H1), (H2), and (H3) we gather

$$\begin{aligned} &\|x_{n+1} - x^*\| \\ &\quad \le \|\Phi(y_n, x_{n+1}+y_n-x_n) - \Phi(y_n, x^*+y_n-x_n)\| \\ (3.4) \qquad &\quad\quad + \|F(x_n) + \Phi(y_n,x^*) - F(x^*) - \Phi(y_n,x_n)\| \\ &\quad\quad + M \|x_n-y_n\| \cdot \{\|x^*-y_n\| + \|x_n-y_n\|\} \\ &\quad \le L\|x_{n+1} - x^*\| + K\|x_n-x^*\|^2 + M\|x_n-y_n\|\{\|x^*-y_n\| + \|x_n-y_n\|\} . \end{aligned}$$

Since $y_n = F(x_n)$, we can apply the following inequalities:

$$\|x^* - F(x_n)\| = \|F(x^*) - F(x_n)\| \le L\|x^* - x_n\| ,$$

$$\|x_n - F(x_n)\| \le \|x_n - x^*\| + \|x^*-F(x_n)\| \le (1+L)\|x^*-x_n\| ,$$

and we obtain

$$(3.5) \qquad \|x_{n+1} - x^*\| \le \frac{1}{1-L} [K+M\cdot(1+L)(1+2L)]\cdot\|x^*-x_n\|^2 .$$

From (3.3) and (3.5) inequality (3.2) follows by mathematical induction, and (3.1) implies convergence.

<u>Theorem 3.3.</u> Under the hypotheses of th. 3.2 Stirling's method converges quadratically to x from any point x_o such that

$$(3.6) \qquad \|x_o-x^*\| < \frac{1-L}{K+M(1+L)(1+2L)} \equiv \rho_S .$$

Proof: In virtue of (3.5) mathematical induction leads to

$$\|x_n-x^*\| \le \left\{\frac{1}{\rho_S} \cdot \|x_o-x^*\|\right\}^{2^n-1} \cdot \|x_o-x^*\| ,$$

thus showing the sufficiency of (3.6) for the quadratic convergence of the Stirling iterates.

Let us turn to the local case, where D is a proper subset of X.

Theorem 3.4. Suppose, D contains the ball

$$B = \{x \in X : \|x-x_o\| \le \frac{2\|x_o - F(x_o)\|}{1-L}\},$$

and (3.1) holds. Then the Stirling iterates x_n converge to the unique fixed point x^* of F in D at the rate given by (3.2).

Proof: Again Banach's fixed point theorem guarantees the existence and the uniqueness of x^*. It suffices to show that the sequences $\{x_n\}$ and $\{y_n\}$ remain in B. This can be verified by an inductive argument which is similar to the proof of th. 2.5 (cf. also the proof of th. 4' in [2]).

4. Newton's Method

The same analysis as in section 3 provides analogous assertions for Newton's method. Actually, the choice $y_n = x_n$ simplifies the proofs, and we can dispense with the hypothesis (H3). Instead of repeeting our arguments we perform some additional error analysis, which, of course, could also be carried through for the methods presented above.

Given a sequence $\{\varepsilon_n\}$ of nonnegative real numbers, we solve the auxiliary fixed point problem (2.1) (with $y_n = x_n$) only within ε_n, i.e. z_{n+1} satisfies

$$\|z_{n+1} - \Phi(z_n, z_{n+1})\| \le \varepsilon_n \; . \tag{4.1}$$

Theorem 4.1. Suppose, the conditions (H1) and (H2) hold in X (D = X). If z_o satisfies

$$q_N \equiv \frac{1}{2} K \frac{\|z_o - F(z_o)\|}{(1-L)^2} < 1 \tag{4.2}$$

and the sequence $\{\varepsilon_n\}$ is bounded above by $(1-q_N)\|z_o - F(z_o)\|$ and has limit zero, then any sequence $\{z_n\}$ generated by (4.1) with starting point z_o converges to the unique fixed point x^* of F. Furthermore

$$\|z_{n+1} - x^*\| \le \frac{1}{1-L} (\varepsilon_n + \frac{K}{2} \|z_n - x^*\|^2) \; . \tag{4.3}$$

Proof: Again existence and uniqueness of the fixed point x^* are ensured. By analogy with (3.4) we calculate

$$\| z_{n+1} - x^* \|$$

$$\leq \| z_{n+1} - \Phi(z_n, z_{n+1}) \| + \| \Phi(z_n, z_{n+1}) - \Phi(z_n, x^*) \|$$

$$+ \| \Phi(z_n, x^*) - F(x^*) \|$$

$$\leq \varepsilon_n + L \| z_{n+1} - x^* \| + \frac{K}{2} \| z_n - x^* \|^2 \quad .$$

Now (4.3) is immediate. We already know from (3.3) that the inequality

$$(4.4) \qquad \| z_k - x^* \| \leq \frac{1}{1-L} \| z_0 - F(z_0) \|$$

is valid for $k = 0$. By mathematical induction it follows from (4.3) and from the boundedness of the ε_n that all distances $\| z_k - x^* \|$ are below this threshold. So we can weaken (4.3) to

$$\| z_{n+1} - x^* \| \leq \frac{\varepsilon_n}{1-L} + q_N \quad \| z_n - x^* \| \; .$$

This gives the estimate

$$\| z_{n+1} - x^* \| \leq \frac{1}{1-L} \sum_{k=0}^{n} q_N^{n-k} \varepsilon_k + q_N^{n+1} \| z_1 - x^* \| \; .$$

As $q_N < 1$ and $\varepsilon_k \to 0$, both terms can be seen to converge to zero. This proves the convergence of the sequence $\{z_n\}$ to x^*.

Now let us establish the corresponding result in bounded regions.

Theorem 4.2. Suppose, the ball

$$B = \{ x \in X : \| x - z_0 \| \leq \frac{2 \| z_0 - F(z_0) \|}{1-L} \}$$

is contained in D. If $q_N < 1$ and the sequence $\{\varepsilon_n\}$ satisfies the conditions of th. 4.1, then any sequence $\{z_n\}$ generated by (4.1) starting in z_0 converges to the unique fixed point x^* of F in D at the rate given by (4.3).

Proof: As above in th. 3.4 existence and uniqueness of x^* in D is guaranteed. Using (4.4) one can easily show by an inductive argument that $\{z_n\} \subset B$.

As the differentiable case already indicates [see 2], we have obtained better bounds for Newton's method than for Stirling's method, which is burdened by the additional hypothesis (H3), besides. Nevertheless one can find examples in which Newton's method provides inferior convergence results to those given by Stirling's method. For a discussion of cases which are favorable to the latter method we refer to [2].

5. Newton's Method with Local Approximations

In the preceding discussions the approximations $\Phi(x,\cdot)$ have always been assumed to be Lipschitz continuous, uniformly with respect to x, in the *whole* space X. This drawback is now removed, at least for Newton's method.

As above let D be an open, convex set in the Banach space X. In this section we postulate that the domain of the continuous operator F contains D, and the continuous mapping Φ is defined on $D \times D$ such that for all $x \in D$, $\xi \in D$, $\xi' \in D$ the inequalities

(H1) $$\|\Phi(x,\xi) - \Phi(x,\xi')\| \le L \, \|\xi - \xi'\| ,$$

(H2) $$\|F(\xi) - \Phi(x,\xi)\| \le \frac{K}{2} \|x - \xi\|^2$$

hold with constants $L \in [0,1)$ and $K > 0$.

<u>Theorem 5.1.</u> If there exists a pair $(x_o,\alpha) \in X \times \mathbb{R}$ such that

$$U = \{x \in X: \|x - x_o\| < \alpha\}$$

is contained in D, and

(5.1) $$\alpha < \frac{1-L}{K}$$

(5.2) $$\|F(x_o) - x_o\| \le \frac{\alpha \cdot K}{2} \left(\frac{1-L}{K} - \alpha\right)$$

hold, then the Newton iterates x_n, i.e.

(5.3) $$x_n = \Phi(x_{n-1}, x_n) ,$$

starting in x_o, converge to the unique fixed point $x^* \in D$. Furthermore

(5.4) $$\|x_{n+1} - x^*\| \le \frac{K}{2(1-L)} \|x_n - x^*\|^2 .$$

For the proof of this theorem we need

<u>Lemma 5.2.</u> Let x_o be an arbitrary element in a Banach space X; U, resp. V an open ball in X of center x_o and radius α, resp. β. Let v be a continuous mapping of $U \times V$ into X such that

$$\|v(x,\xi_1) - v(x,\xi_2)\| \le \gamma \cdot \|\xi_1 - \xi_2\|$$

for all $x \in U$, $\xi_1 \in V$, $\xi_2 \in V$, where γ is a constant such that

$0 \le \gamma < 1$. Then, if

$$\| v(x,x_o) - x_o \| < \beta(1-\gamma)$$

for any $x \in U$, there exists a unique mapping f of U into V such that

$$f(x) = v(x,f(x))$$

for any $x \in U$; and f is continuous in U.

This lemma can be derived from the method of successive approximations (see prop. 10.1.1 in [1]) by a translation argument.

Proof of theorem 5.1. Set $v(x,y) = \Phi(x,y)$, $U = \{x: \| x-x_o \| < \alpha\}$ with the α of the theorem, take $\beta = \alpha/_2$. In virtue of (H2) and (5.2) for any $x \in U$

$$\begin{aligned} &\|\Phi(x,x_o) - x_o\| \\ &\le \| F(x_o) - x_o\| + \frac{K}{2} \| x-x_o\|^2 \\ &< \frac{\alpha}{2} (1-L) \end{aligned}$$

is valid. Therefore lemma 5.2 applies, and yields the unique existence of a continuous mapping $N: U \to V$ such that for any $x \in U$

$$N(x) = \Phi(x,N(x)) .$$

Accordingly Newton's method (5.3) reads

$$x_{n+1} = N(x_n) , \tag{5.5}$$

where all iterates x_n remain in $V \subset U \subset \mathit{dom}N$.

In order to simplify the proof we insert the following

<u>Lemma 5.3.</u> Notations and assumptions as above. Suppose, $x \in V$ and $z \in V$ satisfy

$$\| \Phi(x,z) - z \| < \frac{\alpha}{2} (1-L) .$$

Then

$$\|N(x) - z \| \le \frac{1}{1-L} \|\Phi(x,z) - z\|$$

holds.

Proof of lemma 5.3. Define $z^o = z \in U$, $z^\nu = \Phi(x,z^{\nu-1})$ for all $\nu \in \mathbb{N}$.

Since

$$\|z^1 - x_0\| \le \|z^1 - z^0\| + \|z^0 - x_0\|$$
$$< \frac{\alpha}{2}(1-L) + \frac{\alpha}{2} ,$$

z^1 belongs to $U \subset D$. Now assume that $z^1, z^2, \ldots, z^\nu$ exist, and z^1, $z^2, \ldots, z^{\nu-1}$ belong to U. As

$$\|z^\nu - x^0\| = \|\Phi(x, z^{\nu-1}) - x_0\|$$
$$\le \frac{1}{1-L}\|z^1 - z^0\| + \|z^0 - x^0\|$$
$$< \alpha ,$$

z^ν lies in $U \subset D$, and therefore $z^{\nu+1}$ exists. Thus we have proved that all iterates z^ν exist and remain in U. Because $\Phi(x, \cdot)$ is a contraction, the sequence $\{z^\nu\}$ converges to a $z^* = \Phi(x, z^*)$. By similar computations as above we realize that $z^* \in U$. From the uniqueness of the "solution operator" N in U we conclude that $z^* = N(x)$, and finally we get the estimate

$$\|N(x) - z\| = \|z^* - z^0\|$$
$$\le \frac{1}{1-L}\|z^1 - z^0\| = \frac{\|\Phi(x,z) - z\|}{1-L} .$$

Now let us continue the *proof of theorem 5.1*. As the following inequalities

(5.6) $$\|\Phi(x_n, x_n) - x_n\| \le \frac{K}{2}\|x_n - x_{n-1}\|^2 ,$$

(5.7) $$\|x_n - x_{n-1}\| \le \|x_n - x_0\| + \|x_0 - x_{n-1}\| < \alpha ,$$

$$\frac{K}{2}\alpha^2 < \frac{\alpha}{2}(1-L)$$

hold, lemma 5.3 applies to $x = z = x_n = N(x_{n-1}) \in V$ for an arbitrary $n \in \mathbb{N}$. We obtain

$$\|x_{n+1} - x_n\| \le \frac{\|F(x_n) - x_n\|}{1-L} ,$$

and (5.6) implies

(5.8) $$\|x_{n+1} - x_n\| \le \frac{1}{2} \cdot \frac{K}{1-L}\|x_n - x_{n-1}\|^2 .$$

In virtue of (5.1) and (5.7) $\|x_n - x_{n-1}\| < \frac{1-L}{K}$ is valid. So we can relax (5.8) to

$$\|x_{n+1} - x_n\| < \frac{1}{2}\|x_n - x_{n-1}\| .$$

Thus $\{x_n\}$ is Cauchy, hence converges to a $x^* \in U$. From the continuity of N we obtain $x^* = N(x^*) \in V$; this means $F(x^*) = \Phi(x^*,x^*) = x^*$ is the unique fixed point of F in D. Because of the inequalities

$$\|\Phi(x_n,x^*) - x^*\| \leq \frac{K}{2}\|x_n - x^*\|^2 ,$$

$$\|x_n - x^*\| < \alpha < \frac{1-L}{K}$$

lemma 5.3 applies to $x = x_n$, $z = x^*$, and yields the desired relation (5.4).

Remark 5.4. In view of lemma 5.2 it is also possible to consider an α - neighborhood, and an arbitrary β - neighborhood of x_o. But a thorough analysis shows that we require the following assumptions

$$0 < \beta \leq \frac{\alpha}{2}$$

$$\frac{K}{2}\alpha^2 < \beta(1-L)$$

$$\|F(x_o) - x_o\| \leq \beta(1-L) - \frac{K}{2}\alpha^2$$

to obtain the same convergence result. So the choice $\beta = \frac{\alpha}{2}$ is optimal.

Remark 5.5. If the subproblems

$$\Phi(x_{n-1},\xi) = \xi$$

are only solved approximately, i.e. $z_n \in V$ is computed such that $\|z_{n+1} - \Phi(z_n,z_{n+1})\|$ is within a positive ε_n (cf. section 4), then under the conditions of th. 5.1 the sequence $\{z_n\}$ converges to x^*, if $\varepsilon_n \to 0$. Furthermore

$$\|z^{n+1} - x^*\| \leq \frac{\varepsilon_n}{1-L} + \frac{K}{2(1-L)}\|z_n - x^*\|^2 . \tag{5.9}$$

This estimate can be shown by application of lemma 5.3, and by relaxation of (5.9) (cf. proof of th. 5.1) convergence of the sequence $\{z_n\}$ to x^* can be verified.

In comparison with th. 4.1 and th. 4.2 this theorem 5.1 and remark 5.5 provide the same convergence result, the same convergence rate, but need a more restrictive starting condition concerning $\|F(x_o) - x_o\|$ (by a factor of 16, if α is put $\frac{1-L}{2K}$).

Literature

[1] J. Dieudonné: *Foundations of Modern Analysis.* Academic Press, New York, 1960.

[2] L.B. Rall: *Convergence of Stirling's Method in Banach Spaces.* Aequationes Math. 12 (1975), 12-20.

[3] R.A. Tapia: *The Differentiation and Integration of Nonlinear Operators,* Nonlinear Functional Analysis and Applications, L.B. Rall, ed., Academic Press, New York, 1971, pp. 45-102.

ON A METHOD FOR COMPUTING PSEUDOINVERSES

Joachim Hartung
Institut für Angewandte Mathematik
und Informatik
D-53 Bonn, Wegelerstraße 6

Let A be a continuous linear operator with domain of definition $D(A)=X$ and range of values $R(A)=Y$, where X and Y are Hilbert spaces. We are given an element $y \in Y$ and a subset $M \subset X$, and the problem of finding an element $x \in M$ which solves the equation

$$(1) \qquad Ax = y \qquad , y \in Y, x \in M \subset X .$$

If $y \notin R(A_{|M})$, where $A_{|M}$ denotes the restriction of A to the subset M, there exists no solution of (1). Then we consider the problem of finding an element $x(y) \in M$ of minimum norm which gives a minimum value for the discrepancy $||Ax-y||$, $x \in M$. An element $x(y)$ with this property is called a "best approximate solution"(cf.Holmes[4])or "pseudo-solution"(cf.Morozov[6]) of (1). (The norm $||.||$ may be respectively defined by the inner product $<.,.>$: $||.|| = <.,.>^{1/2}$)
Let $S(M,A,y)$ be the set of $x \in M$ for which Ax is a best $R(A_{|M})$-approximation to y, i.e.

$$(2) \qquad x_0 \in S(M,A,y) \; :\leftrightarrows \; x_0 \in M \text{ and } ||Ax_0 - y|| = \min\{||Ax-y|| \,\big|\, x \in M\}.$$

A best approximate solution of (1) is a solution of the following optimization problem

$$(3) \qquad O_p\{M,A,y\} \; : \min\{||x|| \,\big|\, x \in S(M,A,y)\} .$$

If A is normally solvable ($R(A)$ is closed), there exists a linear and bounded operator A^+ called the pseudoinverse of A, which is uniquely determined by producing for every $y \in Y$ the unique solution $x(y)=A^+ y$ of (3) in the case of $M=X$ (cf. Holmes [4]),i.e.

(4) A_y^+ solves $O_p\{X,A,y\}$ for all $y \varepsilon Y$.

Let

(5) $$i(A) := \inf\{ \frac{||Ax||}{||x||} \mid x \neq 0,\ x \varepsilon N(A)^{\perp} \},$$

where N(A) is the nullspace of A and $N(A)^{\perp}$ its orthogonal complement in the domain of definition of A,
then it holds (Petryshyn [7]):

(6) $$i(A)>0 \rightleftarrows R(A) \text{ closed} \rightarrow A^+ \text{ exists and } ||A^+|| = i(A)^{-1} < +\infty.$$

In many applications we are given the case that M=N(B) , where B: X→Z is a bounded linear operator with D(B)=X and R(B) is contained in a Hilbert space Z.
Let $A_B := A_{|N(B)}$, then A_B^+ is called the N(B)-restricted pseudoinverse of A (Minamide-Nakamura [5]).
If A_B^+ exists, we have:

(7) $A_B^+ y$ solves $O_p\{N(B),A,y\}$, for all $y \varepsilon Y$.

A_B^+ is uniquely determined by this property (7).

Quadratic optimization problems with general equality contraints

$$\min \{ \|Ax-y\|^2 \mid x \varepsilon X,\ Bx=z \},\ (y \varepsilon Y, z \varepsilon Z),$$

like for instance approximation problems with prescribed values at some points,
or more general, the two-stage minimization problem

(8) $O_p\{S(X,B,z),A,y\}$, $(y \varepsilon Y, z \varepsilon Z)$,

can be solved explicitely by using pseudoinverses:

(9) $A_B^+ y + (I-A_B^+ A)B^+ z$ is the solution of (8) provided A_B^+, B^+ exist. I denotes the Identy.

(9) was derived by Minamide-Nakamura [5], who also treated as an example of (8) a control problem, where on the second stage the energy ($\sim \| x \|^2$) is to be minimized.

An other application of pseudoinverses arises in the statistical theory of testing a linear hypothesis.
Let be given the usual model:

$y=Ax+e$, $Ee=0$, $\ddagger_e=\sigma^2 I$,
y is normally distributed with mean Ax and Covariance $\sigma^2 I$, $(\sigma^2>0)$: $y \in \mathcal{N}(Ax,\sigma^2 I)$.

We test the linear hypothesis
(H): $Bx=0$ for at least one x.

Such a hypothesis occurs for example in checking whether two different experimental arrangements give consistent Gauss-Markov estimates for a linear form of the unknown parameter x.
Here we have $X=\mathbb{R}^m$, $Y=\mathbb{R}^n$, $Z=\mathbb{R}^l$, $(m,n,l \in \mathbb{N})$
If $R(A_B) \neq R(A)$
the following test characteristic can be derived:

(10) $$F(y) = \frac{n-r}{r-p} \cdot \frac{\| A(A^+y-A_B^+y)\|^2}{\| y - AA^+y \|^2},$$
where $r=\operatorname{rank}A$, $p=\operatorname{rank}A_B$.

Under the Hypothesis (H) then F(y) belongs to a central $\mathcal{F}$-distrbution with (r-p) and (n-r) degrees of freedom:

(11) $$F(y) \underset{(H)}{\in} \mathcal{F}_{r-p,n-r}(0).$$

If Bx is linear unbiased estimable,(H) becomes the usual hypothesis

$(\tilde{H})$: $Bx=0$ for all x .

Then it can be shown, that $r-p=\operatorname{rank} B$

Now we consider a parametric method for computing pseudoinverses, which permits the application of ordinary inversion methods for self-adjoint positive definite operators.
If A is normally solvable, so it holds

(12) $$A^+ = \lim_{s\to+0} (A^*A+sI)^{-1}A^*, (s\in \mathbb{R}), =: \lim_{s\to+0} A(s)$$

where A^* denotes the adjoint operator of A.
The representation(12) traces back to Ben Israel-Charnes[1] for matrices, and for general Hilbert space operators (12) was derived by Morozov[6], who estimates the rate of convergence by

(13) $$||A^+-A(s)|| \leq s||\hat{A}^{-1}||^3, \ s>0,$$

Where $\hat{A}$ is the operator mapping $N(A)^\perp$ onto $R(A)$ and is the same as the operator A on $N(A)^\perp$. Glashoff [2] gives the estimate

(14) $$||A^+-A(s)|| \leq s||(A^*A)^+||\cdot||A^+||, \ s>0 .$$

Similar results are valid for the restricted pseudoinverse.
If A, B, A_B are normally solvable, Glashoff [2], [3] shows

(15) $$A_B^+y = \lim_{t\to+o} \lim_{s\to+o} (B^*B+sA^*A+s\cdot t\cdot I)^{-1} sA^*y, (s,t\in \mathbb{R}), =: \lim_{t\to+o} \lim_{s\to+o} A_B(s,t)y ,$$

for all $y\in Y$, that is,

(16) $$A_B(s,t) \xrightarrow[t\to+o]{} \xrightarrow[s\to+o]{} A_B^+ \text{ strongly.}$$

Estimating the rate of convergence, respectively deriving an error bound, we show now, that in (16) the operators $A_B(s,t)$ are converging in the norm, too. Then the estimate yields also, that the method, given by (15), can be chosen as a one-parametric one.

Let $[A,B]: X \to Y \times Z$ be the productmapping
$x \mapsto (Ax, Bx)$
and $Y \times Z$ be provided with the induced inner product.

(17) Theorem

If $B, [A,B]$ are normally solvable, then we have

(i) A_B^+ exists

(ii) $||A_B^+|| \leq ||[A,B]^+||$

(iii) $||A_B^+ - (B^*B + sA^*A + s\cdot tI)^{-1} sA^*|| \leq$

$\leq s\cdot t^{-1/2}(||A||^2+t)^{1/2}\cdot ||B^+||^2(||A^*A||\cdot||[A,B]^+||+||A||)+$

$+t^{1/2}(1+s\cdot(||A||^2+t)^{1/2}\cdot||B^+||)\cdot||[A,B]^+||^2, s>0, t>0$

Proof:

We have

$$i(A_B) = \inf\left\{\frac{||A_Bx||}{||x||} \,\middle|\, x \neq 0,\ x \in N(A_B)^{\perp}\right\}$$

$$= \inf\left\{\frac{||Ax||}{||x||} \,\middle|\, x \neq 0,\ x \in N(B),\ x \in (N(B) \cap N(A))^{\perp}\right\}$$

$$= \inf\left\{\frac{(||Ax||^2+||Bx||^2)^{1/2}}{||x||} \,\middle|\, x \neq 0,\ x \in N(B), x \in N([A,B])^{\perp}\right\}$$

$$\geq \inf\left\{\frac{||[A,B]x||_{Y\times Z}}{||x||} \,\middle|\, x \neq 0,\ x \in N([A,B])^{\perp}\right\}$$

$$= i([A,B]),$$

such that (6) yields (i) and (ii).
Let $y \in Y$ be arbitrary chosen, but fixed, and $s>0$, $t>0$.
We define

$$f_{s,t}(x) := ||Bx||^2 + s||Ax-y||^2 + s\cdot t||x||^2, \quad x \in X,$$

$$C_{s,t} := B^*B + sA^*A + s\cdot tI.$$

The self-adjoint operator $C_{s,t}$ is positive definite and has an inverse $C_{s,t}^{-1}$. Then

$$x_{s,t} := C_{s,t}^{-1} s\, A^*y = A_B(s,t)\, y$$

minimizes $f_{s,t}(x)$ over X:

$$f_{s,t}(x_{s,t}) = \min_{x \in X} f_{s,t}(x),$$

because of the Frechét differential of $f_{s,t}$ vanishes in $x_{s,t}$.
We have for all $x \in N(B)$

(18) $$s||Ax_{s,t} - y||^2 + s \cdot t||x_{s,t}||^2 \le f_{s,t}(x_{s,t}) - ||Bx_{s,t}||^2$$
$$\le f_{s,t}(x_{s,t})$$
$$\le f_{s,t}(x)$$
$$\le s||Ax-y||^2 + s \cdot t||x||^2 .$$

From this it follows

(19) $$\lim_{s \to +0} ||Bx_{s,t}||^2 = 0 .$$

Taking specially $x = 0$ in (18) we get

(20) $$t||x_{s,t}||^2 \le ||y||^2 ,$$

that is, $\{x_{s,t}\}_{s \to +0}$ is uniformly bounded, and thus it possesses a weak accumulation point x_t .

Let $\{x_{s',t}\}_{s' \to +0} \subset \{x_{s,t}\}_{s \to +0}$ be a subsequence converging weakly to x_t :

$$x_{s',t} \xrightarrow[s' \to +0]{} x_t .$$

Since the funktions above are lower semi-continuous in the weak topology we get from (18)

(21) $$||Ax_t - y||^2 + t||x_t||^2 \le \overline{\lim_{s' \to +0}} (||Ax_{s',t} - y||^2 + t||x_{s',t}||^2)$$
$$\le ||Ax-y||^2 + t||x||^2,$$
for all $x \in N(B)$,

and from (19)

(22) $$||Bx_t|| \le 0 \text{ , i.e } x_t \in N(B) .$$

Thus x_t minimizes $||Ax-y||^2 + t||x||^2$ over $N(B)$.
A minimizing point of a strictly convex funktion is uniquely determined, such that the whole sequence $\{x_{s,t}\}_{s \to +0}$ is converging weakly to x_t.

With the parallelogram law we have

$$||1/2\ A(x_{s,t} + x_t) - y||^2 + t||1/2(x_{s,t} + x_t)||^2 \le$$

$$\le 1/2||Ax_{s,t} - y||^2 + 1/2||Ax_t - y||^2 + t/2||x_{s,t}||^2 +$$

$$+ t/2||x_t||^2 - t||1/2(x_{s,t} - x_t)||^2 .$$

Using (21) with $x=x_t$ we get

(23) $$\overline{\lim_{s\to+0}}\ ||x_{s,t} - x_t|| \le 0 \text{ , i.e. } x_{s,t} \xrightarrow[s\to+0]{} x_t \text{ strongly .}$$

For all $x \varepsilon S(N(B),A,y)$, the minimizing points of $||Ax-y||$ over $N(B)$, the inequality (21) gives

(24) $$t||x_t||^2 \le ||Ax_t-y||^2 + t||x_t||^2 - ||Ax-y||^2$$

$$\le t||x||^2 \text{ , } x \varepsilon S(N(B),A,y) .$$

This causes

$$\lim_{t\to+0} ||Ax_t-y||^2 = ||Ax-y||^2 \text{ , } x\varepsilon S(N(B),A,y).$$

$\{x_t\}_{t\to+0}$ is uniformly bounded, such that for a weak accumulation point $\hat{x}$ we have

(25) $$||A\hat{x}-y||^2 \le ||Ax-y||^2, \ x\varepsilon S(N(B),A,y)$$

Since $x_t \varepsilon N(B)$, also $\hat{x}\varepsilon N(B)$.
(24) and (25) then yield

(26) $$||\hat{x}|| = \min\{||x|| \,\Big|\, x\varepsilon S(N(B),A,y)\} \text{ , that is } \hat{x} = A_B^+ y .$$

Thus $\hat{x}$ is unique and the whole sequence $\{x_t\}_{t\to+0}$ converges weakly to $\hat{x}$.
The parallelogram law yields with (24) and (26)

$$\overline{\lim_{t\to+0}}\ ||\hat{x}_t - x||^2 \le 4||\hat{x}||^2 - \underline{\lim_{t\to+0}}\ ||x_t+\hat{x}||^2 \le 0,$$

and we have

(27) $$\hat{x} = \lim_{t\to+0} x_t = \lim_{t\to+0} \lim_{s\to+0} x_{s,t} = A_B^+ y .$$

Let for some real parameter $p>0, q>0$ be

$$x_{p,q} = C_{p,q}^{-1}\, p\, A^* y,$$

then with the identities

$$(C_{s,t} - C_{p,q})x_{p,q} = (s-p)\, A^*Ax_{p,q} + (s\cdot t - p\cdot q)\, x_{p,q},$$

$$C_{s,t}\, x_{p,q} = (C_{s,t} - C_{p,q})\, x_{p,q} + pA^*y ,$$

we have

(28) $$\begin{aligned} C_{s,t}(x_{p,q} - x_{s,t}) &= (C_{s,t} - C_{p,q})\, x_{p,q} + pA^*y - sA^*y \\ &= pA^*y - pA^*Ax_{p,q} - pqx_{p,q} - sA^*y + \\ &\quad + stx_{p,q} + sA^*Ax_{p,q} \\ &= B^*Bx_{p,q} + s(A^*Ax_{p,q} - A^*y) + stx_{p,q} , \end{aligned}$$

and by (27)

(29) $$C_{s,t}(\hat{x} - x_{s,t}) = B^*B\,\hat{x} + s(A^*A\hat{x} - A^*y) + st\hat{x} .$$

Now it is $\hat{x} \in S(N(B),A,y)$, that is,

(30) $$B\hat{x} = 0,$$

and $$||A\hat{x} - y||^2 = \min_{x\in N(B)} ||Ax - y||^2$$, which implies, that $A^*A\hat{x} - A^*y \in N(B)^{\perp}$.

Because of B is assumed to be normally solvable it is $N(B)^{\perp} = R(B^*)$ and $(B^*)^+$ exists, such that $z_0 := (B^*)^+(A^*A\hat{x} - A^*y)$ satiesfies the equation

(31) $$B^* z_0 = A^*A\hat{x} - A^*y$$

and by (26) has the property

(32) $$||z_0|| \leq ||B^+||\cdot(||A^*A||\cdot||A_B^+|| + ||A||)\cdot||y|| \quad .$$

Let $u_1, u_2 \varepsilon S(N(B),A,y)$, then we have for all $v \varepsilon N(B)$: $\langle v,\ A^*Au_j - A^*y\rangle = 0$
resp. $\langle v,\ A^*A(u_1-u_2)\rangle = 0$ (j=1,2)
Since $u_1, u_2 \varepsilon N(B)$ it follows, that $u_1 - u_2 \varepsilon N(A) \cap N(B)$.
Thus we have the representation

(33) $$S(N(B),A,y) = \hat{x} + N(A) \cap N(B).$$

It is $||\hat{x}|| = \min\{||x|| \mid x \varepsilon S(N(B),A,y)\}$, such that by (33) it follows: $\hat{x} \varepsilon (N(A) \cap N(B))^{\perp}$.
Since $[A,B]$ is assumed normally solvable, it is $(N(A) \cap N(B))^{\perp} = R([A,B]^*)$ and $([A,B]^*)^+$ exists.
Then $(\hat{y},\hat{z}) := ([A,B]^*)^+ \hat{x}$ fulfills

(34) $$\hat{x} = [A,B]\ (\hat{y},\hat{z}) \ , \quad \text{and}$$

(35) $$||(\hat{y},\hat{z})|| \leq ||[A,B]^+||\cdot||A_B^+||\cdot||y|| \ .$$

Using (30),(31),(34) it follows from(29)

(36) $$C_{s,t}(\hat{x} - \hat{x}_{s,t}) = sB^*z_0 + st[A,B]^*(\hat{y},\hat{z}) \ .$$

It is $$||C_{s,t}^{-1}[A,B]^*|| = \sup\left\{\frac{\|C_{s,t}^{-1}[A,B]^*(y,z)\|}{||(y,z)||} \;\middle|\; (y,z)\varepsilon Y\times Z\right\}$$

$$= \sup\left\{\frac{||C_{s,t}^{-1}\ A^*y + C_{s,t}^{-1}B^*z||}{||(y,z)||} \;\middle|\; (y,z)\varepsilon Y\times Z\right\}$$

$$\leq \sup\left\{\frac{||C_{s,t}^{-1}A^*y||}{||(y,z)||} \;\middle|\; (y,z)\varepsilon Y\times Z\right\} \quad +$$

$$+\sup\left\{\frac{||C_{s,t}^{-1}B^*z||}{||(y,z)||} \;\middle|\; (y,z)\varepsilon Y\times Z\right\}$$

$$\leq ||C_{s,t}^{-1}A^*|| + ||C_{s,t}^{-1}B^*|| .$$

With $\hat{x} = A_B^+y$, $x_{s,t} = A_B(s,t)y$ then (36) yields

(37) $$||(A_B^+ - A_B(s,t))y|| \leq s||C_{s,t}^{-1} B^*||\cdot||z_o|| + $$
$$+st(||C_{s,t}^{-1} A^*|| + ||C_{s,t}^{-1} B^*||)\cdot||(\hat{y},\hat{z})|| .$$

From (20) we have : $||x_{s,t}||\leq t^{-1/2}||y||$, and since $x_{s,t}=C_{s,t}^{-1}sA^*y$ it follows

(38) $$||C_{s,t}^{-1} A^*|| \leq s^{-1}\cdot t^{-1/2} .$$

Let $z\varepsilon Z$, $g_{s,t}(u) := ||Bu - z||^2 + s||Au||^2 + st||u||^2$, then $u_{s,t} := C_{s,t}^{-1} B^*z$ minimizes $g_{s,t}(x)$ over X and we have

$$s\cdot t||u_{s,t}||^2 \leq g_{s,t}(u_{s,t}) - ||Bu_{s,t} - z||^2$$
$$\leq g_{s,t}(B^+z) - ||BB^+z - z ||^2$$
$$\leq s\cdot||AB^+z||^2 + s\cdot t||B^+z||^2 \text{ , that is}$$
$$||u_{s,t}||^2 \leq t^{-1}(||A||^2 + t)\cdot||B^+||^2\cdot||z||^2,$$

this implies

(39) $$||C_{s,t}^{-1} B^*|| \leq t^{-1/2}(||A||^2 + t)^{1/2} \cdot||B^+|| .$$

Using (ii),(32),(35),(38),(39) the inequality (37) then gives the estimate (iii) of the theorem.

If $B\equiv 0$ ($\rightarrow B^+\equiv 0$) we get from (iii) an estimate,which is different from (13) and (14):

(40) <u>Corollary:</u> $$||A^+ - A(s)|| \leq s^{1/2}||A^+||^2, s>0 .$$

References:

[1] Ben Israel,A.,Charnes,A.,Contributions to the theory of generalized inverses, J.Soc.Ind.Appl.Math.11,1963

[2] Glashoff,K.,Regularisierung und Penalty-Methoden, Dissertation, Hamburg 1972

[3] -----------Mehrstufige Optimierungsaufgaben,Oper.Res.Verf. (Meth.of Oper.Res.) XVI, 1973

[4] Holmes,R. B., A course on optimization and best approximation, §35: Generalized inverses, Lect.Not. in Math.,Springer,Berlin-Heidelberg-New York 1972

[5] Minamide,N.,Nakamura,K.,A restricted pseudoinverse and its application to constrained minima,Siam J.Appl. Math. 19,1970

[6] Morozov,V. A.,Pseudo-Solutions,Zh.vȳchisl.Mat.mat.Fiz.,9, 6, 1969

[7] Petryshyn,W.,On generalized inverses and on the uniform convergence of $(I-\beta K)^n$ with application to iterative methods,J.Math.Anal.Appl. 18, 1967.

CHARAKTERISIERUNG LOKALER PARETO-OPTIMA

R. Hettich, T.H. Twente, Enschede.

1. Einleitung. In der mathematischen Ökonomie spielen Pareto-optimale Zustände eine grosse Rolle ([3],[8]). Etwas vage formuliert, sind dies Zustände, in denen sich die Situation jedes Konsumenten höchstens auf Kosten anderer verbessern lässt. Insbesondere ist dabei die Frage von Bedeutung, ob sich solche Zustände als Preisgleichgewichte realisieren lassen. Unter einschneidenden Bedingungen (Konvexität der auftretenden Mengen und Funktionen usw.) wird das z.B. in [3] ausführlich diskutiert und bewiesen, dass sich jedes Pareto-Optimum als Preisgleichgewicht realisieren lässt und umgekehrt jedes solche Pareto-optimal ist. In jüngster Zeit wurden auch für den nichtkonvexen Fall derartige Untersuchungen angestellt ([6],[11], [12]), wobei die Strategie die ist, dass man Kriterien für Pareto-Optima einerseits und Preisgleichgewichte andererseits miteinander vergleicht. Gute Charakterisierungen von Pareto-Optima und Preisgleichgewichten sind hierfür natürlich wichtig. Ausser in den genannten Arbeiten findet man Kriterien für Pareto-Optima in [14] und [15]. Angesichts der mathematischen Verwandtschaft der Probleme ist es nicht überraschend, dass die Beweistechniken in allen Fällen denen für entsprechende Sätze in der Optimierungstheorie ähneln. So basiert etwa die Arbeit von Guesnerie [6] auf der von Dubovitskii und Miljutin [5].

In der vorliegenden Arbeit wird die in den letzten Jahren entwickelte Theorie der Charakterisierung optimaler Punkte von Optimierungsproblemen mit Nebenbedingungen (vgl. etwa [10],[7]) direkt benutzt, um notwendige und hinreichende Bedingungen erster und zweiter Ordnung für Pareto-Optima unter Nebenbedingungen abzuleiten. Jedes solche kann nämlich, wie wir in Abschnitt 2 zeigen (vgl. auch [4]), als Lösung eines gewissen Optimierungsproblems aufgefasst werden. In Abschnitt 3 folgt eine Zusammenfassung der in [7] abgeleiteten Kriterien für Optimierungsprobleme, die in Abschnitt 4 umgesetzt werden in Kriterien für Pareto-Optima. Der letzte Abschnitt enthält einige Schlussbemerkungen.

2. Pareto-Optima und Optimierung. Sei $Z \subseteq R^n$ eine beliebige Menge, $u^i(z)$, $i = 1, \ldots, m$, reellwertige, auf Z definierte Funktionen, die wir zu einer vektorwertigen Funktion $u(z) = (u^1(z), \ldots, u^m(z))^T$ zusammenfassen.

Definition 2.1: Ein $z^o \in Z$ heisst Pareto-optimal bzgl. Z, wenn es kein $z \in Z$ gibt mit

$$u(z) \geq u(z^o), \qquad u(z) \neq u(z^o). \tag{2.1}$$

z^o heisst stark Pareto-optimal bzgl. Z, wenn es kein $z \in Z \sim \{z^o\}$ gibt mit

$$u(z) \geq u(z^o). \tag{2.2}$$

z^o heisst lokal (stark) Pareto-optimal, wenn es eine Umgebung U_o von z^o gibt, sodass für kein $z \in U_o \cap Z$ (2.1)((2.2)) gilt, d.h. wenn z^o (stark) Pareto-optimal bzgl. $U_o \cap Z$ ist.

Wir zeigen (Satz 2.1), dass jedem Pareto-optimalen z^o eine optimale Lösung eines gewissen Optimierungsproblems entspricht und umgekehrt (Satz 2.2) jeder stark optimalen Lösung ein stark Pareto-optimaler Punkt.

O.B.d.A. können wir uns auf den Fall beschränken, dass im betrachteten Punkt $z^o \in Z$ alle Funktionswerte $u^i(z^o)$, $i = 1, \ldots, m$, negativ sind, da die Addition ein und derselben Konstanten zu jeder der Funktionen u^i an der Pareto-Optimalität von z^o nichts ändert.

Satz 2.1: z^o sei Pareto-optimal bzgl. Z, $u^i(z^o) < 0$, $i = 1, \ldots, m$. Sei $w_i = -(u^i(z^o))^{-1}$, $i = 1, \ldots, m$. Dann ist z^o, $v^o = \min_{i=1,\ldots,m} w_i u^i(z^o) = -1$ optimale Lösung des Optimierungsproblems:

(O) Mache $F(z,v) = v$ maximal unter den Nebenbedingungen $z \in Z$, $v - w_i u^i(z) \leq 0$, $i = 1, \ldots, m$.

Beweis: Angenommen, es gebe $v^* > -1$, $z^* \in Z$ mit $v^* - w_i u^i(z^*) \le 0$, $i = 1, \ldots, m$. D.h. aber, es wäre

$$-1 < v^* \le -(u^i(z^o))^{-1} u^i(z^*), \quad i = 1, \ldots, m,$$

also $u^i(z^o) < u^i(z^*)$, $i = 1, \ldots, m$, und z^o wäre nicht Pareto-optimal.

Satz 2.2: Seien $w_i > 0$, $i = 1, \ldots, m$, gegebene Konstanten. z^o, v^o sei stark optimale Lösung des Optimierungsproblems (0). Dann ist z^o stark Pareto-optimal bzgl. Z.

Beweis: Angenommen, es gebe ein $z^* \in Z \sim \{z^o\}$ mit $u^i(z^*) \ge u^i(z^o)$, $i = 1, \ldots, m$. Wegen $w_i > 0$ gilt dann auch $w_i u^i(z^*) \ge w_i u^i(z^o)$, $i = 1, \ldots, m$. Es sei $v^* = \min_{i=1,\ldots,m} w_i u^i(z^*)$. z^*, v^* ist dann eine zulässige Lösung von (0) und es gilt

$$v^* = \min_{i=1,\ldots,m} w_i u^i(z^*) \ge \min_{i=1,\ldots,m} w_i u^i(z^o) \ge v^o$$

im Widerspruch zur Voraussetzung, dass z^o, v^o eine stark optimale Lösung des Optimierungsproblems (0) ist.

Bemerkung: Ist z^o, v^o stark optimale Lösung von (0) und ist $v^o - w_k u^k(z^o) < 0$ für ein bestimmtes k, so bleibt z^o, v^o stark optimale Lösung, wenn man w_k ersetzt durch $w_k' = v^o/u^k(z^o)$ (es ist $u^i(z^o) < 0$, $i=1,\ldots,m$). Wir können daher o.B.d.A. annehmen, dass z^o, v^o die zusätzliche Eigenschaft hat, dass $v^o - w_i u^i(z^o) = 0$, $i = 1, \ldots, m$. Bei der in Satz 2.1 getroffenen Wahl der w_i gilt dies ebenfalls.

3. Extremalkriterien für Optimierungsaufgaben. Beweise der in diesem Abschnitt behandelten Sätze findet man, sofern sie hier nicht gegeben werden, in [7].

Es seien I und J endliche Indexmengen, F, f^i, $i \in I$, und g^j, $j \in J$, auf dem $\mathbb{R}^n$ definierte, reellwertige, stetig differenzierbare Funktionen. Ferner sei $X \subseteq \mathbb{R}^n$ die Menge

$$X = \{x \in \mathbb{R}^n \mid f^i(x) \le 0, \ i \in I, \ g^j(x) = 0, \ j \in J\}. \tag{3.1}$$

I und J dürfen dabei auch leer sein.

Wir betrachten das folgende Optimierungsproblem:

Bestimme $x^o \in X$, sodass $F(x^o) \geq F(x)$ für alle $x \in X$. (3.2)

Ein $x^o \in X$, das (3.2) genügt, heisst optimale Lösung. Ist in (3.2) Gleichheit nur für $x = x^o$ möglich, so nennen wir x^o stark optimal. Gilt (3.2) nur für Punkte x aus einer Umgebung $U_o \subset X$ von x^o (X sei mit der Relativtopologie versehen), so heisst x^o (stark) lokal optimal. Die folgenden Sätze geben notwendige und hinreichende Kriterien dafür, dass ein $x^o \in X$ (lokal) optimale Lösung ist.

Wir führen zunächst einige Bezeichnungen ein.

Im folgenden ist x^o stets ein fester Punkt in X. Es sei

$$I_o = \{i \in I \mid f^i(x^o) = 0\}. \tag{3.3}$$

Den Gradienten einer Funktion von x bezeichnen wir durch einen tiefgestellten Index x. Ein doppelter unterer Index x bezeichnet die Matrix der zweiten Ableitungen. Fehlt das Argument der Funktion, so ist der Funktionswert im Punkt x^o gemeint. So ist beispielsweise F_x der Gradient von F im Punkt x^o.

P sei die Menge der Lösungen ξ des folgenden Systems linearer Relationen

$$\xi^T F_x \geq 0, \tag{3.4}$$

$$-\xi^T f_x^i \geq 0, \; i \in I_o, \tag{3.5}$$

$$\xi^T g_x^j = 0, \; j \in J. \tag{3.6}$$

Wegen $0 \in P$ ist $P \neq \emptyset$. Weiter sei

$$I_1 = \{i \in I_o \mid \xi^T f_x^i = 0 \text{ für alle } \xi \in P\}. \tag{3.7}$$

Die Ungleichungen $-\xi^T f_x^i \geq 0$, $i \in I_1$, heissen singulär. Ferner sei

$$P_o = \{\xi \in P \mid \xi^T F_x = 0\} \tag{3.8}$$

und

$$P_s = \{\xi \in P \mid -\xi^T f_x^i > 0, \; i \in I_o \sim I_1\}. \tag{3.9}$$

Es ist einfach zu sehen, dass P_s nicht leer ist.

Zum Beweis von Satz 3.1 benötigen wir die folgende Verallgemeinerung des Lemmas von Farkas (vgl. etwa [13]).

Hilfssatz 3.1: Sei $d \in R^n$, A eine m x n- und B eine ℓ x n-Matrix. Dann gilt

$$d^T x > 0, \qquad -Ax \geq 0, \qquad Bx = 0$$

ist genau dann nicht lösbar, wenn es $u \in R^m$, $w \in R^\ell$, $u \geq 0$, gibt mit

$$d = A^T u + B^T w.$$

Der folgende Satz gibt hinreichende Kriterien erster Ordnung für lokal stark optimale Lösungen.

Satz 3.1: (a) Ist $P = \{0\}$, so ist x^o eine lokal stark optimale Lösung.
(b) Gibt es unter den Vektoren F_x, f_x^i, $i \in I_o$, g_x^j, $j \in J$, n linear unabhängige, so ist $P = \{0\}$ äquivalent mit:
Es gibt Zahlen $\lambda_i > 0$, $i \in I_o$, und μ_j, $j \in J$, sodass

$$F_x = \sum_{i \in I_o} \lambda_i f_x^i + \sum_{j \in J} \mu_j g_x^j. \tag{3.10}$$

Beweis: (a) wurde in [7] bewiesen.
(3.10) sei erfüllt mit $\lambda_i > 0$, $i \in I_o$. Sei $\xi \in P$. Dann gilt

$$0 \leq \xi^T F_x = \sum_{i \in I_o} \lambda_i \xi^T f_x^i + \sum_{j \in J} \mu_j \xi^T g_x^j \leq 0,$$

also wegen $\lambda_i > 0$

$$\xi^T f_x^i = 0, \; i \in I_o, \; \xi^T g_x^j = 0, \; j \in J, \; \xi^T F_x = 0.$$

Da n dieser Gleichungen linear unabhängig sind, folgt $\xi = 0$ und daher $P = \{0\}$.

Sei umgekehrt $P = \{0\}$. Für jedes $k \in I_o$ ist dann das System

$$\xi^T F_x \geq 0, \qquad -\xi^T f_x^i \geq 0, \; i \in I_o \sim \{k\}$$

$$\xi^T g_x^j = 0, \; j \in J, \qquad -\xi^T f_x^k > 0$$

unlösbar und ebenso das System

$$\xi^T F_x > 0, \qquad -\xi^T f_x^i \geq 0,\ i \in I_o\ , \qquad \xi^T g_x^j = 0,\ j \in J.$$

Nach Hilfssatz 3.1 gibt es Zahlen q_j, q_j^k, $r_k \geq 0$, $p_i \geq 0$ und $p_i^k \geq 0$, sodass

$$F_x = \sum_{i\in I_o} p_i f_x^i + \sum_{j\in J} q_j g_x^j$$

$$r_k F_x = f_x^k + \sum_{i\in I_o \sim \{k\}} p_i^k f_x^i + \sum_{j\in J} q_j^k g_x^j, \qquad k \in I_o.$$

Summation dieser Gleichungen und Division durch $r = 1 + \sum_{k\in I_o} r_k$ ergibt (3.10) mit $\lambda_k = \frac{1}{r}(1+\sum_{i\in I_o} p_k^i + p_k) > 0$. Damit ist Satz 3.1 bewiesen.

<u>Bemerkung</u>: Dem Beweis ist zu entnehmen, dass (3.10) mit $\lambda_i > 0$ aus $P = \{0\}$ auch ohne die Voraussetzung der linearen Unabhängigkeit von n der Vektoren F_x, f_x^i, g_x^j folgt.

Ein notwendiges Kriterium für eine optimale Lösung gibt der bekannte Satz von John [9]:

<u>Satz 3.2</u>: x^o sei optimale Lösung. Dann gibt es Zahlen $\lambda \geq 0$, $\lambda_i \geq 0$, $i \in I_o$, μ_j, $j \in J$, nicht alle gleich Null, sodass

$$\lambda F_x = \sum_{i\in I_o} \lambda_i f_x^i + \sum_{j\in J} \mu_j g_x^j. \tag{3.11}$$

Es gibt eine Reihe von Bedingungen ("constraint qualifications"), unter denen man in (3.11) $\lambda = 1$ setzen kann. Die folgende wird sich später für Pareto-Optima als geeignet erweisen.

(Q1). Zu jedem $\xi \in P_s$ gebe es ein $t_o > 0$ und eine in $[0, t_o]$ stetig differenzierbare vektorwertige Funktion $x(t)$ mit den Eigenschaften

$$x(0) = x^o \tag{3.12}$$

$$\frac{dx(0)}{dt} = \xi \tag{3.13}$$

$$x(t) \in X \text{ für } t \in [0,t_o]. \tag{3.14}$$

Satz 3.3: x^o sei optimale Lösung und (Q1) sei erfüllt. Dann gibt es $\lambda_i \geq 0$, $i \in I_o$, und μ_j, $j \in J$, sodass (3.10) gilt.

Wir wenden uns nun Kriterien zweiter Ordnung zu und setzen für den Rest dieses Abschnitts voraus, dass die Funktionen F, f^i und g^j zweimal stetig differenzierbar sind. Wir beginnen mit einem hinreichenden Kriterium.

Satz 3.4: Zu jedem $\xi \in P_o$ gebe es $\lambda_i \geq 0$, $i \in I_1$, und μ_j, $j \in J$, sodass (3.10) gilt und, falls $\xi \neq 0$,

$$q(\lambda_i,\mu_j,\xi) = \xi^T(F_{xx} - \sum_{i \in I_1} \lambda_i f^i_{xx} - \sum_{j \in J} \mu_j g^j_{xx})\xi < 0. \qquad (3.15)$$

Dann ist x^o eine lokal stark optimale Lösung.

Für das entsprechende notwendige Kriterium brauchen wir wieder eine zusätzliche Voraussetzung.

(Q2) Sei $\hat{P} \subseteq P$, $\hat{P} \neq \emptyset$. Für jedes $\xi \in \hat{P}$ sei das System

$$\eta^T f^i_x \leq - \xi^T f^i_{xx}\xi, \qquad i \in I_o, \qquad (3.16)$$

$$\eta^T g^j_x = - \xi^T g^j_{xx}\xi, \qquad j \in J, \qquad (3.17)$$

lösbar und zu jeder Lösung η gebe es ein $t_o > 0$ und eine in $[0,t_o]$ stetig differenzierbare vektorwertige Funktion x(t) mit den Eigenschaften (3.12), (3.13), (3.14) und

$$\frac{d^2x(t)}{dt^2} \text{ existiert für } t = 0 \text{ und es ist } \frac{d^2x(0)}{dt^2} = \eta. \qquad (3.18)$$

Satz 3.5: x^o sei optimale Lösung und (Q2) sei erfüllt. Dann gibt es zu jedem $\xi \in \hat{P}$ Zahlen $\lambda_i \geq 0$, $i \in I_1$, und μ_j, $j \in J$, sodass (3.10) gilt und $q(\lambda_i,\mu_j,\xi) \leq 0$ (vgl. (3.15)).

Da die Beweisführung in [7] bei diesem Satz von der sonst bei solchen Sätzen üblichen abweicht, wollen wir den Beweis kurz skizzieren.

Beweis: Sei $\xi \in P$, η eine Lösung von (3.16), (3.17) und x(t) wie in (Q2). Die Funktion $\psi(t) = F(x(t))$ hat in $t = 0$ ein Maximum. $\psi(t)$ besitzt für $t = 0$ eine zweite Ableitung und es gilt $\frac{d\psi(0)}{dt} = \xi^T F_x = 0$, $\frac{d^2\psi(0)}{dt^2} = \eta^T F_x + \xi^T F_{xx}\xi \leq 0$. Somit ist der Zielwert (d.i. der maximale Wert der Zielfunktion) der folgenden linearen Optimierungsaufgabe

(ξ fest, variabel ist η) nicht positiv:

Mache $\eta^T F_x + \xi^T F_{xx}\xi$ maximal unter den Nebenbedingungen (3.16), (3.17).

Die duale Aufgabe ist:

Mache $\xi^T(F_{xx} - \sum_{i\in I_o} \lambda_i f^i_{xx} - \sum_{j\in J} \mu_j g^j_{xx})\xi$ minimal unter den Nebenbedingungen $\lambda_i \geq 0$, $i \in I_o$, und (3.10).

Aus dem Dualitätssatz der linearen Optimierung (vgl. etwa [2]) folgt, dass dieses Problem lösbar ist und der Zielwert gleich dem der primalen Aufgabe ist. Multiplikation von (3.10) mit ξ^T von links zeigt, dass $\lambda_i = 0$ für $i \in I_o \sim I_1$, was den Beweis beendet.

Bemerkung: Da die λ_i und μ_j in den Sätzen 3.4 und 3.5 von ξ abhängen können, und damit auch die quadratische Form q, sind die Bedingungen keine Definitheitsbedingungen im üblichen Sinn. Dass die Definitheit eine i.a. zu starke Forderung ist, wird in [7] an einem Beispiel gezeigt. Dort wird ausserdem eine Bedingung angegeben, unter der Definitheit folgt.

4. Kriterien für Pareto-Optima.

Wir betrachten das Problem aus Abschnitt 2, wobei wir annehmen, dass Z gegeben sei durch

$$Z = \{z \in \mathbb{R}^n \mid h^j(z) \leq 0,\ j=1,\ldots,m_h;\ g^\ell(z) = 0,\ \ell=1,\ldots,m_g\} \quad (4.1)$$

mit im R^n stetig differenzierbaren Funktionen h^j und g^ℓ. Die $u^i(z)$, $i = 1, \ldots, m$ seien ebenfalls stetig differenzierbar. Im Hinblick auf die Sätze in Abschnitt 2 betrachten wir zu gegebenen $w_i > 0$, $i = 1, \ldots, m$, das Optimierungsproblem:

Mache $F(x) = x_{n+1}$, $x = \begin{pmatrix} z \\ x_{n+1} \end{pmatrix}$, $z \in \mathbb{R}^n$, $x_{n+1} \in \mathbb{R}$, maximal unter den Nebenbedingungen

$$x_{n+1} - w_i u^i(z) \leq 0, \qquad i = 1, \ldots, m \quad (4.2)$$

$$h^j(z) \leq 0, \qquad j = 1, \ldots, m_h \quad (4.3)$$

$$g^\ell(z) = 0, \qquad \ell = 1, \ldots, m_g. \quad (4.4)$$

Es sei

$$X = \{x \in \mathbb{R}^{n+1} \mid x = \begin{pmatrix} z \\ x_{n+1} \end{pmatrix} \text{ erfüllt (4.2), (4.3), (4.4)}\}. \qquad (4.5)$$

$x^o = \begin{pmatrix} z^o \\ \\ x^o_{n+1} \end{pmatrix} \in X$ sei ein fester Punkt. Entsprechend der Bemerkung am Ende von Abschnitt 2 nehmen wir zusätzlich an, dass
$x^o_{n+1} - w_i u^i(z^o) = 0, \quad i = 1, \ldots, m.$

Wir definieren

$$I^h_o = \{i \in \{1, \ldots, m_h\} \mid h^i(z^o) = 0\}. \qquad (4.6)$$

Mit $\xi = \begin{pmatrix} \zeta \\ \xi_{n+1} \end{pmatrix} \in \mathbb{R}^{n+1}$ geht das System (3.4), (3.5), (3.6) über in

$$\xi_{n+1} \geq 0 \qquad (4.7)$$

$$-\xi_{n+1} + w_i \zeta^T u^i_z \geq 0, \; i = 1, \ldots, m \qquad (4.8)$$

$$-\zeta^T h^j_z \geq 0, \; j \in I^h_o \qquad (4.9)$$

$$\zeta^T g^\ell_z = 0, \; \ell = 1, \ldots, m_g. \qquad (4.10)$$

P sei wieder die Menge der Lösungen ξ dieses Systems und

$$I^u_1 = \{i \in \{1, \ldots, m\} \mid w_i \zeta^T u^i_z = \xi_{n+1} \text{ für alle } \xi \in P\}, \qquad (4.11)$$

$$I^h_1 = \{i \in I^h_o \mid \zeta^T h^i_z = 0 \text{ für alle } \xi \in P\}. \qquad (4.12)$$

P_o und P_s werden entsprechend (3.8), (3.9) definiert.

Der Beweis des folgenden Hilfssatzes ist trivial.

Hilfssatz 4.1: (a) Genau dann ist $P = \{0\}$, wenn das folgende System nur die Lösung $\zeta = 0$ besitzt:

$$\begin{aligned} &\zeta^T u^i_z \geq 0, \; i = 1, \ldots, m, \qquad -\zeta^T h^j_z \geq 0, \; j \in I^h_o \\ &\zeta^T g^\ell_z = 0, \; \ell = 1, \ldots, m_g. \end{aligned} \qquad (4.13)$$

(b) Unter den Vektoren $F_x = (0,\ldots,0,1)^T, \begin{pmatrix} u^i_z \\ -1 \end{pmatrix}, \; i = 1, \ldots, m,$

$\begin{pmatrix} h_z^j \\ 0 \end{pmatrix}$, $j \in I_o^h$, und $\begin{pmatrix} g_z^\ell \\ 0 \end{pmatrix}$, $\ell = 1, \ldots, m_g$, gibt es genau dann n + 1 linear unabhängige, wenn es unter den u_z^i, h_z^j, g_z^ℓ n linear unabhängige gibt.

Hiermit und mit Satz 2.2 erhält man aus Satz 3.1 unmittelbar den folgenden.

Satz 4.1: (a) $\bar{P}$ sei die Menge der Lösungen von (4.13). Ist $\bar{P} = \{0\}$, so ist z^o lokal stark Pareto-optimal.

(b) Gibt es unter den Vektoren u_z^i, $i = 1, \ldots, m$, h_z^j, $j \in I_o^h$, und g_z^ℓ, $\ell = 1, \ldots, m_g$, n linear unabhängige, so ist $\bar{P} = \{0\}$ äquivalent mit:

Es gibt Zahlen $\lambda_i > 0$, $i = 1, \ldots, m$, $\rho_j > 0$, $j \in I_o^h$, μ_ℓ, $\ell = 1, \ldots, m_g$, sodass

$$\sum_{i=1}^{m} \lambda_i u_z^i = \sum_{j \in I_o^h} \rho_j h_z^j + \sum_{\ell=1}^{m_g} \mu_\ell g_z^\ell . \tag{4.14}$$

Aus Satz 3.2 und Satz 2.1 folgt ein Satz, der sich auch bei Smale [12] findet:

Satz 4.2: z^o sei Pareto-optimal. Dann gibt es $\lambda_i \geq 0$, $i = 1, \ldots, m$, $\rho_j \geq 0$, $j \in I_o^h$, und μ_ℓ, $\ell = 1, \ldots, m_g$, (nicht alle λ_i, ρ_j, μ_ℓ gleich Null), sodass (4.14) gilt.

Als nächstes wollen wir (Q1) auf eine Bedingung an Z reduzieren. Es sei

$$\bar{P}_s = \{\zeta \in \bar{P} \mid \zeta^T u_z^i > 0,\ i \in \{1, \ldots, m\} \sim I_1^u,\ -\zeta^T h_z^j > 0,$$
$$j \in I_o^h \sim I_1^h\}.$$

(A). Zu jedem $\zeta \in \bar{P}_s$ gebe es ein $t_o > 0$ und eine in $[0,t_o]$ stetig differenzierbare vektorwertige Funktion z(t) mit den Eigenschaften

$$z(0) = z^o \tag{4.15}$$

$$\frac{dz(0)}{dt} = \zeta \tag{4.16}$$

$$z(t) \in Z \text{ für } t \in [0,t_o]. \tag{4.17}$$

<u>Hilfssatz 4.2</u>: Gilt (A), so ist (Q1) für das zu Beginn dieses Abschnitts formulierte Optimierungsproblem erfüllt.

<u>Beweis</u>: Sei $\xi = \begin{pmatrix}\zeta \\ \xi_{n+1}\end{pmatrix} \in P_s$. Dann ist $\zeta \in \bar{P}_s$. $z(t)$ sei wie in (A). Wir sind fertig, wenn wir zeigen können, dass es eine in $[0,t_o]$ stetig differenzierbare Funktion $x_{n+1}(t)$ gibt mit $x_{n+1}(0) = x^o_{n+1}$, $\frac{dx_{n+1}(0)}{dt} = \zeta_{n+1}$ und $x_{n+1}(t) \le w_i u^i(z(t))$ in $[0,t_o]$, $i = 1, \ldots, m$.

Die Funktion $q(t) = \min \{\xi_{n+1}, w_i \frac{d}{dt} u^i(z(t)) \mid i = 1, \ldots, m\}$ ist stetig und es ist $q(0) = \xi_{n+1}$ wegen $\xi \in P$.

$x_{n+1}(t) = x^o_{n+1} + \int_0^t q(\tau)d\tau$ erfüllt $x_{n+1}(0) = x^o_{n+1}$ und $\frac{dx_{n+1}(0)}{dt} = \xi_{n+1}$.

Wegen $x_{n+1}(0) \le w_i u^i(z(0))$ $(x^o \in X)$ und

$\frac{dx_{n+1}(t)}{dt} = q(t) \le \frac{d}{dt} w_i u^i(z(t))$ in $[0,t_o]$ ist für $i = 1, \ldots, m$

$x_{n+1}(t) \le w_i u^i(z(t))$ in $[0,t_o]$.

Damit ist der Hilfssatz bewiesen.

Aus Satz 3.3 folgt nun:

<u>Satz 4.3</u>: z^o sei Pareto-optimal und (A) sei erfüllt. Dann gibt es $\lambda_i \ge 0$, $i = 1, \ldots, m$, $\sum_{i=1}^m \lambda_i = 1$, $\rho_j \ge 0$, $j \in I^h_o$, und μ_ℓ, $\ell = 1, \ldots, m_g$, sodass (4.14) gilt.

Zum Beweis bemerken wir nur, dass Satz 3.3 unmittelbar die Existenz von $\bar{\lambda}_i$, $\bar{\rho}_j$, $\bar{\mu}_\ell$, $\sum_{i=1}^m \bar{\lambda}_i$ liefert, sodass

$$\sum_{i=1}^m \bar{\lambda}_i w_i u^i_z = \sum_{j\in I^h_o} \bar{\rho}_j h^j_z + \sum_{\ell=1}^{m_g} \bar{\mu}_\ell g^z_\ell .$$

Mit $p = (\sum_{i=1}^m \bar{\lambda}_i w_i)^{-1}$ haben dann $\lambda_i = pw_i\bar{\lambda}_i$, $\rho_j = p\bar{\rho}_j$ und $\mu_\ell = p\bar{\mu}_\ell$ die behaupteten Eigenschaften.

Um die Kriterien zweiter Ordnung übertragen zu können, setzen wir für den Rest dieses Abschnitts voraus, dass die Funktionen u^i, h^j und g^ℓ zweimal stetig differenzierbar sind. Beachtet man, dass $\zeta \in \bar{P}$ wenn $\xi = \begin{pmatrix}\zeta \\ \xi_{n+1}\end{pmatrix} \in P_o$ und dass ξ_{n+1} in (3.15) keinen Beitrag liefert, so erhält man aus Satz 3.4 den folgenden Satz.

Satz 4.4: Zu jedem $\zeta \in \bar{P}$ gebe es $\lambda_i \geq 0$, $i \in I_1^u$, $\sum_{i \in I_1^u} \lambda_i = 1$, $\rho_j \geq 0$, $j \in I_1^h$, und μ_ℓ, $\ell = 1, \ldots, m_g$, sodass (4.14) gilt und, falls $\zeta \neq 0$ ist,

$$q(\lambda_i, \rho_j, \mu_\ell, \zeta) = \zeta^T \Big(\sum_{i \in I_1^u} \lambda_i u_{zz}^i - \sum_{j \in I_1^h} \rho_j h_{zz}^j - \sum_{\ell=1}^{m_g} \mu_\ell g_{zz}^\ell \Big) \zeta < 0. \quad (4.18)$$

Dann ist z^o lokal stark Pareto-optimal.

Ähnlich wie (Q1) reduzieren wir (Q2) auf eine Bedingung an Z.

(B). Sei $\hat{\hat{P}} \subseteq \bar{P}$, $\hat{\hat{P}} \neq \emptyset$. Für jedes $\zeta \in \hat{\hat{P}}$ sei das System

$$\nu^T h_z^j \leq - \zeta^T h_{zz}^j \zeta, \quad j \in I_o^h \quad (4.19)$$

$$\nu^T g_z^\ell = - \zeta^T g_{zz}^\ell \zeta, \quad \ell = 1, \ldots, m_g, \quad (4.20)$$

lösbar und zu jeder Lösung ν gebe es $t_o > 0$ und eine in $[0, t_o]$ zweimal stetig differenzierbare vektorwertige Funktion $z(t)$ mit den Eigenschaften (4.15), (4.16) und (4.17) aus (A) und

$$\frac{d^2 z(0)}{dt^2} = \nu. \quad (4.21)$$

Hilfssatz 4.3: Gilt (B), so ist (Q2) für das zu Beginn dieses Abschnitts formulierte Optimierungsproblem erfüllt für

$$\hat{P} = \{\xi = \begin{pmatrix} \zeta \\ \xi_{n+1} \end{pmatrix} \mid \zeta \in \hat{\hat{P}}, \ \xi_{n+1} = 0\} \subseteq P_o.$$

Beweis: Sei $\xi \in \hat{P}$. Mit $\eta = \begin{pmatrix} \nu \\ \eta_{n+1} \end{pmatrix}$ geht das System (3.16), (3.17) über in die Ungleichungen (4.19), (4.20) und

$$\eta_{n+1} - w_i \nu^T u_z^i \leq w_i \zeta^T u_{zz}^i \zeta, \quad i = 1, \ldots, m. \quad (4.22)$$

Wegen $\xi = \begin{pmatrix} \zeta \\ \xi_{n+1} \end{pmatrix} \in \hat{P}$ ist $\zeta \in \hat{\hat{P}}$ und es gibt stets ein ν, sodass (4.19) und (4.20) erfüllt sind. Für $\eta_{n+1} \leq \min \{w_i(\nu^T u_z^i + \zeta^T u_{zz}^i \zeta) \mid i = 1, \ldots, m\}$ ist dann auch (4.22) erfüllt; das System (4.19), (4.20), (4.22) ist also für jedes $\xi \in \hat{P}$ lösbar.

Sei $\eta = \begin{pmatrix} \nu \\ \eta_{n+1} \end{pmatrix}$ eine Lösung, $z(t)$ wie in (B). Wir sind fertig, wenn wir eine zweimal stetig differenzierbare Funktion $x_{n+1}(t)$ angeben können mit den Eigenschaften $x_{n+1}(0) = x^o_{n+1}$, $\frac{dx_{n+1}(0)}{dt} = \xi_{n+1} = 0$,

$\frac{d^2x_{n+1}(0)}{dt^2} = \eta_{n+1}$ und $x_{n+1}(t) \le w_i u^i(z(t))$, $t \in [0,t_o]$, $i = 1, \ldots, m$.

$s(t) = \min \{\eta_{n+1}, \frac{d^2}{dt^2} w_i u^i(z(t)) \mid i = 1, \ldots, m\}$ ist wegen der zweimaligen stetigen Differenzierbarkeit von z und u^i, $i = 1, \ldots, m$, stetig in $[0,t_o]$.

$x_{n+1}(t) = x^o_{n+1} + \int_0^t [\int_0^\tau s(\sigma)d\sigma]d\tau$ erfüllt $x_{n+1}(0) = x^o_{n+1}$ und $\frac{dx_{n+1}(0)}{dt} = 0$.

Wegen $\frac{d^2}{dt^2} w_i u^i(z(t))\Big|_{t=0} = w_i \nu^T u^i_z + w^i \zeta^T u^i_{zz}\zeta \ge \eta_{n+1}$ ist

$\frac{d^2x_{n+1}(0)}{dt^2} = s(0) = \eta_{n+1}$.

Für $i = 1, \ldots, m$ gilt:

$$x_{n+1}(0) \le w_i u^i(z(0)) \qquad (x^o \in X)$$

$$\frac{dx_{n+1}(0)}{dt} = 0 = \xi_{n+1} \le w_i \zeta^T u^i_z = \frac{d}{dt} w_i u^i(z(t))\Big|_{t=0} \qquad (\xi \in P_o)$$

und $$\frac{d^2x_{n+1}(t)}{dt^2} = s(t) \le \frac{d^2}{dt^2} w_i u^i(z(t)) \text{ in } [0,t_o].$$

Hieraus folgt $x_{n+1}(t) \le w_i u^i(z(t))$, $t \in [0,t_o]$, für $i = 1, \ldots, m$, womit der Hilfssatz bewiesen ist.

Aus Satz 3.5 erhalten wir somit:

<u>Satz 4.5</u>: z^o sei Pareto-optimal und (B) sei erfüllt. Dann gibt es zu jedem $\zeta \in \hat{\hat{P}}$ Zahlen $\lambda_i \ge 0$, $i \in I^u_1$, $\sum_{i \in I^u_1} \lambda_i = 1$, $\rho_j \ge 0$, $j \in I^h_1$, und μ_ℓ, $\ell = 1, \ldots, m_g$, sodass (4.14) gilt und (vgl. 4.18) $q(\lambda_i, \rho_j, \mu_\ell, \zeta) \le 0$.

Wir schliessen diesen Abschnitt mit dem Nachweis, dass die Bedingungen (A) und (B) für Gebiete Z, die durch lineare Nebenbedingungen definiert sind, stets gelten.

Hilfssatz 4.4: Sind die $h^j(z)$ und $g^\ell(z)$ lineare Funktionen: $h^j(z) = a_j^T z + b_j$, $g^\ell(z) = c_\ell^T z + d_\ell$, so sind (A) und (B) für $\hat{\bar{P}} = \bar{P}$ erfüllt.

Beweis: Wir zeigen (B) für $\hat{\bar{P}} = \bar{P}$, womit zugleich auch (A) bewiesen ist. Sei $\zeta \in \bar{P}$. Wegen $h_{zz}^j = g_{zz}^\ell = 0$ (0-Matrizen) wird aus (4.19), (4.20)

$$\nu^T a_j \le 0, \quad j \in I_o^h, \qquad \nu^T c_\ell = 0, \quad \ell = 1, \ldots, m_g,$$

ein System, das stets die Lösung $\nu = 0$ hat, also stets lösbar ist. Sei ν eine solche Lösung. Setzt man

$$z(t) = z^o + t\zeta + \frac{1}{2}t^2\nu,$$

so gelten (4.15), (4.16) und (4.21). Zu zeigen bleibt, dass es ein $t_o > 0$ gibt mit $z(t) \in Z$ für $t \in [0,t_o]$.
Es ist $h^j(z(t)) = a_j^T z_o + b_j + t\zeta^T a_j + \frac{1}{2}t^2\nu^T a_j$.
Ist $j \notin I_o^h$, so ist $a_j^T z_o + b_j < 0$ und somit gibt es ein $t^j > 0$, sodass $h^j(z(t)) \le 0$ für $t \in [0,t^j]$. Setze

$$t_o = \min\{1, t^j \mid j = 1, \ldots, m_h, j \notin I_o^h\}.$$

Dann ist $h^j(z(t)) \le 0$, $t \in [0,t_o]$, für $j \notin I_o^h$.
Ist $j \in I_o^h$, so wird $h^j(z(t)) = t\zeta^T a_j + \frac{1}{2}t^2\nu^T a_j \le 0$ für alle $t \ge 0$.
Ferner gilt $g^j(z) = c_j^T z_o + d_j + t\zeta^T c_j + \frac{1}{2}t^2\nu^T c_j = 0$ für alle $t \ge 0$.
Also ist (4.21) und damit (B) erfüllt und der Hilfssatz ist bewiesen.

5. Schlussbemerkungen. Die Ergebnisse, die wir in Abschnitt 4 erhalten haben, verallgemeinern die von Smale und von Wan. Weinberger kommt zu ähnlichen Resultaten, jedoch ist bei ihm sowohl bei den notwendigen als bei den hinreichenden Bedingungen der Fall, dass alle λ_i gleich Null sind, nicht ausgeschlossen. Bei den hinreichenden Bedingungen wird dafür angenommen, dass (4.14) erfüllbar ist mit $\rho_j > 0$, $j \in I_o^h$.
Die Theorie lässt sich ohne Schwierigkeiten auf semi-infinite Probleme ausdehnen. Abschnitt 2 bleibt voll gültig, sodass die Kriterien aus [16] oder [7] unmittelbar übertragbar sind.

Literatur.

[1] L. Collatz, W. Krabs, Approximationstheorie, Teubner Studienbücher Mathematik, Teubner, Stuttgart, 1973.

[2] L. Collatz, W. Wetterling, Optimierungsaufgaben, Heidelberger Taschenbücher Bd. 15, Springer, Berlin-Heidelberg-New York, 1966.

[3] G. Debreu, Theory of Value, Wiley, New York, 1959.

[4] Dixon, The theory of joint maximization, North-Holland Publ. Comp., Amsterdam-Oxford, Amer. Elsevier Publ. Comp., Inc., New York, 1975.

[5] A.J. Dubovitskii, A.A. Miljutin, Extremum Problems in the Presence of Restrictions, Zh. Vychisl. Mat. Fiz., 5, 395-453, 1965; U.S.S.R. Comp. Math. and Math. Physics, 5, 1-80.

[6] R. Guesnerie, Pareto Optimality in Non-Convex Economics, Econometrica, 43, 1-29, 1975.

[7] R. Hettich, Extremalkriterien für Optimierungs- und Approximationsaufgaben, Technische Hogeschool Twente, Enschede, Dissertation, 1973.

[8] M. Intriligator, Mathematical Optimization and Economic Theory, Eaglewood Cliffs, N.J., Prentice-Hall, 1971.

[9] F. John, Extremum problems with inequalities as subsidiary conditions, Studies and Essays, Courant Anniversary Volume, Interscience, New York, 187-204, 1948.

[10] G.P. McCormick, Second order conditions for constrained minima, SIAM J. Appl. Math., 15, 641-652, 1967.

[11] S. Smale, Global analysis and economics, III, Pareto optima and price equilibria, to appear in the J. of Math. Economics.

[12] S. Smale, Global analysis and economics, V. Pareto theory with constraints, to appear.

[13] J. Stoer, Ch. Witzgall, Convexity and optimization in finite dimensions I, Springer, Berlin-Heidelberg-New York, 1970.

[14] S. Wan, On lokal Pareto optima, J. of Math. Economics, 2, 35-42, 1975.

[15] H.F. Weinberger, Conditions for a lokal Pareto optimum, Preprint, University of Minnesota, 1975.

[16] W. Wetterling, Definitheitsbedingungen für relative Extrema bei Optimierungs- und Approximationsaufgaben, Numer. Math., 15, 122-136, 1970.

On the Exact Evaluation of Finite Activity Networks with Stochastic Durations of Activities

Höpfinger, E., U. Steinhardt
Universität Karlsruhe

Summary: The networks treated here contain two different types of deterministic nodes or events. The durations of the activities are assumed to be independent random variables. Two procedures for calculating the common distribution of the time points of the realizations of the networks' terminals are described.

1. Basic Definitions:

The networks treated here can be considered as generalizations of PERT (Program Evaluation and Review Technique[1]) since they admit two types of nodes or events and arbitrary distribution functions of the durations of the activities.

Definition 1: Let E and V be finite disjoint sets. The elements of E are called activities, the elements of V events. Then (E,V,A,B,d,F) is called an activity network if

(1) $A:E \to V$ and $B:E \to V$ are two maps from E into V. A(e) is considered as the event preceding e, B(e) as the event following e. We assume that the "directed graph" (E,V,A,B) has no cycles, i.e., there is no finite sequence $e_1,\ldots,e_n$ such that $B(e_i) = A(e_{i+1})$ $(i = 1,\ldots,n-1)$ and $B(e_n) = A(e_1)$.

(2) d is a map $d:V \to \{0,1\}$ denoting the type of the node or event. If $d(v) = 0$ then $v \varepsilon S$ or $|\{e|B(e) = v\}| = 1$ or v occurs as soon as one preceding activity $e \varepsilon \{e'|B(e') = v\}$ is realized. If $d(v) = 1$ then v occurs as soon as all preceding activities $\{e|B(e) = v\}$ are realized and $|\{e|B(e) = v\}| \geq 2$.

(3) F is a map from $E \times \mathbb{R}$ into the set of all one-dimensional distribution functions. For each $e \varepsilon E$ let D_e denote the duration of the activity e and let Y_v denote the time of occurrence of v $(v \varepsilon V)$. Then $F(e,t',.)$ is considered as the distribution function of the random variable D_e if e is started at t':

$P(D_e \leq t | Y_{A(e)} = t') = F(e,t',t)$ $(t \varepsilon \mathbb{R})$.

$F(e,t',.)$ is assumed to satisfy $F(e,t',-0) = 0$ and $F(e,t',\infty) = 1$, $t' \to F(e,t',t)$ is supposed to be measurable for every $t \varepsilon \mathbb{R}$.

We admit several sources and terminals.

Definition 2: A subnetwork (E',V',A',B',d',F') of (E,V,A,B,d,F) is an activity network such that E',V' are subsets of E,V respectively and A',B',d',F' are the restrictions of A,B,d,F to E',E',V and $E' \times \mathbb{R}$ respectively. $v \varepsilon V'$ is called a terminal of (E',V',A',B',d',F') if there is no $e \varepsilon E'$ with $A(e) = v$ or there is an $e \varepsilon E \setminus E'$ with $A(e) = v$.

Our aim is the exact calculation of the conditional common distribution function

$(t_v)_{v\epsilon T} \to P(Y_v \leqq t_v(v\epsilon T)|Y_v = t'_v(v\epsilon S))$ $(t'_v\epsilon \mathbb{R}$ $(v\epsilon S))$

where S,T are the sets of the network's sources and terminals respectively.

<u>First Procedure</u> (moving backwards from T to S)

<u>First iteration step:</u> The first subnetwork U^1 is assumed to be $(\emptyset,T,d^1,F^1)$. The set of sources S^1 of U^1 is $S^1 = T$. The first distribution function is given by

$$P(Y_v \leqq t_v(v\epsilon T)|Y_v = t'_v(v\epsilon T)) = \begin{cases} 1 \text{ if } t_v \geqq t'_v \text{ for all } v\epsilon T \\ 0 \text{ if } t_v < t'_v \text{ for one } v\epsilon T \text{ at last.} \end{cases}$$

<u>i-th iterations step:</u> After (i-1) iterations $(i \geqq 2)$ we have obtained a subnetwork $U^{i-1} = (E^{i-1}, V^{i-1}, d^{i-1}, F^{i-1})$ with the set $S^{i-1} \subseteq E^{i-1}$ of sources. Furthermore we have the conditional distribution function

$(t_v)_{v\epsilon T} \to P(Y_v \leqq t_v(v\epsilon T)|Y_v = t'_v(v\epsilon S^{i-1}))$ $(t'_v\epsilon \mathbb{R}(v\epsilon S^{i-1}))$

First one can show that $S^{i-1} = S$ if and only if U^{i-1} is the original network. Then the procedure stops. In case of $S^{i-1} \neq S$ on account of the absence of cycles a $v^i \epsilon S^{i-1} \setminus S$ exists such that $\{e|A(e) = v\} \subseteq E^{i-1}$, i.e. all emanating activities are in E^{i-1}. Let $E(v^i) := \{e\epsilon E|B(e) = v^i\}$ and $V(v^i) := \{A(e)|e\epsilon E(v^i)\}$. Then the subnetwork $U^i = (E^i,V^i,d^i,F^i)$ is defined by

$E^i := E^{i-1} \cup E(v^i)$
$V^i := V^{i-1} \cup V(v^i)$.

The set S^i of sources of U^i is equal $S^i = (S^{i-1} \cup V(v^i)) \setminus \{v^i\}$. The distribution function of T belonging to U^i is given by

$$(t_v)_{v\epsilon T} \to P(Y_v \leqq t_v(v\epsilon T)|Y_v = t'_v(v\epsilon S^i)) =$$

$$\int_{-\infty<\tau<\infty} P(Y_v \leqq t_v(v\epsilon T)|Y_v = t'_v(v\epsilon S^{i-1} \setminus \{v^i\}), Y_{v^i} = \tau) \times$$

$$\times P(\tau \leqq Y_{v^i} \leqq \tau + d\tau|Y_v = t'_v(v\epsilon V(v^i)).$$

The last term is determined by

$$P(Y_{v^i} \leqq \tau|Y_v = t'_v(v\epsilon V(v^i)) = 1 - \prod_{e\epsilon E(v^i)} \{1 - F(e,t'_{A(e)},\tau-t'_{A(e)})\}$$

in case of $d(v^i) = 0$ and equal to

$$\prod_{e\epsilon E(v^i)} F(e,t'_{A(e)},\tau-t'_{A(e)}) \text{ in case of } d(v^i) = 1.$$

The procedure stops after finitely many iteration steps.

<u>Second Procedure</u> (moving forward from the set S of sources to the set T of terminal events).

<u>First iteration step:</u> First subnetwork: $U^1 = (E^1,V^1,d^1,F^1) = (\emptyset,S,d^1,F^1)$.

$$(t_v)_{v\in S} \to P(Y_v \leq t_v(v\in S) | Y_v = t'_v(v\in S)) = \begin{cases} 1 & t_v \geq t'_v \quad \text{for all } v\in S \\ 0 & \text{else} \end{cases}$$

Set of terminals $T^1 = S$.

<u>i-th iterations step:</u> After (i-1) iterations ($i \geq 2$) we have obtained a subnetwork $S^{i-1} = (E^{i-1}, V^{i-1}, d^{i-1}, F^{i-1})$ with a set T^{i-1} of terminals and the common distribution function

$$(t_v)_{v\in T^{i-1}} \to P(Y_v \leq t_v(v\in T^{i-1}) | Y_v = t'_v(v\in S)) \quad (t'_v \in \mathbb{R}(v\in S)).$$

One can easily verify that $T^{i-1} = T$ if and only if U^{i-1} is equal to (E,V,d,F). Then we have finished. Now assume $T^{i-1} \neq T$. Since there are no cycles there exists an $v^i \in V \setminus V^{i-1}$ such that each $e\in E$ with $B(e) = v^i$ satisfies $A(e) \in T^{i-1}$, i.e., all events preceding v^i belong to V^{i-1}. Let $E(v^i) := \{e\in E | B(e) = v^i\}$ and let $V(v^i) := \{v\in T^{i-1} | A(e) = v \text{ implies } e\in E^{i-1} \cup E(v^i)\}$ be the set of all events losing the property of a terminal. U^{i-1} is replaced by $U^i = (E^i, V^i, d^i, F^i)$ where $E^i := E^{i-1} \cup E(v^i)$ and $V^i = V^{i-1} \cup V(v^i)$. The set of terminals of U^i is given by $T^i = (T^{i-1} \setminus V(v^i)) \cup \{v^i\}$. A simple derivation yields

$$\begin{aligned}(t_v)_{v\in T^i} &\to P(Y_v \leq t_v(v\in T^i) | Y_v = t'_v(v\in S)) \\ &= \int_{\substack{-\infty<\tau_v \leq t_v (v\in T^{i-1}\setminus V(v^i)) \\ -\infty<\tau_v<\infty \ (v\in V(v^i))}} \cdots \int P(Y_{v^i} \leq t_{v^i} | Y_v = \tau_v \ (v\in T^{i-1})) \times \\ &\qquad \times P(\tau_v \leq Y_v \leq \tau_v + d\tau_v \ (v\in T^{i-1}) | Y_v = t'_v \ (v\in S)).\end{aligned}$$

The procedure stops after finitely many steps with the desired result.

In case of numerical calculations one can replace the given distributions by distributions belonging to lattice random variables. Then the given two integrals can be written as sums. These aspects together with a generalized description of the procedures are scheduled for publication.

[1] Elmaghraby, Salah E., "Some Models in Management Science, Springer Verlag, Berlin 1970.

APPROXIMATION OF A PARABOLIC BOUNDARY CONTROL PROBLEM BY THE LINE METHOD

M. Köhler
Institut für Operations Research
der Universität Zürich

Abstract.

Applying the line method approximation to a parabolic boundary control problem a sequence of ordinary control problems is generated. It is shown that the line method is a consistent and stable discretization. The convergence of the extreme values of the ordinary control problems to the extreme value of the parabolic control problem is proved. Finally, error estimations are given.

1. Introduction and Problem Statement.

We consider a parabolic boundary control problem of the following type:

$$\text{minimize} \quad \int_0^T\int_0^1 g(y(t,x),u(t),x)\,dx\,dt \tag{1.1}$$

subject to

$$a(t,x)y_{xx}+b(t,x)y_x+c(t,x)y+d(t,x)-\frac{\partial}{\partial t}y=0 \quad \text{in } B:=(0,T]\times(0,1), \tag{1.2}$$

$$y(0,x)=y_0(x) \quad \text{for } x \in (0,1), \tag{1.3}$$

$$y_x(t,0)+\alpha(t)y(t,0)=\beta(t)u_1(t) \quad \text{for } t \in [0,T], \tag{1.4}$$

$$y_x(t,1)+\gamma(t)y(t,1)=\delta(t)u_2(t) \quad \text{for } t \in [0,T], \tag{1.5}$$

$$u:=\begin{pmatrix}u_1\\u_2\end{pmatrix} \in Q_U \subset C^2[0,T]. \tag{1.6}$$

$C^m[0,T]$ denotes the Banach space of continuous functions from $[0,T]$ into R^m. Further, $C(\overline{B})$ denotes the Banach space of continuous functions from $[0,T]\times[0,1]$ into R. The norms in the appropriate spaces are given by

$$\|u\|_{C^m[0,T]} := \sup_{t\in[0,T]} \Big(\sum_{i=1}^m |u_i(t)|^2\Big)^{\frac{1}{2}} \quad \text{for } u \in C^m[0,T], \text{ and}$$

$$\|a\|_{C(\overline{B})} := \sup_{(t,x)\in\overline{B}} |a(t,x)| \quad \text{for } a \in C(\overline{B}).$$

In addition, y_x (y_{xx}) denotes the first (second) partial derivative with respect to x.

Throughout this paper, let the following assumptions hold:

<u>A1</u>. $a, b, c, d \in C(\overline{B}), \alpha, \beta, \gamma, \delta \in C[0,T]$.

<u>A2</u>. For each control $u \in Q_U$, there exists one and only one solution $y \in C(\overline{B})$ with $\frac{\partial}{\partial t}y, y_x, y_{xx} \in C(\overline{B})$ satisfying the constraints (1.2)-(1.5).

<u>A3</u>. $g: R\times R^2\times R\to R$ is continuous for all $(s,r,x) \in R\times R^2\times R$.

<u>A4</u>. There is a $K\in R$, $K<\infty$, satisfying $\|u\|_{C^2[0,T]}\leq K$ for all $u \in Q_U$.

In order to apply the theory presented in [3], let us define:

$Y:=C(\overline{B})$ with $\|\cdot\cdot\|_Y:=\|\cdot\cdot\|_{C(\overline{B})}$,

$D:=\{y\in Y\mid \frac{\partial}{\partial t}y,\ y_x,\ y_{xx}\in Y\}$,

$U:=C^2[0,T]$ with $\|\cdot\cdot\|_U:=\|\cdot\cdot\|_{C^2[0,T]}$,

$P:=Y$.

Furthermore, define the operator $T:Y\times U\to P$ by (1.7)

$$T(y,u)(t,x):\begin{cases}\left\{\begin{array}{l} a(t,x)y_{xx}(t,x)+b(t,x)y_x(t,x)+c(t,x)y(t,x)\\ \qquad +d(t,x)-\frac{\partial}{\partial t}y(t,x) \quad \text{in } B,\\ y(0,x)-y_0(x) \quad \text{for } x\in(0,1),\\ y_x(t,0)+\alpha(t)y(t,0)-\beta(t)u_1(t),\ x=0,\ t\in[0,T]\\ y_x(t,1)+\gamma(t)y(t,1)-\delta(t)u_2(t),\ x=1,\ t\in[0,T]\end{array}\right\} & \text{if } y\in D,\\ \left\{+\infty \qquad \text{in } \overline{B}\right\} & \text{if } y\notin D,\end{cases}$$

and $F:Y\times U\to R$ by

$$F(y,u):=\int_0^T\int_0^1 g(y(t,x),u(t),x)\,dx\,dt\ . \tag{1.8}$$

Hence, the control problem (1.1)-(1.6) is equivalent to the following problem $\mathcal{P}$:

$\mathcal{P}$: Min! $F(y,u)$

$$T(y,u)=0, \tag{1.9}$$

$$(y,u)\in Q:=Y\times Q_U\ . \tag{1.10}$$

Because of A2, there is a map $\Gamma:Q_U\to Y$ defined by

$$\Gamma u:=y \quad \text{with } T(y,u)=0\ . \tag{1.11}$$

2. Line Method Approximation.

We consider the line method approximation as a special case of the general discretization method presented in [3] (cf. also [5]).

Definition 2.1. A discretization method $\mathcal{V}$ applicable to $\mathcal{P}$ consists of an infinite sequence $\{Y_n\times U_n,\xi_n,\phi_n,\tau_n,P_n,\pi_n\}_{n\in N'}$, where

(i) N' is an infinite subset of N;

(ii) Y_n, U_n, P_n are Banach spaces;

(iii) $\xi_n:=\chi_n\times\partial_n$ with linear $\chi_n:Y\to Y_n$ and linear $\partial_n:U\to U_n$ (i.e. $\xi_n:Y\times U\to Y_n\times U_n$ is a map defined by $\xi_n(y,u):=(\chi_n(y),\partial_n(u))$) and $\pi_n:P\to P_n$ are linear mappings with $\lim_{n\to\infty}\|\xi_n(y,u)\|_{Y_n\times U_n}:=\lim_{n\to\infty}(\|\chi_n(y)\|_{Y_n}+\|\partial_n(u)\|_{U_n})=\|(y,u)\|_{Y\times U}:=\|y\|_Y+\|u\|_U$ for each fixed $(y,u)\in Y\times Q_U$, and $\lim_{n\to\infty}\|\pi_n(p)\|_{P_n}=\|p\|_P$ for each fixed $p\in P$;

(iv) $\tau_n:\{Y\times U\to P\}\to\{Y_n\times U_n\to P_n\}$ are mappings with T in the domain of all τ_n,
$\phi_n:\{Y\times U\to R\}\to\{Y_n\times U_n\to R\}$ are mappings with F in the domain of all ϕ_n.

Now, let us define for each $n\in N':=\{n\in N\mid n\geq 2\}$:

$$x_\nu:=\nu h,\ \nu=-1,0,1,2,\dots,n+1,\ \text{with } h:=\frac{1}{n}, \tag{2.1}$$

$$I_n:=\{x_\nu\mid \nu=0,1,\dots,n\}, \tag{2.2}$$

$$I_n':=I_n\cup\{x_{-1},x_{n+1}\}, \tag{2.3}$$

$$Y_n:=\{y^n:[0,T]\times I_n'\to R\mid (t,x_\nu)\to y^n(t,x_\nu),\ y^n(.,x_\nu)\in C[0,T],\ \nu=-1,0,\dots,n+1\},$$
$$\text{with } \|y^n\|_{Y_n}:=\sup_{(t,x_\nu)\in[0,T]\times I_n'}|y^n(t,x_\nu)|, \tag{2.4}$$

$$D_n:=\{y^n\in Y_n\mid \tfrac{\partial}{\partial t}y^n(.,x_\nu)\in C[0,T] \text{ for } \nu=-1,0,\dots,n+1\}, \tag{2.5}$$

$\chi_n:Y\to Y_n$ by $\chi_n(y):[0,T]\times I_n'\to R$ with

$$\chi_n(y)(t,x_\nu):=\begin{cases} y(t,x_\nu),\ \nu=0,1,\dots,n \\ \left.\begin{cases}\frac{1}{2}h^2y_{xx}(t,0)-hy_x(t,0)+y(t,0) & \text{if } y\in D,\\ +\infty & \text{if } y\notin D,\end{cases}\right\}\ \nu=-1, \\ \left.\begin{cases}\frac{1}{2}h^2y_{xx}(t,1)+hy_x(t,1)+y(t,1) & \text{if } y\in D,\\ +\infty & \text{if } y\notin D,\end{cases}\right\}\ \nu=n+1,\end{cases} \tag{2.6}$$

$$U_n:=U, \tag{2.7}$$

$$\partial_n:=\text{id (id is the identity map)}, \tag{2.8}$$

$$P_n:=\{p^n:[0,T]\times I_n\to R\mid (t,x_\nu)\to p^n(t,x_\nu),\ p^n(.,x_\nu)\in C[0,T],\ \nu=0,1,\dots,n\}, \tag{2.9}$$
$$\text{with } \|p^n\|_{P_n}:=\sup_{(t,x_\nu)\in[0,T]\times I_n}|p^n(t,x_\nu)|,$$

$$\pi_n:P\to P_n \text{ by } \pi_n(p):[0,T]\times I_n\to R \text{ with } \pi_n(p)(t,x_\nu):=p(t,x_\nu),\ \nu=0,1,\dots,n. \tag{2.10}$$

Further, define $T_n:=\tau_n(T):Y_n\times U\to P_n$ by

$$T_n(y^n,u)(t,x_\nu):= \tag{2.11}$$
$$:=\left.\begin{cases} a(t,x_\nu)\frac{1}{h^2}(y^n(t,x_{\nu+1})+y^n(t,x_{\nu-1})-2y^n(t,x_\nu)) \\ \quad +b(t,x_\nu)\frac{1}{2h}(y^n(t,x_{\nu+1})-y^n(t,x_{\nu-1}))+c(t,x_\nu)y^n(t,x_\nu) \\ \quad +d(t,x_\nu)-\frac{\partial}{\partial t}y^n(t,x_\nu) \text{ for } (t,x_\nu)\in[0,T]\times I_n, \\ y^n(0,x_\nu)-y_0(x_\nu) \quad \text{for } \nu=0,1,\dots,n, \\ \frac{1}{h}(\frac{1}{2h}(y^n(t,x_1)-y^n(t,x_{-1}))+\alpha(t)y^n(t,0)-\beta(t)u_1(t)),\ t\in[0,T],\ \nu=0, \\ \frac{1}{h}(\frac{1}{2h}(y^n(t,x_{n+1})-y^n(t,x_{n-1}))+\gamma(t)y^n(t,1)-\delta(t)u_2(t)),\ t\in[0,T],\ \nu=n,\end{cases}\right\}\text{if } y^n\in D_n,$$

$\left\{+\infty \right\}$ if $y^n \notin D_n$,

and $F_n := \phi_n(F): Y_n \times U \to R$ by

$$F_n(y^n,u) := \frac{1}{n}\sum_{\nu=0}^{n-1}\int_0^T g(y^n(t,x_\nu),u(t),x_\nu)dt \quad . \qquad (2.12)$$

For every $u \in Q_U$, the operator T_n represents the application of the line method to the parabolic boundary value problem (1.2)-(1.5) (cf. [6]). According to Definition 2.1., (2.2)-(2.12) defines a discretization method $\mathcal{V}=\{Y_n \times U, \xi_n, \phi_n, \tau_n, P_n, \pi_n\}_{n\in N}$, applicable to $\mathcal{P}$, where $\xi_n: Y\times U \to Y_n \times U$ is given by $\xi_n := \chi_n \times id$, i.e. $\xi_n(y,u)=(\chi_n(y),u)$ for $(y,u) \in Y\times U$. Hence, there exists a sequence of problems $\{\mathcal{P}_n\}_{n\in N}$, generated by $\mathcal{V}$, where $\mathcal{P}_n$ denotes a control problem of the following type:

$$\mathcal{P}_n: \qquad \text{Min! } F_n(y^n,u)$$

$$T_n(y^n,u)=0, \qquad (2.13)$$

$$(y^n,u) \in Q_n := Y_n \times Q_U \quad . \qquad (2.14)$$

Now, define for every $t \in [0,T]$

$y_\nu^n(t) := y^n(t,x_\nu)$, $\nu=-1,0,1,\dots\dots,n+1$,

$a_\nu(t) := a(t,x_\nu)$, $b_\nu(t) := b(t,x_\nu)$, $c_\nu(t) := c(t,x_\nu)$, $d_\nu(t) := d(t,x_\nu)$,

$y_{0\nu} := y_0(x_\nu)$ for $\nu=0,1,\dots\dots,n$, and

$g_\nu(s,r) := g(s,r,x_\nu)$ for $\nu=0,1,\dots\dots,n-1$.

Further, define for every $t \in [0,T]$ the $(n+1)\times(n+1)$-matrix

$A^n(t)=(a_{ij}^n(t))$ by (2.15)

$a_{00}^n(t) := 2a_0(t)(1-h\alpha(t))+h^2 b_0(t)\alpha(t)-h^2 c_0(t)$,

$a_{01}^n(t) := -2a_0(t)$,

$a_{nn-1}^n(t) := -2a_n(t)$,

$a_{nn}^n(t) := 2a_n(t)(1+h\gamma(t))+h^2 b_n(t)\gamma(t)-h^2 c_n(t)$,

$a_{\nu\nu-1}^n(t) := -(a_\nu(t)-\frac{h}{2}b_\nu(t))$, $\nu=1,2,\dots\dots,n-1$,

$a_{\nu\nu}^n(t) := 2a_\nu(t)-h^2 c_\nu(t)$, $\nu=1,2,\dots\dots,n-1$,

$a_{\nu\nu+1}^n(t) := -(a_\nu(t)+\frac{h}{2}b_\nu(t))$, $\nu=1,2,\dots\dots,n-1$,

$a_{ij}^n(t) := 0$ for $i,j \in \{0,1,\dots\dots,n\}$, $|i-j| \geq 2$

(i.e. $A^n(t)$ is a tridiagonal matrix), and the $(n+1)\times 2$-matrix $B^n(t)=(b_{jk}^n(t))$ by

(2.16)

$b_{jk}^n(t) := 0$, $j=1,2,\dots\dots,n-1$, $k=0,1$,

$b_{00}^n(t):=-\frac{1}{h}(2a_0(t)-hb_0(t))\beta(t)$, $b_{01}^n(t):=0$,

$b_{n1}^n(t):=\frac{1}{h}(2a_n(t)+hb_n(t))\delta(t)$, $b_{n0}^n(t):=0$.

Finally, define for every $t \in [0,T]$

$$\hat{d}^n(t):=\begin{pmatrix} d_0(t) \\ \vdots \\ d_n(t) \end{pmatrix}, \hat{y}^n(t):=\begin{pmatrix} y_0^n(t) \\ \vdots \\ y_n^n(t) \end{pmatrix}, \hat{y}_0^n:=\begin{pmatrix} y_{00} \\ \vdots \\ y_{0n} \end{pmatrix}. \tag{2.17}$$

Making use of (2.15),(2.16),(2.17) we obtain from (2.13) by elimination of y_{-1}^n and y_{n+1}^n that the problem $\mathcal{P}_n$ is equivalent to the ordinary control problem

$$\text{Min!} \quad \frac{1}{n}\sum_{\nu=0}^{n-1}\int_0^T g_\nu(y^n(t),u(t))dt \tag{2.18}$$

$$\frac{d}{dt}\hat{y}^n(t)=-\frac{1}{h^2}A^n(t)\hat{y}^n+B^n(t)u(t)+\hat{d}^n(t), \tag{2.19}$$

$$\hat{y}^n(0)=\hat{y}_0^n, \tag{2.20}$$

$$u \in Q_U. \tag{2.21}$$

3. Convergence of the Approximation.

Let the space R^{n+1} be normed by $\|..\|_{n+1}$, where

$$\|s\|_{n+1}:=\max_{\nu=0,1,..,n}|s_\nu| \quad \text{for } s=\begin{pmatrix} s_0 \\ \vdots \\ s_n \end{pmatrix} \in R^{n+1}, \tag{3.1}$$

let us denote by $\|..\|_{(n+1)\times(n+1)}$ the matrix norm given by

$$\|A\|_{(n+1)\times(n+1)}:=\max_{i=0,....,n}\sum_{j=0}^{n}|a_{ij}| \tag{3.2}$$

for every $(n+)\times(n+1)$-matrix $A=(a_{ij})$, and let us denote by $\|..\|_{C^{n+1}}$ the norm

$$\|y\|_{C^{n+1}}:=\sup_{t\in[0,T]}\|y(t)\|_{n+1} \quad \text{for } y \in C^{n+1}[0,T].$$

Because of A1, there exists for every $u \in Q_U$ one and only one $\hat{y}^n \in C^{n+1}[0,T]$ satisfying (2.19) and (2.20). Let $\Phi^n(t,0)$ with $\Phi^n(0,0)=E^n$ (E^n is the $(n+1)\times(n+1)$ identity-matrix) be the fundamental matrix of (2.19),(2.20), i.e.

$$\frac{d}{dt}\Phi^n(t,0)=-\frac{1}{h^2}A^n(t)\Phi^n(t,0).$$

<u>Lemma 3.1</u>. Suppose that there exists a constant $L \in R$, $L<\infty$, and an $n^*\in N'$ satisfying

$$\sup_{t\in[0,T]}\{\sup_{\tau\in[0,t]}\|\Phi^n(t,\tau)\|_{(n+1)\times(n+1)}\}\leq L \quad \text{for all } n\geq n^*.$$

Then the discretization method $V=\{Y_n\times U,\xi_n,\phi_n,\tau_n,P_n,\pi_n\}_{n\in N'}$, is stable, that is, there exists an $S>0$ (S independent of n) such that for all $n\in N'$, $n\geqq n^*$,

$$\|(y^n,u)-(y^{*n},u)\|_{Y_n\times U} \leqq S\|T_n(y^n,u)-T_n(y^{*n},u)\|_{P_n}$$

holds for all $(y^n,u),(y^{*n},u)\in Q_n$.

<u>Proof</u>. Defining

$$w_\nu^n(t):=T_n(y^n,u)(t,x_\nu)-T_n(y^{*n},u)(t,x_\nu),\ \nu=0,1,\dots\dots,n, \text{ and} \tag{3.3}$$

$$v_\nu^n(t):=y^n(t,x_\nu)-y^{*n}(t,x_\nu),\ \nu=-1,0,1,\dots\dots,n+1, \tag{3.4}$$

for arbitrarily given $(y^n,u),(y^{*n},u)\in Q_n$, we have by (2.11) and (2.15) that

$$\frac{d}{dt}\hat{v}^n(t)=-\frac{1}{h^2}A^n(t)\hat{v}^n(t)-\hat{w}^{*n}(t), \tag{3.5}$$

$$\hat{v}^n(0)=0 \tag{3.6}$$

holds, where

$$\hat{v}^n(t):=\begin{pmatrix}v_0^n(t)\\ \vdots\\ v_n^n(t)\end{pmatrix},\quad \hat{w}^n(t):=\begin{pmatrix}w_0^n(t)\\ \vdots\\ w_n^n(t)\end{pmatrix},\ \text{and}$$

$$\hat{w}^{*n}(t):=\hat{w}^n(t)+\begin{pmatrix}(2a_0(t)-hb_0(t))w_0^n(t)\\ 0\\ \vdots\\ 0\\ -(2a_n(t)+hb_n(t))w_n^n(t)\end{pmatrix}. \tag{3.7}$$

Now, the solution of the initial value problem (3.5),(3.6) is given by

$$\hat{v}^n(t)=-\int_0^t\Phi^n(t,0)(\Phi^n(\tau,0))^{-1}\hat{w}^{*n}(\tau)d\tau=-\int_0^t\Phi^n(t,\tau)\hat{w}^{*n}(\tau)d\tau\ . \tag{3.8}$$

Hence, we have

$$\|\hat{v}^n(t)\|_{n+1}\leqq\int_0^t\|\Phi^n(t,\tau)\|\,\|\hat{w}^{*n}(\tau)\|_{n+1}d\tau \quad \text{implying}$$

$$\|\hat{v}^n(t)\|_{n+1}\leqq t\sup_{\tau\in[0,T]}\|\Phi^n(t,\tau)\|\sup_{\tau\in[0,T]}\|\hat{w}^{*n}(\tau)\|_{n+1}$$

(the index of the norm (3.2) is omitted in the sequel), and

$$\|\hat{v}^n\|_{C^{n+1}}\leqq TL\|\hat{w}^{*n}\|_{C^{n+1}}\ . \tag{3.9}$$

Since

$$\|\hat{w}^{*n}\|_{C^{n+1}}\leqq\|\hat{w}^n\|_{C^{n+1}}+L_0\sup_{t\in[0,T]}\{\max\{|w_0^n(t)|,|w_n^n(t)|\}\}\leqq(1+L_0)\|\hat{w}^n\|_{C^{n+1}}$$

with $L_0:=2\|a\|_{C(\overline{B})}+\|b\|_{C(\overline{B})}$, we have by (3.9)

$$\|\hat{v}^n\|_{C^{n+1}} \leq TL(1+L_0)\|\hat{w}^n\|_{C^{n+1}} \ . \tag{3.10}$$

Defining

$\alpha^*:=\max(\|\alpha\|_{C[0,T]}, \|\gamma\|_{C[0,T]})$ we have by (2.11) and (3.10) that

$$\sup_{t\in[0,T]} \{\max(|v^n_{-1}(t)|, |v^n_{n+1}(t)|)\} \leq (1+2\alpha^*)\|\hat{v}^n\|_{C^{n+1}} + 2\|\hat{w}^n\|_{C^{n+1}}$$
$$\leq (2+(1+2\alpha^*)(1+L_0)TL)\|\hat{w}^n\|_{C^{n+1}} \ . \tag{3.11}$$

From (3.10) and (3.11) we obtain

$$\|y^n - y^{*n}\|_{Y_n} = \|(y^n,u)-(y^{*n},u)\|_{Y_n \times U} \leq S\|T_n(y^n,u)-T_n(y^{*n},u)\|_{P_n} \tag{3.12}$$

with $S:=2+(1+L_0)(1+2\alpha^*)TL$, and the assertion follows.

Lemma 3.2. Suppose that for all $t \in [0,T]$ $\alpha(t)=\alpha<0$, $\gamma(t)=\gamma>0$, $a(t,x)=a(x)$, $b(t,x)=b(x)$, $c(t,x)=c(x)$, and that, in addition, $a(x)>0$ for all $x \in [0,1]$. Then there exists an $n^*\in N'$ such that

$$\sup_{t\in[0,T]} \{ \sup_{\tau\in[0,t]} \|\Phi^n(t,\tau)\|_{(n+1)\times(n+1)}\} \leq \exp(T\|c\|_{C(\bar{B})})$$

holds for all $n \geq n^*$.

Proof. Since $A^n(t)=A^n$ for every $t \in [0,T]$, we have

$$\Phi^n(t,\tau)=\exp(-(t-\tau)n^2A^n) \ . \tag{3.13}$$

Furthermore,

$$\exp(-(t-\tau)n^2A^n)=\exp(-(t-\tau)n^2s)\exp((t-\tau)n^2(sE^n-A^n)) \tag{3.14}$$

holds for every $s \in R$. Defining

$$s:=2\|a\|_{C(\bar{B})}+\|c\|_{C(\bar{B})}+\alpha^*(2\|a\|_{C(\bar{B})}+\|b\|_{C(\bar{B})}) \quad (\alpha^* \text{ from above}) \tag{3.15}$$

and

$$n^*:=\min\{n\in N' \mid n > \frac{\|b\|_{C(\bar{B})}}{\inf_{x\in[0,1]}\{2a(x)\}}\} \ , \tag{3.16}$$

we obtain

$$sE^n - A^n \geq 0 \text{ for all } n \geq n^* . \tag{3.17}$$

Hence, by (2.15), (3.16) and the assumption of the lemma

$$\|sE^n-A^n\| = \max\{|s-a^n_{00}|+|a^n_{01}|, |s-a^n_{nn}|+|a^n_{nn-1}|,$$
$$\max_{\nu=1,\dots,n-1} (|a^n_{\nu\nu-1}|+|s-a^n_{\nu\nu}|+|a^n_{\nu\nu+1}|)\}$$
$$=\max\{s+2ha_0\alpha-h^2b_0\alpha+h^2c_0, s-2ha_n\gamma-h^2b_n\gamma+h^2c_n,$$
$$\max_{\nu=1,\dots,n-1} (s+h^2c_\nu)\} = \max_{\nu=0,\dots,n} (s+h^2c_\nu)$$
$$\leq s+h^2\|c\|_{C(\bar{B})} \tag{3.18}$$

holds for $n \geq n^*$. Because of (3.18),

$$\|\Phi^n(t,\tau)\| = \|\exp(-(t-\tau)n^2s)\exp((t-\tau)n^2(sE^n-A^n))\| \leq$$

$$\leq \exp(-(t-\tau)n^2 s)\exp((t-\tau)n^2(\|sE^n - A^n\|))$$

$$\leq \exp(-(t-\tau)n^2 s)\exp((t-\tau)n^2(s+h^2\|c\|_{C(\overline{B})})) = \exp((t-\tau)\|c\|_{C(\overline{B})})$$

holds for all $n \geq n^*$, and we obtain

$$\sup_{t\in[0,T]}\{\sup_{\tau\in[0,t]}\|\Phi^n(t,\tau)\|\} \leq \exp(T\|c\|_{C(\overline{B})}).$$

(The index of the matrix-norm (3.2) has been omitted.)

Lemma 3.3. For $i=0,1,2$, let the subsets

$$\{\frac{\partial^i}{\partial x^i}\Gamma u \mid u\in Q_U\} \subset C(\overline{B}) \text{ be compact.}$$

Then the following condition holds: For every $\varepsilon>0$, there is an $n_0\in N'$ such that $n\geq n_0$ implies

$$|F(y,u)-F_n(\xi_n(y,u))| < \varepsilon$$

and

$$\|T_n(\xi_n(y,u))\|_{P_n} < \varepsilon \qquad \text{for all } (y,u)\in T^{-1}(0)\cap Q$$

(That is, V is consistent on $T^{-1}(0)\cap Q$, uniformly with respect to $u\in Q_U$, cf. [3], Definition 2.2.)

Proof. Because of A2, we have for $y=\Gamma u$, $u\in Q_U$, that

$$T_n(\xi_n(y,u))(t,x_\nu) = T_n(\xi_n(y,u))(t,x_\nu) - \pi_n(T(y,u))(t,x_\nu) =$$

$$= \begin{cases} a(t,x_\nu)(\frac{1}{h^2}(y(t,x_{\nu+1})+y(t,x_{\nu-1})-2y(t,x_\nu))-y_{xx}(t,x_\nu)) \\ +b(t,x_\nu)(\frac{1}{2h}(y(t,x_{\nu+1})-y(t,x_{\nu-1}))-y_x(t,x_\nu)) \quad \text{on } [0,T]\times I_n, \\ 0 \quad \text{for } t=0,\ x_\nu\in I_n, \\ \frac{1}{h}(\frac{1}{2h}(y(t,x_1)-y(t,x_{-1}))-y_x(t,0)) \quad \text{for } t\in[0,T],\ \nu=0, \\ \frac{1}{h}(\frac{1}{2h}(y(t,x_{n+1})-y(t,x_{n-1}))-y_x(t,1)) \quad \text{for } t\in[0,T],\ \nu=n. \end{cases} \tag{3.19}$$

Further, assumption A2 and (2.6) imply that, for $\nu=0,1,\dots,n$,

$$y(t,x_{\nu+1})+y(t,x_{\nu-1})-2y(t,x_\nu) = \frac{h^2}{2}y_{xx}(t,\zeta_\nu)+\frac{h^2}{2}y_{xx}(t,\zeta_{\nu-1}) \tag{3.20}$$

and

$$y(t,x_{\nu+1})-y(t,x_{\nu-1}) = 2hy_x(t,x_\nu)+\frac{h^2}{2}(y_{xx}(t,\zeta_\nu)-y_{xx}(t,\zeta_{\nu-1})) \tag{3.21}$$

$$\text{with } \zeta_\nu\in[x_\nu,x_{\nu+1}],\ \nu=0,1,\dots,n-1,\ \zeta_{-1}=0 \text{ and } \zeta_n=1$$

hold. Together with (3.20) and (3.21) we obtain from (3.19) that

$$|T_n(\xi_n(y,u))(t,x_\nu)| \leq$$

$$\frac{1}{2}|a(t,x_\nu)|(|y_{xx}(t,\zeta_\nu)-y_{xx}(t,x_\nu)|+|y_{xx}(t,\zeta_{\nu-1})-y_{xx}(t,x_\nu)|)+$$

$$+\frac{1}{4}(1+|b(t,x_\nu)|)(|y_{xx}(t,\zeta_\nu)-y_{xx}(t,\zeta_{\nu-1})|) \tag{3.22}$$

holds for $t\in[0,T]$, $\nu=0,1,\ldots\ldots,n$. For $y=\Gamma u$, $u\in Q_U$, the assumption of the lemma implies that, for every $\varepsilon>0$, there exists an $\delta(\varepsilon)>0$ ($\delta(\varepsilon)$ independent of u) such that

$$|x-x^*| < \delta(\varepsilon),\ x,x^*\in[0,1],\ \text{implies}$$
$$\left|\frac{\partial^i}{\partial x^i}y(t,x)-\frac{\partial^i}{\partial x^i}y(t,x^*)\right| < \varepsilon \quad \text{for } i=0,1,2, \text{ and all } t\in[0,T]. \tag{3.23}$$

Now, fix $\varepsilon>0$ arbitrarily. Defining

$\varepsilon^* := \dfrac{4\varepsilon}{1+4\|a\|_{C(\bar{B})}+\|b\|_{C(\bar{B})}}$ we obtain from (3.22) and (3.23) that for all $n > \dfrac{2}{\delta(\varepsilon^*)}$

$$|T_n(\xi_n(y,u))(t,x_\nu)| \leqq |a(t,x_\nu)|\varepsilon^*+\frac{1}{4}\varepsilon^*(1+|b(t,x_\nu)|) \tag{3.24}$$

holds on $[0,T]\times I_n$. Hence,

$$\|T_n(\xi_n(y,u))\|_{P_n} \leqq \frac{1}{4}(1+4\|a\|_{C(\bar{B})}+\|b\|_{C(\bar{B})})\varepsilon^*=\varepsilon \quad \text{for } n>\frac{2}{\delta(\varepsilon^*)} \quad . \tag{3.25}$$

Furthermore, we have

$$\int_0^T\int_0^1 g(y(t,x),u(t),x)\,dx\,dt=\int_0^T\Big(\sum_{\nu=0}^{n-1}\int_{\frac{\nu}{n}}^{\frac{\nu+1}{n}} g(y(t,x),u(t),x)\,dx\Big)dt, \tag{3.26}$$

and

$$\frac{1}{n}\sum_{\nu=0}^{n-1} g(y(t,x_\nu),u(t),x_\nu)=\sum_{\nu=0}^{n-1}\int_{\frac{\nu}{n}}^{\frac{\nu+1}{n}} g(y(t,x_\nu),u(t),x_\nu)\,dx \tag{3.27}$$

for every $t\in[0,T]$.

Thus, by (3.26) and (3.27) we have

$$\Big|\int_0^T\int_0^1 g(y(t,x),u(t),x)\,dx\,dt-\frac{1}{n}\sum_{\nu=0}^{n-1}\int_0^T g(y(t,x_\nu),u(t),x_\nu)\,dt\Big| \leqq$$
$$\leqq \int_0^T\Big(\sum_{\nu=0}^{n-1}\int_{\frac{\nu}{n}}^{\frac{\nu+1}{n}}|g(y(t,x),u(t),x)-g(y(t,x_\nu),u(t),x_\nu)|\,dx\Big)dt. \tag{3.28}$$

Since $\{\Gamma u\,|\,u\in Q_U\}$ is compact in $C(\bar{B})$, there is an $K_0\in R$, $K_0<\infty$, such that

$$\|\Gamma u\|_{C(\bar{B})} \leqq K_0 \text{ for all } u\in Q_U.$$

Hence by A3 and A4, we obtain that $g: R\times R^2\times R\to R$ is uniformly continuous on $[-K_0,K_0]\times[-K,K]^2\times[0,1] \subset R\times R^2\times R$. Therefore, there exists for every $\varepsilon>0$ an$\delta^*(\varepsilon)>0$ such that

$$|s-s^*|+|x-x^*| < \delta^*(\varepsilon) \quad \text{implies}$$
$$|g(s,r,x)-g(s^*,r,x^*)| < \frac{1}{T}\varepsilon\ , \tag{3.29}$$

whenever $(s,r,x),(s^*,r,x^*) \in [-K_0,K_0]\times[-K,K]^2\times[0,1]$.
(3.23) and (3.29) imply that

$$|y(t,x)-y(t,x_\nu)|+|x-x_\nu| < \delta^*(\varepsilon) \text{ for } x\in[x_\nu,x_{\nu+1}],\ \nu=0,1,\ldots,n-1,$$

whenever

$$n > \kappa:=\max\left\{\frac{1}{\frac{\delta^*(\varepsilon)}{2}},\frac{1}{\delta(\frac{\delta^*(\varepsilon)}{2})}\right\} .$$

Therefore, by (3.29)

$$\int_0^T\Big(\sum_{\nu=0}^{n-1}\int_{\frac{\nu}{n}}^{\frac{\nu+1}{n}}|g(y(t,x),u(t),x)-g(y(t,x_\nu),u(t),x_\nu)|dx\Big)dt \leqq Tnh\frac{1}{T}\varepsilon=\varepsilon \tag{3.30}$$

holds for all $n\geqq\kappa$. Defining $n_0:=\min\{n\in N'\,|\,n>\max(\kappa,\frac{2}{\delta(\varepsilon^*)})\}$
we obtain from (3.25),(3.28) and (3.30) the assertion.

In order to give sufficient conditions for A2 and the assumption of Lemma 3.3., we refer to the Hölder spaces

$$H^{l+2}[0,1],\ H^{l,\frac{l}{2}}(\overline{B}),\ H^{l+1,\frac{l+1}{2}}[0,T]$$

with the appropriate norms $\|\cdot\cdot\|_{l+2}$, $\|\cdot\cdot\|_{l,\frac{l}{2}}$, $\|\cdot\cdot\|_{l+1,\frac{l+1}{2}}$
presented in [4], pp. 7-8.
We obtain

<u>Lemma 3.4</u>. For $0<l<1$ let

$$y_0 \in H^{l+2}[0,1],\ a,\ b,\ c,\ d \in H^{l,\frac{l}{2}}(\overline{B}),\ \alpha,\ \beta,\ \gamma,\ \delta \in H^{l+1,\frac{l+1}{2}}[0,T].$$

Suppose, in addition, that $a(t,x)>0$ for all $(t,x)\in\overline{B}$, that

$$\frac{d}{dx}y_0(0)+\alpha(0)y_0(0)=0 \quad \text{if } \beta(0)=0,$$

and that

$$\frac{d}{dx}y_0(1)+\gamma(0)y_0(1)=0 \quad \text{if } \delta(0)=0.$$

Further, assume that

$$Q_U:=\left\{u\in C^2[0,T]\ \middle|\ \begin{array}{l} u_i\in H^{l+1,\frac{l+1}{2}}[0,T],\ \|u_i\|_{l+1,\frac{l+1}{2}} < K<\infty \text{ for } i=1,2, \\ \left\{\begin{array}{ll} u_1(0)=\frac{1}{\beta(0)}(\frac{d}{dx}y_0(0)+\alpha(0)y_0(0)) & \text{if } \beta(0)\neq 0, \\ u_2(0)=\frac{1}{\delta(0)}(\frac{d}{dx}y_0(1)+\gamma(0)y_0(1)) & \text{if } \delta(0)\neq 0. \end{array}\right. \end{array}\right\}$$

Then assumptions A1, A2, A4 and the assumptions of Lemma 3.3. are satisfied.

<u>Proof</u>. The assertion follows directly from Theorem 5.3., p. 320 in [4].

Finally, if we define the extreme value of the problem (1.1)-(1.6) by

$$e(P):=\inf\{F(y,u)\,|\,(y,u)\in T^{-1}(0)\cap Q\}\ ,$$

and the extreme value of the problem (2.18)-(2.21) by

$$e(P_n):=\inf\{F_n(y^n,u)\,|\,(y^n,u)\in T_n^{-1}(0)\cap Q_n\}\ ,$$

we obtain the following approximation theorem:

Theorem 3.1. Suppose that assumptions A1, A2, A3, A4 hold. Assume, in addition, that the assumptions of Lemma 3.1. and Lemma 3.3. are satisfied. Then

$$\lim_{n\to\infty} e(P_n) = e(P)$$

holds.

Proof. Applying Satz 3.2 in [3] the assertion follows.

4. Error Estimations.

First, we give an estimation of the global discretization error. Therefore, we make the following additional assumption.

A5. There exists a continuous function $\omega:[0,1]\to R$ with $\lim_{\delta\to 0}\omega(\delta)=0$ such that $|x'-x''|<\delta$, $x',x''\in[0,1]$, implies

$$\left|\frac{\partial^2}{\partial x^2}\Gamma u(t,x'')-\frac{\partial^2}{\partial x^2}\Gamma u(t,x')\right| \leqq \omega(\delta)$$

for all $u\in Q_U$ and all $t\in[0,T]$.

For every $n\in N'$, the global discretization error of V is defined for every $(y,u)\in T^{-1}(0)\cap Q$ by

$$\varepsilon_n(u):=(y^n,u)-\xi_n(y,u) \quad \text{with } T_n(y^n,u)=0 \quad \text{(cf. [3])}.$$

Theorem 4.1. Let assumptions A1-A5 hold. Assume, in addition, that the assumption of Lemma 3.1. is satisfied. Defining

$$S_0:=\frac{1}{4}(1+4\,\|a\|_{C(\bar{B})}+\|b\|_{C(\bar{B})})(2+(1+L_0)(1+2\alpha^*)TL)$$

$\varepsilon_n(u)$ permits an estimate

$$\|\varepsilon_n(u)\|_{Y_n\times U} \leqq S_0\,\omega\!\left(\frac{2}{n}\right)$$

for all $n\geqq n^*$, uniformly in $u\in Q_U$.

Proof. Lemma 3.1. implies that

$$\|\varepsilon_n(u)\|_{Y_n\times U} \leqq S\,\|T_n(\xi_n(y,u))\|_{P_n} \tag{4.1}$$

holds for all $n\geqq n^*$ with $S=2+(1+L_0)(1+2\alpha^*)TL$. From (3.22) and A5 we obtain that

$$\|T_n(\xi_n(y,u))\|_{P_n} \leqq \frac{1}{4}(1+4\,\|a\|_{C(\bar{B})}+\|b\|_{C(\bar{B})})\,\omega\!\left(\frac{2}{n}\right) \tag{4.2}$$

holds. (4.1) and (4.2) imply the assertion.

In order to give an estimation of $|e(\mathcal{P})-e(\mathcal{P}_n)|$ we make one further assumption.

<u>A6</u>. There exist constants L_x, L_s such that
$|g(s^*,r,x)-g(s,r,x)| \leq L_s|s^*-s|$ and
$|g(s,r,x^*)-g(s,r,x)| \leq L_x|x^*-x|$

holds for all $(s,r,x),(s^*,r,x),(s,r,x^*)\in[-K_0,K_0]\times[-K,K]^2\times[0,1]$ (K_0 from p. 9, K from A4).

We obtain

<u>Theorem 4.2</u>. Let assumptions A1-A6 hold. Assume, in addition, that the assumptions of Lemma 3.1. and Lemma 3.3. are satisfied. Then

$$|e(\mathcal{P})-e(\mathcal{P}_n)| \leq T(L_sK_1+L_x)\frac{1}{n}+TL_sS_0\omega(\frac{2}{n})$$

holds for all $n\geq n^*$.

(According to Lemma 3.3., K_1 is a constant satisfying $\|\frac{\partial}{\partial x}\Gamma u\|_{C(B)}\leq K_1$ for all $u\in Q_U$.)

<u>Proof</u>. The assumption of Lemma 3.3. implies that

$$|\Gamma u(t,x')-\Gamma u(t,x'')|\leq K_1|x'-x''| \text{ holds for all } x',x''\in[0,1]. \qquad (4.3)$$

From (3.28) we obtain, using (4.3) and A6, that for every $n\in N'$

$$|F(y,u)-F_n(\chi_n(y),u)|\leq\int_0^T\Big(\sum_{\nu=0}^{n-1}\int_{\frac{\nu}{n}}^{\frac{\nu+1}{n}}|g(y(t,x),u(t),x)-g(y(t,x_\nu),u(t),x_\nu)|dx\Big)dt$$

$$\leq T(L_sK_1+L_x)h \text{ holds for all } (y,u)\in T^{-1}(0)\cap Q. \qquad (4.4)$$

On the other hand, we have for every $n\geq n^*$, using Theorem 4.1. and A5, that

$$|F_n(\chi_n(y),u)-F_n(y^n,u)|\leq\frac{1}{n}\sum_{\nu=0}^{n-1}\int_0^T|g(y(t,x_\nu),u(t),x_\nu)-g(y^n(t,x_\nu),u(t),x_\nu)|dt$$

$$\leq\frac{1}{n}\sum_{\nu=0}^{n-1}\int_0^T L_s|y(t,x_\nu)-y^n(t,x_\nu)|dt \leq TL_s\omega(\frac{2}{n})S_0 \qquad (4.5)$$

holds for every $(y,u)\in T^{-1}(0)\cap Q$, $(y^n,u)\in T_n^{-1}(0)\cap Q_n$.

Now, for arbitrarily given $(y,u)\in T^{-1}(0)\cap Q$, we deduce from (4.4) and (4.5) that for all $n\geq n^*$,

$$e(\mathcal{P})\leq F(y,u)\leq F_n(\chi_n(y),u)+T(L_sK_1+L_x)h\leq F_n(y^n,u)+T(L_sK_1+L_x)h+TL_sS_0\omega(\frac{2}{n})$$

holds with $T_n(y^n,u)=0$. Consequently, we have

$$e(\mathcal{P})\leq e(\mathcal{P}_n)+T(L_sK_1+L_x)\frac{1}{n}+TL_sS_0\omega(\frac{2}{n}) \quad \text{for all } n\geq n^*. \qquad (4.6)$$

Conversely, we obtain in the same manner that

$$e(\mathcal{P}_n)\leq e(\mathcal{P})+T(L_sK_1+L_x)\frac{1}{n}+TL_sS_0\omega(\frac{2}{n}) \qquad (4.7)$$

holds for all $n\geq n^*$. Combining (4.6) and (4.7) we are done.

In view of Lemma 3.4. a characterization of the function ω required in A5 is possible.

Lemma 4.1. Let the assumptions of Lemma 3.4. hold. Then there exists a constant C such that, for all $t \in [0,T]$,

$$\left|\frac{\partial^2}{\partial x^2}\Gamma u(t,x') - \frac{\partial^2}{\partial x^2}\Gamma u(t,x'')\right| \leq C|x'-x''|^{l}$$

holds for all $x',x'' \in [0,1]$ and all $u \in Q_U$. Further, C is given by

$$C := c_0(\|d\|_{l,\frac{l}{2}} + \|y_0\|_{l+2} + K(\|\beta\|_{C[0,T]} + \|\beta\|_{l+1,\frac{l+1}{2}} + \|\delta\|_{C[0,T]} + \|\delta\|_{l+1,\frac{l+1}{2}}))$$

(with another constant c_0) and satisfies

$$\left\|\frac{\partial}{\partial x}\Gamma u\right\|_{C(\overline{B})} \leq C \quad \text{for all } u \in Q_U.$$

Proof. Theorem 5.3., p. 320 in [4].

Applying Lemma 4.1. we get the following estimates:

Theorem 4.3. Let assumptions A1-A4 and A6 hold. Assume, in addition, that the assumptions of Lemma 3.1. and Lemma 3.4. hold. Then $\varepsilon_n(u)$ permits an estimation

$$\|\varepsilon_n(u)\|_{Y_n\times U} \leq S_0 C 2^{l} \cdot \frac{1}{n^{l}} \quad \text{for all } n \geq n^*, \text{ uniformly in } u \in Q_U.$$

Further,

$$|e(\mathcal{P}) - e(\mathcal{P}_n)| \leq S_1\left(\frac{1}{n} + \frac{1}{n^{l}}\right)$$

holds for all $n \geq n^*$, where $S_1 := T\max\{L_s C + L_x, L_s S_0 C 2^{l}\}$.

Proof. From Lemma 4.1. we obtain that A5 is satisfied. Defining $K_1 := C$ and applying Theorem 4.1. and Theorem 4.2. we are done.

References.

[1] Butkovskiy, A.G.: Distributed Control Systems. Elsevier, New York-London-Amsterdam (1969).

[2] Glashoff, K.; Krabs, W.: Konvergenz der Linienmethode bei einem parabolischen Rand-Kontrollproblem. ZAMM 54 (1974), 551-555.

[3] Köhler, M.: Approximation optimaler Prozesse unter Verwendung stabiler und konsistenter Diskretisierungsverfahren. Operations Research Verfahren 20, Verlag Anton Hain·Meisenheim (1975), 49-65.

[4] Ladyzenskaja, O.A.; Solonnikov, V.A.; Ural'ceva, N.N.: Linear and Quasilinear Equations of Parabolic Type. Translation of Mathematical Monographs 23, American Mathematical Society, Providence (1968).

[5] Stetter, H.J.: Analysis of Discretization Methods for Ordinary Differential Equations. Springer Tracts in Natural Philosophy 23, Springer Verlag, Berlin-Heidelberg-New York (1973).

[6] Walter, W.: Differential and Integral Inequalities. Springer Verlag, Berlin-Heidelberg-New York (1970).

Regularisation of Optimization Problems and Operator Equations

Peter Kosmol, Kiel

The first part of this article deals with Tihonov's regularisation of optimization problems and with Browder's regularisation of operator equations. In the second part an iterative procedure is introduced which solves equations involving monotone operators.

1. Uniformly Convex Functions and Strong Solvability of Convex Programs

Let X be a Banach Space and let C be a convex and closed subset of X. A continuous function $g: C \to \mathbb{R}$ is called uniformly convex on C ([11]) if there exists a monotonic function $\tau: \mathbb{R}_+ \to \mathbb{R}_+$ such that $\tau(o) = o$, $\tau(s) > o$ for $s > o$ and

(1.1) $$g\left(\frac{x+y}{2}\right) \le \frac{1}{2} g(x) + \frac{1}{2} g(y) - \tau(\|x-y\|) \quad \text{for all } x, y \in X .$$

For the following we shall need a weaker concept of uniform convexity.

So let a continuous function $g: C \to \mathbb{R}$ be called level uniformly convex on C (or l-uniformly convex on C) if for every $r \in \mathbb{R}$ there exists a $\tau_r: \mathbb{R}_+ \to \mathbb{R}_+$ with the above properties, but with (1.1) required only for all $x, y \in l_g(r) := \{x \in C \mid g(x) \le r\}$.

Lemma: The level sets $l_g(r)$ of the l-uniformly convex function $g: C \to \mathbb{R}$ are bounded.

Proof: We may assume that $o \in C$. On the set $B = \{x \in C \mid \|x\| \le 1\}$ the function g is bounded from below. Let $g(x) \ge \alpha$ for $x \in B$. Suppose that $l_g(r_o)$ is not bounded for some $r_o > g(o)$. Then there exists a sequence (x_n) where $x_n \in C$, $\|x_n\| = 1$ and $nx_n \in l_g(r_o)$. Hence

$$g(nx_n) \ge 2g((n-1)x_n) - g((n-2)x_n) + 2\tau_{r_o}(2) .$$

Let $\beta = 2\tau_{r_o}(2)$. For $2 \le k \le n$ we have

$$g(nx_n) \ge kg((n-k+1)x_n) - (k-1)g((n-k)x_n) + \frac{k(k-1)}{2}\beta$$

Equating $k = n$ results in the contradiction

$$r_o \ge g(nx_n) \ge ng(x_n) - (n-1)g(o) + \frac{n(n-1)}{2}\beta \ge n\left[\alpha - g(o) + \frac{(n-1)\beta}{2}\right] + g(o) \xrightarrow[n\to\infty]{} \infty .$$

q.e.d.

In the finite dimensional case g is 1-uniformly convex iff g is strictly convex and has bounded level sets $l_g(r)$.

One can also prove that a norm $\|\cdot\|$ is uniformly convex iff $\|\cdot\|^\lambda (\lambda>1)$ is 1-uniformly convex. If a 1-uniformly convex function exists on all of X then X has to be reflexive (even uniformly normable). 1-uniformly convex functions g on X have the following important property. For every closed and convex subset S of X the convex program (g,S) is strongly solvable. This means ([8] p.150), (g,S) has a unique solution x_o and $g(x_n) \to g(x_o)$ implies $x_n \to x_o$. The characterization of functions with this property led to the following concept ([10]).

<u>Definition:</u> A continuous function $f: X \to \mathbb{R}$ is called <u>locally uniformly convex</u> if

(1.2) $\forall x \in X$, $\forall x^* \in \delta f(x)$, $\exists \tau: \mathbb{R}_+ \to \mathbb{R}_+$ monotonic such that $\tau(o) = o$, $\tau(s) > o$ for $s > o$, $\lim\limits_{s\to\infty} \frac{\tau(s)}{s} = \infty$ and $f(x+y) \geq f(x) + \langle x^*, y\rangle + \tau(\|y\|)$ for all $y \in X$ where $\delta f(x)$ is the subdifferential of f at x.

The characterization in [10] concerns only bounded functions, that is functions which map bounded sets into bounded sets:

Let X be a reflexive Banach Space. Then a nonnegative bounded convex function $f: X \to \mathbb{R}$ has a strong minimum on every closed convex subset of X iff f^2 is locally uniformly convex.

<u>2. Regularisation of Optimization Problems</u>

Let C be a set and $f: C \to \mathbb{R}$. Tihonov's ([14]) regularisation method selects for the given problem (f,C) a function $g: C \to \mathbb{R}$ and solves a sequence of auxiliary problems (f_n, C), where $f_n = f + \alpha_n g$ and $o < \alpha_n \to o$.

We denote $M(f,C) := \{x \in C | x$ is solution of $(f,C)\}$. The result is

Theorem (Tihonov): Let C be a topological space,

$f,g: C \to \mathbb{R}$ lower semicontinuous and bounded from below, $x_n \in M(f_n, C)$ and $H[x_n] = \{x | x$ is a cluster point of $(x_n)\} \neq \emptyset$. Then

(a) $H[x_n] \subseteq M(g, M(f,C))$

(b) $g(x_n) \to \min\{g(x) | x \in M(f,C)\}$

(c) $\dfrac{f(x_n) - \min\{f(x) | x \in C\}}{\alpha_n} \to o$.

Is $S \subseteq C$, $f(x) = 0$ for $x \in S$ but $f(x) > 0$ for $x \notin S$ then the above approach corresponds to the penalty method.

Additional assumptions about f, g and C yield

Theorem (Levitin - Poljak [11]). Let C be a convex and closed subset of a reflexive Banach Space X, let $f: C \to \mathbb{R}$ be a convex continuous function such that $M(f,C) \neq \emptyset$ and let $g: C \to \mathbb{R}$ be uniformly convex. Then the solutions of $(f+\alpha_n g, C)$ converge $(\alpha_n \to 0)$ to the solution of $(g, M(f,C))$.

Note: If the function f is not differentiable then $f_n = f + \alpha_n g$ will normally not be differentiable either. If f is to be approximated by differentiable functions the following construction may be employed.

Let C be a topological space and let $f: C\times[0,a] \to \mathbb{R}$ be given for some $a > 0$.

Denoting $f^{(n)}(x,\alpha) := \frac{\partial^n}{\partial\alpha^n} f(x,\alpha)$ and $M_\alpha = M(f(\cdot,\alpha), C)$ one has ([9])

Theorem 1: Let f fulfil the following conditions

(1) $f(x,\cdot): [0,a] \to \mathbb{R}$ is twice continuously differentiable for all x.

(2) $f^{(1)}(\cdot,0)$ and $f(\cdot,\alpha)$ are lower semicontinuous and bounded from below for all α.

(3) $f^{(2)}(\cdot,\cdot)$ is bounded from below.

Now, if $\alpha_n \to 0$, $x_n \in M_{\alpha_n}$ and $H[x_n] \neq \emptyset$ then

(a) $H[x_n] \subseteq M(f^{(1)}(\cdot,0), M_0)$

(b) $f^{(1)}(x_n,0) \to \min\{f^{(1)}(x,0) \mid x \in M_0\}$

(c) $(f(x_n,0) - \min\{f(x,0) \mid x \in C\})\alpha_n^{-1} \to 0$.

In this connection the Levitin-Poljak Theorem may be generalized as follows

Theorem 2: Let C be a convex and closed subset of a reflexive Banach Space X and let $f: C\times[0,a] \to \mathbb{R}$ satisfy

(1) $f(x,\cdot) \in C^{(2)}[0,a]$ for all x

(2) $f(\cdot,\alpha)$ convex and continuous and $M_\alpha \neq \emptyset$ for all α

(3) $f^{(1)}(\cdot,0)$ 1-uniformly convex

(4) $f^{(2)}(\cdot,\cdot)$ bounded from below.

Now, if $\alpha_n \to 0$ and $x_n \in M_{\alpha_n}$ then (x_n) converges to the solution x_0 of $(f^{(1)}(\cdot,0), M_0)$.

Proof: Let $\overline{x} \in M_o$. Then (4) yields a $\lambda \in \mathbb{R}$ which is independent of x and α such that

$$f(x,\alpha) - f(\overline{x},\alpha) \geq f(x,o) - f(\overline{x},o) + \alpha[f^{(1)}(x,o) - f^{(1)}(\overline{x},o)] + \alpha^2\lambda .$$

This implies

$$f^{(1)}(x_n,o) \leq f^{(1)}(\overline{x},o) + a^2|\lambda| .$$

Since $f^{(1)}(\cdot,o)$ is 1-uniformly convex the sequence (x_n) is bounded (Lemma). Hence by Theorem 1, (x_n) converges weakly to x_o and $f^{(1)}(x_n,o) \to f^{(1)}(x_o,o)$. Let $f^{(1)}(x_n,o) \leq r$. Then

$$\tau_r(\|x_n - x_o\|) \leq \frac{1}{2} f^{(1)}(x_n,o) + \frac{1}{2} f^{(1)}(x_o,o) - f^{(1)}\left(\frac{x_n+x_o}{2}, o\right) .$$

Since $f^{(1)}(\cdot,o)$ is weakly lower semicontinuous the strong convergence of (x_n) follows. q.e.d.

As an illustration ([9]) consider a finite measure space (T,Σ,μ), an n-dim subspace X of $L_2(\mu)$ and a convex closed subset C of X. Let

$$f(x,\alpha) := \int_T |x|^{1+\alpha} d\mu \qquad o \leq \alpha \leq 1 .$$

Then the solutions x_α of $(f(\cdot,\alpha),C)$ converge to the solution x_o of $(f(\cdot,o),C)$, which is determined uniquely by the additional requirement

$$\int_T |x_o| \log|x_o| d\mu \leq \int_T |x| \log|x| d\mu$$

for all other solutions x of $(f(\cdot,o),C)$.

Because one has

$$f^{(1)}(x,\alpha) = \int_T |x|^{1+\alpha} \log|x| d\mu$$

and

$$f^{(2)}(x,\alpha) = \int_T |x|^{1+\alpha} \log^2|x| d\mu \geq o \quad \text{for all } x \text{ and } \alpha.$$

Note that $f^{(1)}(\cdot,o)$ is 1-uniformly convex on the positive cone in X but not uniformly convex.

3. Regularisation of Operator Equations

If the subset C in Tihonov's regularisation method is the entire Banach Space X and if the functions f and g are convex, continuous and Gateaux differentiable, then the solutions of (f_α,C) may be obtained as the solutions of the Operator Equation $$f'(x) + \alpha g'(x) = o .$$

The Operator $f': X \to X^*$ is monotone, which means $<f'(x) - f'(y), x-y> \geq o$ for all $x,y \in X$.

Moreover f' is demicontinuous ([4] and [13]), that is the strong convergence $x_n \to x_o$ implies the weak convergence $f'(x_n) \rightharpoonup f'(x_o)$.

For the purpose of solving the Operator Equation $(I-A)\,x = o$ ($A: X \to X$ nonexpansive, X a Hilbert Space) Browder ([2]) introduced the regularisation

$$A_\alpha = (I-A) + \alpha I .$$

He showed that the solutions of $A_\alpha x = o$ converge ($\alpha \to o$) to the solution x_o of $(I-A)\,x = o$ which satisfies $\|x_o\| \leq \|x\|$ for all other solutions x.

Cruceanu ([6]) obtained similar statements in respect to regularisations $A_\alpha = A + \alpha B$, where B a uniformly monotone potential operator and A monotone.

<u>Definition:</u> The Operator $B: X \to X^*$ is called locally uniformly monotone if for every $x \in X$ there is a τ_x as in (1.2) such that

$$<Bx - By,\ x - y> \geq \tau_x(\|x-y\|) \qquad \text{for all} \quad y \in X .$$

A modification of the proof from the Theorem 1 in [1] leads to the following

Theorem: Let X be a reflexive Banach Space, let $A: X \to X^*$ be monotone and demicontinuous and $B: X \to X^*$ demicontinuous and locally uniformly monotone. If $S = \{x \mid Ax = o\} \neq \emptyset$ then the solutions x_α of $A_\alpha x := Ax + \alpha Bx = o$ converge to the solution x^* of S which satisfies

$$<Bx^*, x> \geq <Bx^*, x^*> \qquad \text{for all} \quad x \in S . \tag{3.1}$$

Proof: Let $A_\alpha = A + \alpha B$. Then A_α is strictly monotone, demicontinuous and coercive. This follows from

$$\frac{<A_\alpha y, y>}{\|y\|} = \frac{<Ay-A(o),y> + <A(o),y> + \alpha<By-B(o),y> + \alpha<B(o),y>}{\|y\|} \geq \alpha\,\frac{\tau_o(\|y\|)}{\|y\|} -$$

$$- \ \alpha\,\|B(o)\| - \|A(o)\| .$$

According to Browder and Minty ([15] p.222), A_α has a unique solution x_α. For $x \in S$ one has

$$\langle Ax_\alpha - Ax, x_\alpha - x\rangle + \alpha\langle Bx_\alpha, x_\alpha - x\rangle = o$$

and therefore

(3.2) $$\langle Bx_\alpha, x_\alpha - x\rangle \leq o$$

since A is monotone. We conclude from (3.2) that the set $\{x_\alpha\}_1^\infty$ is bounded.

$$\frac{\tau_x(\|x_\alpha - x\|)}{\|x_\alpha - x\|} \leq \frac{\langle Bx_\alpha - Bx, x_\alpha - x\rangle}{\|x_\alpha - x\|} = \frac{\langle Bx_\alpha, x_\alpha - x\rangle}{\|x_\alpha - x\|} + \frac{\langle Bx, x_\alpha - x\rangle}{\|x_\alpha - x\|} \leq \|Bx\| .$$

Consequently x_α contains a weakly convergent subsequence x_β with the limit x^*. We show that $Ax^* = o$. For an arbitrary $y \in X$ one has

$$\langle Ay - Ax_\beta, y - x_\beta\rangle + \beta\langle Bx_\beta, x_\beta - y\rangle = \langle Ay, y - x_\beta\rangle .$$

From $\beta \to o$ and

$$\langle Bx_\beta, x_\beta - y\rangle = \langle Bx_\beta - By, x_\beta - y\rangle + \langle By, x_\beta - y\rangle \geq \langle By, x_\beta - y\rangle$$
$$\geq -\|By\| \, \|x_\beta - y\|$$

we conclude

$$\langle Ay, y - x^*\rangle \geq o .$$

Let $y = x^* + tz$, then

$$\langle Ax^*, z\rangle \geq o$$

since A is demicontinuous. But this inequality can hold for arbitrary z only if $Ax^* = o$. Therefore (3.2) yields

$$\tau_{x^*}(\|x^* - x_\beta\|) \leq \langle Bx^* - Bx_\beta, x^* - x_\beta\rangle = \langle Bx^*, x^* - x_\beta\rangle + \langle Bx_\beta, x_\beta - x^*\rangle$$
$$\leq \langle Bx^*, x^* - x_\beta\rangle \to o .$$

Hence x_β converges strongly to x^* and

$$\langle Bx^*, x^* - x\rangle \leq o \qquad \forall x \in S .$$

Since B is strictly monotone x^* is uniquely determined by (3.1). Therefore x_α converges strongly to x^*. q.e.d.

Remark 1: In the special case of B being the derivative of a convex function $g: X \to \mathbb{R}$ (X-Hilbert Space) (3.1) is equivalent to $g(x^*) \le g(x)$ for all $x \in X$. Because S is convex ([3]) and (3.1) is in this case the characterization of minimal solutions of convex programs ([8] p. 31).

4. An Iterative Procedure

The regularisation methods of Tihonov and Browder have two important properties. One is the convergence of the solutions of the regularised problems to a two-stage solution. Second, the regularised mappings f_α (or A_α) are uniformly convex (or uniformly monotone). However, the modulus of convexity $\alpha\tau$ decreases towards o as $\alpha \to o$. This may cause numerical difficulties on account of rounding errors. An iterative procedure with constant modules of convexity seems to be desirable and will now be described.

Let X be a Hilbert Space and $A: X \to X$ monotone and demicontinuous, and again $S = \{x | Ax = o\} \neq \emptyset$.

Starting with an arbitrary x_o let x_n be the solution of $A_{n-1}x := Ax+x-x_{n-1} = o$. Obviously $\langle A_n x - A_n y, x-y \rangle \ge \|x-y\|^2$ for all $n \in \mathbb{N}$ and $x, y \in X$. For solving equations with uniformly monotone operators see [7] p. 110.

We shall now prove

Theorem 4: The sequence (Ax_n) converges strongly to o and (x_n) converges weakly to a solution of $Ax = o$.

Proof: Let $T: X \to X$ be defined by $A(Tx) + Tx - x = o$. Obviously $Tx = x$ iff $Ax = o$. The map T is monotone and nonexpansive ($\|Tx-Ty\| \le \|x-y\|$). From

$$A(Tx) + Tx - x = o \quad \text{and} \quad A(Ty) + Ty - y = o$$

follows that

$$\langle A(Tx) - A(Ty) + Tx - Ty - x + y, Tx - Ty \rangle = o \quad . \tag{1}$$

Since A is monotone we have

(2) $$<Tx-Ty,\ x-y> \geq <Tx-Ty,\ Tx-Ty> \geq o$$

and therefore

$$\|Tx-Ty\|\ \|x-y\| \geq \|Tx-Ty\|^2$$

or

$$\|x-y\| \geq \|Tx-Ty\|$$

Now let $Ap = o$ and $Ax \neq o$. Then (1) implies

(3) $$o \geq <Tx-x,\ Tx-p> = <Tx-x,\ Tx-x> + <Tx-x,\ x-p> \quad .$$

Furthermore (2) yields

(4) $$\|Tx-p\|^2 \leq <Tx-p,\ x-p> = <x-p,\ x-p> + <Tx-x,\ x-p> \ .$$

Combining (3) and (4) we have

$$\|Tx-p\|^2 \leq \|x-p\|^2 - \|Tx-x\|^2 \quad .$$

It follows that T is asymptotically regular: $x_{n+1} - x_n \to o$. According to [12] the sequence (x_n) converges weakly to a solution of $Ax = o$. Furthermore $Ax_n = x_{n-1} - x_n$ converges strongly to o. q.e.d.

<u>Corollary:</u> If $A = A_1 + \alpha B_1$, A_1- monotone and B_1- locally uniformly monotone, then (x_n) converges strongly to the solution of $Ax = o$.

Because $Ap = o$ implies

$$\alpha\tau_p(\|x_n-p\|) \leq \alpha <B_1x_n - B_1p,\ x_n-p> \leq <Ax_n-Ap,\ x_n-p> \to o \ .$$

<u>Remark 2:</u> If A is Lipschitz continuous, that is if $\|Ax-Ay\| \leq L\ \|x-y\|$ then $U_t = I - t(A+I-x_n)$ and $t < 2/L^2$ is strictly contractive ([5]).

Its fixed point is the solution of $A_nx = Ax + x - x_n = o$.

The above procedure may also be applied to constrained convex programs in a Hilbert Space X. Let C be a closed convex subset of X and $f: C \to \mathbb{R}$ be continuous and convex. Then for every $y \in C$ the function $f_y(\cdot) := f + \frac{\|(\cdot)-y\|^2}{2}$ is uniformly convex on C. Now let the map $T: C \to C$ be defined pointwise by Ty being the solution of (f_y, C). Starting with an arbitrary x_o let $x_n = Tx_{n-1}$. Then provided that

$M(f,C) \neq \emptyset$ the following statement is true.

Theorem 5: The sequence $f(x_n)$ converges monotonically to the minimum of f on C and (x_n) converges weakly to a solution of (f, C).

Proof:

$$\text{Let } \overline{f}(x) = \begin{cases} f(x) & \text{if } x \in C \\ \infty & \text{if } x \in X \setminus C \end{cases} .$$

Tx is the solution of (f_x, C) and Ty is the solution of (f_y, C) iff subgradients $g \in \delta\overline{f}(Tx)$ and $h \in \delta\overline{f}(Ty)$ exist and ([8] p. 25 and p. 30)

$$g + Tx - x = o , \quad h + Ty - y = o .$$

This implies $Tx = x$ iff x is a solution of (f, C). In exactly the same way as in the proof of Theorem 4 it is shown that $T: C \to C$ is nonexpansive and asymptotically regular. Again according to [12] the weak convergence of the sequence (x_n) to a solution of (f, C) follows. Furthermore the sequence $g_n \in X$ exists such that

$$g_n \in \delta\overline{f}(x_n) \quad \text{and} \quad g_n = x_{n-1} - x_n .$$

The sequence (x_n) is bounded and hence if $Tp = p$ we have

$$o \le f(x_n) - f(p) \le \langle g_n, x_n - p\rangle \to o .$$

Finally it is clear that $f(x_n)$ is monotonically decreasing: the definition of x_n yields

$$f(x_n) \ge f(x_{n+1}) + \frac{\|x_{n+1} - x_n\|^2}{2} . \qquad \text{q.e.d.}$$

References

[1] J.P. Alber: On the solvability of nonlinear equations with monotone operators in Banach spaces (Russian). Sibirsk. Mat. Ž. 16, No. 1 (1975), 3 - 12.

[2] F.E. Browder: Convergence of approximations to fixed points of nonexpansive nonlinear mappings in Banach spaces. Arch. Rat. Mech. and Anal. (1967), 82 - 90.

[3] F.E. Browder: Nonlinear mappings of nonexpansive and accretive type in Banach spaces. Bull. Amer. Math. Soc. 73 (1967), 875 - 882.

[4] F.E. Browder: Nonlinear operators and nonlinear equations of evolution in Banach spaces. In: Nonlinear Functional Analysis - Proceedings of Symposia in Pure Mathematics, Vol. 18, Part 2. American Mathematical Society, 1972.

[5] F.E. Browder and W.V. Petryshyn: Construction of fixed points of nonlinear mappings in Hilbert space. J. Math. Anal. and Appl. 20 (1967), 197 - 228.

[6] S. Cruceanu: Rêgularisation pour les problêmes â operateurs monotones et la mêthode de Galerkine. Comment. Math. Univ. Carol. 12 No. 1 (1971), 1 - 13.

[7] K. Deimling: Nichtlineare Gleichungen und Abbildungsgrade. Springer-Verlag, Berlin - Heidelberg - New York, 1974.

[8] R.B. Holmes: A course on optimization and best approximation. Lect. Notes in Math. 257. Springer-Verlag, Berlin - Heidelberg - New York, 1972.

[9] P. Kosmol: Optimierung konvexer Funktionen mit Stabilitäts-Betrachtungen. To appear in Dissertationes Math.

[10] P. Kosmol and M. Wriedt: Starke Lösbarkeit von Optimierungsaufgaben. To appear.

[11] E.S. Levitin and B.T. Polyak: Convergence of minimizing sequences in conditional extremum problems. Soviet. Math. Dokl. 7 (1966), 764 - 767.

[12] Z. Opial: Weak convergence of the sequence of successive approximation for nonexpansive mappings. Bull. Amer. Math. Soc. 73 (1967), 591 - 597.

[13] R.T. Rockafellar: Characterization of the subdifferentials of convex functions. Pacific. J. Math. 17 (1966), 497 - 510.

[14] A.N. Tihonov: Solution of incorrectly formulated problems and the regularization method. Soviet. Math. Dokl. 4 (1963), 1035 - 1038.

[15] M.M. Vainberg: Variational method and method of monotone Operators in the theory of nonlinear equations. John Wiley & Sons, New York - Toronto, 1973.

Mathematisches Seminar
der Universität Kiel
23 Kiel
Olshausenstr. 40-60

Some extensions of linearly constrained nonlinear programming

Vera Kovacevic, Belgrad

1. Introduction

It is a purpose of this paper to give a general framework for feasible direction methods that solve linearly constrained nonlinear programming problems. In Sections 5 and 6 the definitions of admissibility of the sequence of directions $\{s_j\}$ and the sequence of stepsizes $\{\sigma_j\}$ are introduced. The antizigzagging precaution defined in Section 7 uses only information from the constraints active at the point at which it is applied. Section 8 contains convergence results. Examples of some well known methods that fit into our framework are given in Section 9.

2. Formulation of the problem and notation

Let $x \in R^n$, $F(x): R^n \to R^1$. If $F(x)$ is differentiable at a point x_j, we denote its gradient at x_j by $\nabla F(x_j)$ or g_j. For any column vector x and any matrix X we denote the transpose by x' and X', respectively. The symbol $\|x\|$ is used for the Euclidean norm of x.

Let $a_i \in R^n$, $i \in I$. Define $R = \{x \in R^n \mid a_i' x \leq b_i,\ i \in I\}$. We assume $\text{int}(R) \neq \emptyset$. Consider the problem of determining $z \in R$ such that $F(z) \leq F(x)$ for all $x \in R$.

Throughout the paper we shall make the following

Assumption 1: The algorithm starts with $x_0 \in R$ such that there is a convex compact set S with the property

$$\{x \in R \mid F(x) \leq F(x_0)\} \subset S$$

and $F(x)$ is continuously differentiable on some open set containing S.

An $x \in R$ is said to be a stationary point if there are real numbers λ_i such that

$$\nabla F(x) = \sum_{i=1}^{n} \lambda_i a_i,$$

$$\lambda_i (a_i' x - b_i) = 0, \quad \lambda_i \leq 0, \quad i \in I.$$

By the Kuhn-Tucker theorem, every local minimizer $z \in R$ of a continuously differentiable function $F(x)$ is a stationary point. We shall describe the general conditions under which a minimization algorithm generates a sequence of points $\{x_j\}$ with the property that every cluster point of $\{x_j\}$ is a stationary point.

3. Definition of the algorithm

A general cycle of the algorithm consists of three steps which are described below. At the beginning of the j-th cycle the following data are available: $x_j \in R$, g_j, $I_j = \{i \in I \mid a_i' x_j = b_i\}$ and an indicator $\beta_j \in \{0,1\}$.

Step 1: Choose the direction s_j such that the following two conditions are satisfied

$$\text{i) } a_i' s_j \geq 0, \; i \in I_j$$
$$\text{ii) } g_j' s_j > 0$$

Step 2: Choose the stepsize σ_j such that the following two conditions are satisfied

$$\text{i) } 0 < \sigma_j \leq \sigma_j^*$$
$$\text{ii) } F(x_j - \sigma_j s_j) < F(x_j),$$

where $\sigma_j^* = \min \{\frac{a_i' x_j - b_i}{a_i' s_j} \mid a_i' s_j < 0, \; i \in I\}$.

If $\sigma_j = \sigma_j^*$ set $\beta_{j+1} = 1$; otherwise set $\beta_{j+1} = 0$.

Step 3: Set $x_{j+1} = x_j - \sigma_j s_j$. Compute g_{j+1}. Set $I_{j+1} = \{i \in I \mid a_i' x_{j+1} = b_i\}$.

Replace j by j+1 and go to Step 1.

Stopping criterion: Stop if no s_j satisfying i) and ii) in Step 1 exists. The following two propositions state that the algorithm is well defined. The proofs are simple and will be omitted.

Proposition 3.1. If there is no s_j satisfying i) and ii) in Step 1 then x_j is a stationary point.

Proposition 3.2. If there is an s_j satisfying i) and ii) in Step 1 then there exists a σ_j satisfying i) and ii) in Step 2.

4. Definition of forcing functions

Let
$$L_\nu = \{x \in R^n \mid a_i' x = b_i, \; i \in I_\nu\}$$
$$T_\nu = \{x \in R^n \mid a_i' x = 0, \; i \in I_\nu\}$$
$$K_\nu = \{x \in R^n \mid x = \sum_{i \in I_\nu} \lambda_i a_i, \lambda_i \leq 0\},$$

where $I_\nu \subset I$ is a set of indices. Then for arbitrary $x \in R^n$ the representation $x = x_1 + x_2$, $x_1 \in T_\nu$, $x_2 \in T_\nu^\perp$ is unique. Let $t_\nu(x) = \|x_1\|$, $v_\nu(x) = d(x_2, K_\nu)$.

The family of mappings $\{\varphi_{\nu\beta} : R^n \to R^+, \beta \in N\}$ is called P-forcing with respect to L_ν if for any convergent sequence $\{x_j\}$ the following two conditions hold:

i) $\varphi_{\nu j}(x_j)\to 0$ implies $t_\nu(x_j)\to 0$, $j\to\infty$

ii) $t_\nu(x_j)\to 0$ implies $\varphi_{\nu j}(x_j)\to 0$, $j\to\infty$

The family of mappings $\{\Psi_{\nu\beta}:R^n\to R^+,\beta\in N\}$ is called <u>C-forcing with respect to L_ν</u> if for any convergent sequence $\{x_j\}$ the following condition holds:

iii) $\varphi_{\nu j}(x_j)\to 0$ and $\Psi_{\nu j}(x_j)\to 0$ imply $v_\nu(x_j)\to 0$, $j\to\infty$.

5. Definition of an admissible sequence of directions

Let $T_j = \{x\in R^n | a_i' x=0, i\in I_j\}$. Let $\tilde{I}_j\subset I_j$ be the set of indices of the constraints dropped from the set of active constraints at the cycle j. Let φ_{jj} and Ψ_{jj} be elements of the families $\{\varphi_{j\beta},\beta\in N\}$ and $\{\Psi_{j\beta},\beta\in N\}$ respectively which are P and C forcing with respect to T_j.

The sequence of directions $\{s_j\}$ is <u>admissible</u> if the following conditions hold for every infinite set of indices J:

i) If there is $\varepsilon>0$ such that $\varphi_{jj}(g_j)\geq\varepsilon$ and $\tilde{I}_j=\emptyset, j\in J$, then there is $\delta>0$ such that $g_j' s_j\geq\delta$, $j\in J$.

ii) If there is $\varepsilon>0$ such that $\Psi_{jj}(g_j)\geq\varepsilon$ and $\tilde{I}_j\neq\emptyset, j\in J$ then there is $\delta>0$ such that $g_j' s_j\geq\delta, j\in J$.

iii) If $\dfrac{g_j' s_j}{\|s_j\|}\to 0, j\to\infty$, $j\in J$ then $\|s_j\|\to 0, j\to\infty$, $j\in J$.

iv) The sequence $\{\|s_j\|\}$ is bounded.

6. Definition of an admissible sequence of stepsizes

Let σ_j^* be as defined in Section 3. The sequence of stepsizes $\{\sigma_j\}$ is <u>admissible</u> if the following conditions hold for every infinite set of indices J:

i) If there is $\varepsilon>0$ such that $g_j' s_j\geq\varepsilon$ and $\sigma_j^*\geq\varepsilon$, $j\in J$, then there is $\delta>0$ such that $F(x_j-\sigma_j s_j)\leq F(x_j)-\delta, j\in J$.

ii) If there is $\varepsilon>0$ such that $g_j' s_j\geq\varepsilon$ and $\sigma_j\to 0, j\to\infty$, $j\in J$, then there is $j_0\in J$ such that $\sigma_j=\sigma_j^*$, $j\geq j_0, j\in J$.

iii) The sequence $\{\sigma_j\}$ is bounded.

7. Definition of the antizigzagging precaution

Let I_j be as defined in Section 3 and $T_j = \{x\in R^n | a_i' x=0, i\in I_j\}$. Let $\tilde{I}_j$ be as defined in Section 5 and $\tilde{T}_j=\{x\in R^n | a_i' x=0, i\in I_j\setminus\tilde{I}_j\}$. Let φ_{jj} and Ψ_{jj} be as defined in Section 5.

The antizigzagging precaution is satisfied at the j-th cycle of the algorithm if the following conditions hold:

i) $\tilde{I}_j \begin{cases} =\emptyset \text{ if } \varphi_{jj}(g_j) \geq \Psi_{jj}(g_j) \text{ or } (\beta_j=1 \text{ and } \varphi_{jj}(g_j) \neq 0) \\ \neq \emptyset \text{ otherwise} \end{cases}$

ii) If $\tilde{I}_j \neq \emptyset$ then $\dim(T_j^{\perp}) = \dim(\tilde{T}_j^{\perp})+1$.

8. Convergence results

Throughout this section we shall make the following assumption: If $x \in R$ then the vectors a_i such that $a_i'x = b_i$ are linearly independent. This assumption is made in order to make the proofs simpler. In the case when it is not satisfied it is possible to prove all the results with a somewhat more complicated, antizigzagging precaution. The details will be reported in a forthcoming paper.

The following lemma, which we state without proof, characterizes the distance of cluster points of an arbitrary sequence $\{x_j\} \subset T_\nu^{\perp}$ from the cone K_ν. The proof is straightforeward when the fact that K_ν is closed is used.

Lemma 8.1. Let I_ν, T_ν and K_ν be as in Section 4. Let $\{x_j\}$ be any sequence in $T_\nu^{\perp}$. Then $d(x_j,K_\nu) \to 0$, $j \to \infty$, if and only if $\max\{0,\lambda_{ij},\ i \in I_\nu\} \to 0$, $j \to \infty$, where $x_j = \sum_{i \in I_\nu} \lambda_{ij} a_i$.

In the lemmas that follow we assume that the sequence $\{x_j\}$ is generated by the algorithm in Section 3 and that the sequences $\{s_j\}$ and $\{\sigma_j\}$ are admissible in the sense of the definitions in Sections 5 and 6.

Lemma 8.2. Let z be a cluster point of $\{x_j\}$ and let $\{x_j, j \in J\}$ be a subsequence of $\{x_j\}$ converging to z with the property $I_j = I_\nu$, $j \in J$

i) If $P_\nu \nabla F(z) \neq 0$ then there is $\delta > 0$ such that, for $j \in J$ large enough, $\varphi_{\nu j}(g_j) \geq \delta$.

ii) If $\nabla F(z) = \sum_{i \in I_\nu} \lambda_i a_i$ and $\max\{\lambda_i |\ i \in I_\nu\} > 0$ then $\varphi_{\nu j}(g_j) \to 0$, $j \to \infty$, $j \in J$,

and there is $\delta > 0$ such that, for $j \in J$ large enough, $\Psi_{\nu j}(g_j) \geq \delta$.

Proof: i) As $\{x_j, j \in J\}$ converges to z and $\nabla F(x)$ is continuous it follows that $P_\nu g_j \to P_\nu \nabla F(z)$, $j \to \infty$, $j \in J$, and hence there is $\varepsilon > 0$ such that $t_\nu(g_j)$ satisfies the inequality

$$t_\nu(g_j) \geq \varepsilon \text{ for } j \in J \text{ large enough.}$$

By the part i) of the definition in Section 4 it follows that an $\delta>0$ exists such that

$$\varphi_{\nu j}(g_j)\geq\delta \text{ for } j\in J \text{ sufficiently large.}$$

ii) Write g_j in the form

$$g_j = P_\nu g_j + \sum_{i\in I_\nu}\lambda_{ij}a_i.$$

Since $g_j\to\nabla F(z)$, $j\to\infty$, $j\in J$, it follows that $P_\nu g_j\to P_\nu\nabla F(z)$ and

$$\sum_{i\in I_\nu}\lambda_{ij}a_i\to\sum_{i\in I_\nu}\lambda_i a_i,\ j\to\infty,\ j\in J$$

The linear independency of the a_i implies $\lambda_{ij}\to\lambda_i$, $j\to\infty$, $j\in J$. Hence there is $\varepsilon>0$ such that $\lambda_{\tilde{\imath}j}=\max\{\lambda_{ij},\ i\in I_\nu\}\geq\varepsilon$ for $j\in J$ sufficiently large. By Lemma 8.1, there is $\delta>0$ such that

(1) $v_\nu(g_j)\geq\delta$ for $j\in J$ sufficiently large.

Furthermore $P_\nu\nabla F(z) = 0$ implies $t_\nu(g_j)\to 0$, $j\to\infty$, $j\in J$ and by the part ii) of Definition 4,

(2) $\varphi_{\nu j}(g_j)\to 0$, $j\to\infty$, $j\in J$.

Finally relations (1) and (2) imply by the part iii) of Definition 4 that there is $\delta>0$ such that $\Psi_{\nu j}(g_j)\geq\delta$ for $j\in J$ sufficiently large.

Lemma 8.3. Let z be a cluster of the sequence $\{x_j\}$ and let $\{x_j, j\in J_\nu\}$ be a subsequence of $\{x_j\}$ converging to z with the property $I_j = I_\nu$, $j\in J$.

i) If $P_\nu\nabla F(z) \neq 0$ there is $\varepsilon>0$ such that $g_j's_j\geq\varepsilon$ for $j\in J$ sufficiently large.

ii) If $\nabla F(z) = \sum_{i\in I_\nu}\lambda_i a_i$, $\max\{\lambda_i \mid i\in I_\nu\}>0$ and $\tilde{I}_j\neq\emptyset, j\in J$ then there is $\varepsilon>0$ such that $g_j's_j\geq\varepsilon$ for $j\in J$ large enough.

Proof: i) By Lemma 8.2 there exists $\delta>0$ such that $\varphi_{\nu j}(g_j)\geq\delta$, $j\in J$ large enough. Let $J_1\subset J$ be the set of indices such that $\tilde{I}_j=\emptyset$, $j\in J_1$. From the property i) of the admissibility definition 5 it follows that there is $\varepsilon_1>0$ such that

(1) $g_j's_j \geq \varepsilon_1$ for $j\in J_1$ sufficiently large.

Let $J_2=J\setminus J_1$. Then $\tilde{I}_j\neq\emptyset, j\in J_2$ and by the antizigzagging precaution 7 we have $\varphi_{\nu j}(g_j)<\Psi_{\nu j}(g_j)$, $j\in J_2$ and hence $\Psi_{\nu j}(g_j)\geq\delta$ for $j\in J_2$ large enough. From the property ii) of the admissibility definition 6 it follows that there is $\varepsilon_2>0$ such that

(2) $g_j's_j\geq\varepsilon_2$ for $j\in J_2$ sufficiently large.

The conclusion of the Lemma follows from (1) and (2) with $\varepsilon=\min\{\varepsilon_1,\varepsilon_2\}$.

ii) By Lemma 8.2 there exists $\delta>0$ such that $\Psi_{\nu j}(g_j)\geq\delta$ for $j\in J$ large enough. From the property ii) of the admissibility definition 5 follows then the existence of an $\varepsilon>0$ such that

$$g_j's_j\geq\varepsilon \text{ for } j\in J \text{ sufficiently large.}$$

Lemma 8.4. Let I_ν, $I_\rho\subset I$ be such that $I_\nu\setminus I_\rho\neq\emptyset$. Suppose

$$\nabla F(z) = \sum_{i\in I_\nu}\lambda_i a_i,\ \max\{\lambda_i, i\in I_\nu\}>0.$$

If there is a subsequence $\{x_j, j\in J\}$ converging to z with the properties that $I_j=I_\nu, I_{j+1}=I_\rho$, $j\in J$, and $\{x_{j+1}, j\in J\}$ converges to z then $P_\rho\nabla F(z)\neq 0$.

Proof: Using Lemma 8.3. we conclude from the assumptions stated in the Lemma that there is $\varepsilon>0$ such that

$$(1)\ g_j's_j\geq\varepsilon>0 \text{ for } j\in J \text{ sufficiently large.}$$

Suppose that $P_\rho\nabla F(z)=0$, i.e. $\nabla F(z) = \sum_{i\in I}\lambda_i a_i$. As both $\{x_j, j\in J\}$ and $\{x_{j+1}, j\in J\}$ converge to z it follows that $I_\nu\subset I(z)$ and $I_\rho\subset I(z)$, where $I(z)=\{i\in I\mid a_i'z=b_i\}$.
Hence the $a_i, i\in I_\nu\cup I_\rho$, are linearly independent. Then $\nabla F(z)=\sum_{i\in I_\nu\cap I_\rho}\lambda_i a_i$ and therefore

$$(2)\ \lambda_i = 0,\qquad i\in I_\nu\setminus I_\rho .$$

Consider the representation $g_j=P_\nu g_j + \sum_{i\in I_\nu}\lambda_{ij}a_i$. As $g_j\to\nabla F(z)$, $j\to\infty$, $j\in J$, and the a_i, $i\in I_\nu$, are linearly independent it follows that $P_\nu g_j\to 0$ and $\lambda_{ij}\to\lambda_i, j\to\infty, j\in J$. Relation (2) implies $\lambda_{ij}\to 0$, $j\to\infty$, $j\in J$, $i\in I_\nu\setminus I_\rho$. Now

$$g_j's_j=(P_\nu g_j)'s_j + \sum_{i\in I_\nu\setminus I_\rho}\lambda_{ij}a_i's_j,\ j\in J$$

and the boundedness of $\{\|s_j\|\}$ implies $g_j's_j\to 0, j\to\infty, j\in J$, which contradicts (1).

Lemma 8.5. Let I_ν, $I_\rho\subset I$ be such that $I_\nu\setminus I_\rho|\neq\emptyset$, $I_\rho\setminus I_\nu\neq\emptyset$.
Suppose that $\nabla F(z)=\sum_{i\in I_\nu}\lambda_i a_i$. If there is a subsequence $\{x_j, j\in J\}$ converging to z with the properties that $I_j=I_\nu, I_{j+1}=I_\rho, j\in J$, and $\{x_{j+1}, j\in J\}$ converges to z then $\tilde{I}_{j+1}=\emptyset$ for $j\in J$ large enough.

Proof: Using Lemma 8.4 and the assumptions of the lemma we conclude that $P_\rho\nabla F(z)\neq 0$. Hence by Lemma 8.2., there is $\varepsilon>0$ such that $\varphi_{j+1,j+1}(g_{j+1})= \varphi_{\rho,j+1}(g_{j+1})$ satisfies the inequality

$$(1)\ \varphi_{\rho,j+1}(g_{j+1})\geq\varepsilon \text{ for } j \text{ sufficiently large.}$$

Furthermore, $I_\rho\setminus I_\nu\neq\emptyset$ implies

$$(2)\ \beta_{j+1}=1,\ j\in J.$$

It follows from (1) and (2) and the antizigzagging precaution 7 that $\tilde{I}_{j+1}=\emptyset$ for $j\in J$ large enough.

In the following we need

Definition 8.1. Let z be a cluster point of $\{x_j\}$. Let I(z) be such that $a_i'z=b_i$, $i\in I(z)$, $a_i'z<b_i$, $i\notin I(z)$. The set L_ν is said to be maximal if it has the following property: The set L_ν contains a subsequence of $\{x_j\}$ which converges to z and for $i\in I_\nu$ the set $L_\nu\cap\{x\in R^n | a_i'x=b_i\}$ contains no such subsequence.

It is clear that $I_\nu\subset I(z)$, since $\{x_j, j\in J\}$ converges to z and, for j large enough, no constraint not active at z can be active at x_j. Let $\Omega(z)$ be a set of subsets $I_\nu\subset I(z)$ such that for each $I_\nu\in\Omega(z)$ the corresponding intersection L_ν is maximal.

Lemma 8.6. Let $I_\nu\in\Omega(z)$ and let $\{x_j, j\in J\}\subset L_\nu$ be a subsequence of $\{x_j\}$ converging to z. If $\nabla F(z) = \sum_{i\in I_\nu}\lambda_i a_i$ and $\max\{\lambda_i, i\in I_\nu\}>0$ then there is an infinite subsequence $\{x_j, j\in J_1\}\subset L_\nu$ converging to z such that $\tilde{I}_j\neq\emptyset$, $j\in J_1$.

Proof: By Lemma 8.2, $\varphi_{\nu j}(g_j)\to 0$, $j\to\infty$, $j\in J$, and $\Psi_{\nu j}(g_j)\geq\varepsilon>0$ for $j\in J$ sufficiently large.
If there is some infinite subset $J_1\subset J$ such that $\tilde{I}_j\neq\emptyset$, $j\in J_1$, the statement of the lemma is true. Otherwise, it follows from the antizigzagging precaution that $\beta_j=1$ and $\varphi_{\nu j}(g_j)\neq 0$ for all $j\in J$ large enough. Then $s_j\in T_\nu$ and the assumptions of the lemma imply

$$\frac{g_j's_j}{\|s_j\|}\to 0,\ j\to\infty\ ,\ j\in J.$$

By the admissibility definition 5, $\|s_j\|\to 0$, $j\to\infty$, $j\in J$. As by the admissibility definition 6, $\{\sigma_j\}$ is bounded it follows that $\|x_{j+1}-x_j\|\to 0$, $j\to\infty$, $j\in J$, i.e. $\{x_{j+1}, j\in J\}$ converges to z. As $\{x_{j+1}, j\in J\}\subset L_\nu$, by the definition of L_ν, $\sigma_j<\sigma_j^*$, i.e. $\beta_{j+1}=0$ for $j\in J$ large enough. Let $J_1=\{j | j-1\in J\}$. Then it follows from the antizigzagging precaution 7 that $\tilde{I}_j\neq\emptyset$, $j\in J_1$.

Lemma 8.7. Let $I_\nu\in\Omega(z)$ and $\{x_j, j\in J\}\subset L_\nu$ be any subsequence of $\{x_j\}$ converging to z. If $P_\nu\nabla F(z)\neq 0$ and $\tilde{I}_j=\emptyset$ for $j\in J$, then there exists $\varepsilon>0$ such that $g_j's_j\geq\varepsilon$ and $\sigma_j^*\geq\varepsilon$ for $j\in J$ large enough.

Proof: By Lemma 8.3, there is $\varepsilon_1>0$ such that

(1) $g_j's_j\geq\varepsilon_1$ for $j\in J$ large enough.

If there is an infinite set $J_1\subset J$ such that $\sigma_j\to 0$, $j\to\infty$, $j\in J_1$ it follows from the admissibility definition 6 that $\sigma_j=\sigma_j^*$ for $j\in J_1$ sufficiently large. The boundedness of $\{\|s_j\|\}$ implies that $\|x_{j+1}-x_j\|\to 0$, $j\to\infty$, $j\in J_1$, i.e. $\{x_{j+1}, j\in J_1\}$ converges to z. Since the number of constraints is finite, it follows that there is $k\in I_\nu$ such that $x_{j+1}'a_k=b_k$ infinitely often. Therefore, there exists a subsequence of $\{x_j\}$ converging to z contained in the intersection $L_\nu\cap\{x\in R^n \mid a_k'x=b_k\}$. This contradicts the definition of L_ν. Hence there is $\varepsilon_2>0$ such that

(2) $\sigma_j^*\geq\sigma_j\geq\varepsilon_2$ for $j\in J$ sufficiently large.

The statement of the lemma follows from (1) and (2) with $\varepsilon=\min\{\varepsilon_1,\varepsilon_2\}$.

Lemma 8.8. Let $I_\nu\in\Omega(z)$ be a set satisfying $\mathrm{card}(I_\nu)=\max\{\mathrm{card}(I_\theta), I_\theta\in\Omega(z)\}$ and let $\{x_j, j\in J\}\subset L_\nu$ be any subsequence converging to z. If $\nabla F(z)=\sum_{i\in I_\nu}\lambda_i a_i$, $\max\{\lambda_i \mid i\in I_\nu\}>0$, there is an infinite subset $J_2\subset J$ and an $\varepsilon>0$ such that either $g_j's_j\geq\varepsilon$ and $\sigma_j^*\geq\varepsilon$, $j\in J_2$ or there exists $I_\rho\in\Omega(z)$ such that $\{x_{j+1}, j\in J_2\}\subset L_\rho$ converges to z and $g_{j+1}'s_{j+1}\geq\varepsilon$ and $\sigma_{j+1}^*\geq\varepsilon$, $j\in J_2$.

Proof: By Lemma 8.6 the assumptions of the lemma imply that there is an infinite subsequence $\{x_j, j\in J_1\}\subset L_\nu$ converging to z such that $\tilde{I}_j\neq\emptyset, j\in J_1$. By Lemma 8.2, there exists an $\varepsilon_1>0$ such that

(1) $g_j's_j\geq\varepsilon_1$, $j\in J_1$.

If $\sigma_j\not\to 0, j\to\infty$, $j\in J_1$ then there is an infinite subset $J_2\subset J_1$ and an $\varepsilon_2>0$ such that

(2) $\sigma_j^*\geq\sigma_j\geq\varepsilon_2, j\in J_2$,

and the first alternative of the lemma follows from (1) and (2) with $\varepsilon=\min\{\varepsilon_1,\varepsilon_2\}$

Suppose that $\sigma_j\to 0$, $j\to\infty$, $j\in J_1$. It follows from the boundedness of $\{\|s_j\|\}$ that $\|x_{j+1}-x_j\|\to 0, j\to\infty, j\in J_1$.

Hence

(3) $\{x_{j+1}, j\in J_1\}$ converges to z.

By the property ii) of the admissibility definition 6,

(4) $\sigma_j=\sigma_j^*$ for $j\in J_1$ large enough.

Since the number of constraints is finite, there is an infinite subset $J_2\subset J_1$ and a set of indices I_ρ such that

(5) $\{x_{j+1}, j\in J_2\}\subset L_\rho$.

Since by the antizigzagging precaution 7 at most one active constraint is dropped at the step j and by (4) at least one new active constraint is obtained, the assumption $card(I_\nu)=\max\{card(I_\theta)|I_\theta\in\Omega(z)\}$ implies that $I_\rho\in\Omega(z)$. Now the conditions of Lemmas 8.4 and 8.5 are satisfied and, by Lemma 8.5, $\tilde{I}_{j+1}=\emptyset$ for $j\in J_2$ sufficiently large. Furthermore by Lemma 8.7, it follows that there is $\varepsilon>0$ such that $g'_{j+1}s_{j+1}\geq\varepsilon$ and $\sigma^*_{j+1}\geq\varepsilon$ for $j\in J_2$ large enough.

Lemma 8.9. Let $I_1,\ldots,I_k$ be the set of elements of $\Omega(z)$ satisfying $card(I_i)=\max\{card(I_\nu), I_\nu\in\Omega(z)\}$, $i=1,\ldots,k$ and suppose that $P_i\nabla F(z)\neq 0$, $1\leq i\leq K$. Then at least one of the sets $L_1,\ldots,L_k$ contains an infinite subsequence $\{x_j, j\in J\}$ converging to z such that, for some $\varepsilon>0$ and $j\in J$ large enough, either $g'_j s_j\geq\varepsilon$ and $\sigma^*_j\geq\varepsilon, j\in J$

or $g'_{j+1}s_{j+1}\geq\varepsilon$ and $\sigma^*_{j+1}\geq\varepsilon, j\in J$.

Proof: Let $\{x_j, j\in J_1\}\subset L_1$ be an arbitrary sequence converging to z. By Lemma 8.3, there exists an $\varepsilon>0$ such that

(1) $g'_j s_j\geq\varepsilon$ for $j\in J_1$ large enough.

If for infinitely many $j\in J_1$, $\tilde{I}_j=\emptyset$, then the conclusion of the lemma follows from Lemma 8.7.

Suppose that $\tilde{I}_j\neq\emptyset$ for all $j\in J_1$ large enough. If for infinitely many $j\in J_1$, σ_j is greater than or equal to some positive constant, the statement of the lemma is true since $\sigma_j^*\geq\sigma_j$ for all j. Suppose that

(2) $\sigma_j\to 0,\ j\to\infty,\ j\in J_1$.

Then (1) and (2) imply, by the admissibility definition 6,

(3) $\sigma_j=\sigma^*_j$ and $\beta_{j+1}=1$ for $j\in J_1$ large enough.

Since $\{\|s_j\|\}$ is bounded, (2) implies

(4) $\|x_{j+1}-x_j\|\to 0, j\to\infty,\ j\in J_1$.

As the number of constraints is finite, there is an infinite set $J\subset J_1$ and a set of indices I_ρ such that $\{x_{j+1}, j\in J\}\subset L_\rho$. By a similar argument as in Lemma 8.8 it follows that $I_\rho\in\Omega(z)$ and $\rho\in\{2,\ldots,k\}$. By the assumptions of the lemma $P_\rho\nabla F(z)\neq 0$ and by the antizigzagging precaution $\tilde{I}_{j+1}=\emptyset$. Therefore the assumptions of Lemma 8.7. are satisfied and there is $\varepsilon>0$ such that $g'_{j+1}s_{j+1}\geq\varepsilon$ and $\sigma^*_{j+1}\geq\varepsilon$ for $j\in \mathbf{J}$ large enough.

Theorem 8.1. Let the sequences $\{s_j\}$, $\{\sigma_j\}$, and $\{x_j\}$ be generated by the algorithm in Section 3 with the antizigzagging precaution given in Section 7. If the sequences $\{s_j\}$ and $\{\sigma_j\}$ are admissable, then every cluster point of $\{x_j\}$ is a stationary point.

Proof: Suppose z is a cluster point of $\{x_j\}$ and let $\Omega(z)$ be as defined in Definition 8.1. If z is not a stationary point, for each $I_\nu \in \Omega(z)$, either $P_\nu \nabla F(z) \neq 0$ or $P_\nu \nabla F(z)=0$ and $d(\nabla F(z), K_\nu)>0$. Let $I_1,\dots,I_k$ be all elements of $\Omega(z)$ satisfying card $(I_i)=\max\{\mathrm{card}(I_\nu) \mid I_\nu \in \Omega(z)\}$, $1 \leq i \leq k$. If for some $\rho \in \{1,\dots,k\}$, $P_\rho \nabla F(z)=0$ and $d(\nabla F(z), K_\rho)>0$.it follows from Lemma 8.8 that there is an infinite set of indices $\mathcal{J}$ and an $\varepsilon>0$ such that, for $j \in \mathcal{J}$ large enough, either $g_j' s_j \geq \varepsilon$ and $\sigma_j^* \geq \varepsilon$ or $g_{j+1}' s_{j+1} \geq \varepsilon$ and $\sigma_{j+1}^* \geq \varepsilon$. If for all i, $1 \leq i \leq k$, $P_i \nabla F(z) \neq 0$, Lemma 8.9 implies that there is an infinite set of indices J and an $\varepsilon>0$ such that, for $j \in J$ large enough, either $g_j' s_j \geq \varepsilon$ and $\sigma_j^* \geq \varepsilon$ or $g_{j+1}' s_{j+1} \geq \varepsilon$ and $\sigma_{j+1}^* \geq \varepsilon$.

Hence in both cases there exists $\varepsilon>0$ and an infinite set of indices K, such that $K=J$ or $K=\{j+1, j \in J\}$ and

(1) $\quad g_j' s_j \geq \varepsilon$ and $\sigma_j^* \geq \varepsilon$ for $j \in J$ sufficiently large.

By the admissibility definition 6, it follows from (1) that there is $\delta>0$ such that

(2) $\quad F(x_{j+1}) \leq F(x_j)-\delta$ for $j \in K$ large enough.

As $F(x_{j+1})<F(x_j)$ for every j, (2) implies that $F(x_j) \to -\infty$, $j \to \infty$, in contr diction to the fact that F(x) is bounded below on R.

9. Examples

In this section we will illustrate how the theory of the previous sections can be applied to some well-known methods. As the admissibility definitions for the sequence of directions $\{s_j\}$ and the sequence of stepsizes $\{\sigma_j\}$ are independent, we will first give examples of two methods that generate admissible sequences of directions and then an example of a step size procedure that generates an admissible sequence of stepsizes.

9.1. Rosen's gradient projection method [1]. Let I_j be the set of indi ces of the constraints active at the step j, $T_j=\{x \in R^n \mid a_i' x=0, i \in I_j\}$, $A_j=(a_i'), i \in I_j$ and $P_j=I-A_j'(A_j A_j')^{-1} A_j$.

Set $\quad \varphi_{jj}(g_j)=\|P_j g_j\|, \qquad \Psi_{jj}(g_j)=(u_j)_l$,

where $u_j=(A_j A_j')^{-1} A_j g_j$ and l is such that $(u_j)_l=\max\{(u_j)_i\}$.

Set $\quad \tilde{I}_j=\begin{cases} \emptyset & \text{if } \varphi_{jj}(g_j) \geq \Psi_{jj}(g_j) \text{ or } (\beta_j=1 \text{ and } \varphi_{jj}(g_j) \neq 0) \\ \{l\} & \text{otherwise} \end{cases}$

Set $\quad s_j=\begin{cases} P_j g_j & \text{if } \tilde{I}_j=\emptyset \\ \overline{P}_j g_j & \text{if } \tilde{I}_j=\{l\}, \end{cases}$

where $\bar{P}_j = I - \bar{A}_j'(\bar{A}_j\bar{A}_j')^{-1}A_j$, $\bar{A}_j = (a_i')$, $i \in I_j$ and $\bar{I}_j = I_j \setminus \{1\}$.

It is easy to verify that φ_{jj} and Ψ_{jj} are P and C forcing functions and $\{s_j\}$ is admissible in the sense of Definition 5.

9.2. Ritter's method of conjugate directions [2]. Let $D_j' = (d_{1j},\dots,d_{nj})$, $D_j^{-1} = (c_{1j},\dots,c_{nj})$ and $J(x_j) = \{\alpha_{1j},\dots,\alpha_{nj}\}$ be as defined in [2]

Set

$$\varphi_{jj}(g_j) = |(v_j)_k|, \quad \Psi_{jj}(g_j) = (v_j)_1,$$

where $u_j = D_j^{-1} g_j$, $u_j' = ((u_j)_1, \dots (u_j)_n)$

$$(v_j)_i = \frac{(u_j)_i}{\|c_{ij}\|}, \quad v_j = ((v_j)_1, \dots, (v_j)_n)$$

and let l and k be such that

$$(v_j)_1 \geq (v_j)_i \text{ for all i with } \alpha_{ij} > 0$$

$$|(v_j)_k| \geq |(v_j)_i| \text{ for all i with } \alpha_{ij} \leq 0.$$

Set

$$\tilde{I}_j = \begin{cases} \emptyset & \text{if } \varphi_{jj}(g_j) \geq \Psi_{jj}(g_j) \text{ or } (\beta_j = 1 \text{ and } \varphi_{jj}(g_j) \neq 0) \\ \{1\} & \text{otherwise} \end{cases}$$

Set

$$s_j = \begin{cases} c_{kj}(v_j)_k & \text{if } \tilde{I}_j = \emptyset \\ c_{1j}(v_j)_1 & \text{if } \tilde{I}_j = \{1\}. \end{cases}$$

Using a similar argument as in Lemma 2 in [2] it is possible to prove that φ_{jj} and Ψ_{jj} are P and C forcing functions and that $\{s_j\}$ is admissible in the sense of Definition 5.

9.3. Modified Armijo-Goldstein procedure. Let $\delta > 0$ be a given constant.

Let

$$\sigma_j^* = \min\{\frac{a_i'x_j - b_i}{a_i'x_j} \mid a_i's_j < 0\}.$$

Let

$$\bar{\sigma}_j = \min\{1, \sigma_j^*\}. \text{ Let}$$

$$h(x_j, \sigma) = \frac{F(x_j) - F(x_j - \sigma s_j)}{\sigma g_j' s_j}$$

Let ν_j be the smallest nonegative integer such that $h(x_j, (\frac{1}{2})^{\nu_j}\bar{\sigma}_j) \geq \delta$ and set $\sigma_j = (\frac{1}{2})^{\nu_j}\bar{\sigma}_j$.

It is easy to check that the sequence $\{\sigma_j\}$ is admissible in the sense of the Definition 6.

References

[1] J.B.ROSEN The gradient projection method for nonlinear programming, SIAM J.Appl.Math., 8(1960)181-217

[2] K.RITTER A method of conjugate directions for linearly constrained nonlinear programming problems, Mathematical Programming SIAM J.Numer.Anal., 12(1975)273-303

[3] A.GOLDSTEIN Constructive Real Analysis, Harper and Row, New York 1967

[4] G.ZOUTENDIJK Methods of Feasible Directions, Elsevier, Amsterdam 1960

[5] V.L.MANGASARIAN Nonlinear Programming, McGraw-Hill, New York, 1969

Yu-11040 Belgrad
Milovana Marinkovica 5

BOUNDARY CONTROL OF THE HIGHER-DIMENSIONAL WAVE EQUATION

by

W. Krabs

1. Introduction

In [2] we have considered the problem of controlling a vibrating string at the right-hand side where the left end is kept fixed. The results of this paper can be easily extended to vibrating systems of several dimensions which are described by the wave equation. This is the purpose of the present note. In [2] we also discussed the problem of controllability that has been previously studied by Butkovskiy [1], Russell [4], [5], and Roxin [3]. In [6],[7] controllability for hyperbolic equations in several space dimensions has been investigated by Russell too.

Here we are concerned with control problems for the higher-dimensional wave equation and boundary conditions of the first kind. Other boundary conditions, however, could be treated as well in a similar manner. After the statement of the problem in section 2 we show in section 3 that optimal controls exist and in section 4 that they can be characterized by a weak bang-bang-principle. This, however, is not strong enough to guarantee the uniqueness of optimal controls.
In order to obtain approximate solutions to the problem one can proceed in the same way as in [2] by using Fourier's method.

2. Statement of the Problem

In the following we shall make use of the concept of generalized solutions to boundary value problems in the sense of distributions. Since all the details of the corresponding theory can be found in the book [8] of Triebel to the full extent as they are needed here, we shall omit most of the basic definitions and refer the reader to this book without always explicitly mentioning. We shall also adopt the general notations from [8] .

Let Ω be a bounded open domain in $\mathbb{R}^n$ whose boundary $\partial\Omega$ belongs to the class C^1. We assume the motion of the vibrating medium in Ω to be described by the wave equation

$$y_{tt}-\Delta_x y=o \text{ in } \Omega\times(o,T), \tag{2.1}$$

where $T>o$ is a given fixed time for the vibration process and Δ_x

denotes the Laplacian operator in $\mathbb{R}^n$. The initial conditions are given by

$$y(x,o)=y_o(x),\ y_t(x,o)=y_1(x),\ x\in\Omega, \tag{2.2}$$

with $y_o\in \overset{o}{W}{}_2^1(\Omega)$ and $y_1\in L_2(\Omega)$.

The control on the boundary is performed in such a way that the following boundary conditions hold

$$y(x,t)=\varphi(x)v(t),\ x\in\partial\Omega, t\in[o,T], \tag{2.3}$$

where φ is the restriction of a given function $\phi\in W_2^2(\Omega)$ to $\partial\Omega$ and v is allowed to vary in the Hilbert space

$$X=\{v\in C^1[o,T] \mid v'' \text{ exists a.e. and is in } L_2[o,T], \text{ and } v(o)=v'(o)=o\} \tag{2.4}$$

provided with the scalar product

$$\langle v_1,v_2\rangle_X=\int_o^T v_1''(t)v_2''(t)dt,\ v_1,v_2\in X,$$

and the norm

$$\|v\|_X=\langle v,v\rangle_X^{1/2},\ v\in X.$$

Let $\hat{r}\in W_2^2(\Omega)$ be the solution of the Dirichlet problem

$$\Delta_x\hat{r}=o \text{ in } \Omega,\quad \hat{r}=\varphi \text{ on } \partial\Omega. \tag{2.5}$$

By [8] , Satz 43.1, there exists exactly one generalized solution $y=\hat{y}$ of the initial boundary value problem (2.1),(2.2) with homogeneous boundary conditions

$$y(x,t)=o, x\in\partial\Omega, t\in[o,T], \tag{2.6}$$

and $\hat{y}$ can be represented as

$$\hat{y}(x,t)=\sum_{k=1}^{\infty}\left[\langle y_o,z_k\rangle_{L_2(\Omega)}\cos\sqrt{\lambda_k}t+\frac{\langle y_1,z_k\rangle_{L_2(\Omega)}}{\sqrt{\lambda_k}}\sin\sqrt{\lambda_k}t\right]z_k(x), \tag{2.7}$$

where $\langle\cdot,\cdot\rangle_{L_2(\Omega)}$ denotes the $L_2(\Omega)$-scalar-product, (λ_k) is the sequence of the eigenvalues of Δ_x with

$$o<\lambda_1\leq\lambda_2\leq\ldots \text{and} \quad \lim_{k\to\infty}\lambda_k=+\infty$$

and (z_k) is the corresponding sequence of normalized eigenfunctions.

By [8] , Satz 43.3, for all $r\in L_2(\Omega)$ and all $v\in X$ there exists a unique generalized solution $y=y^*(x,t,v,r)$ of the initial boundary value problem

$$y_{tt}-\Delta_x y=r\cdot v'' \text{ in } \Omega\times(o,T),$$
$$y(x,o)=y_t(x,o)=o,\ x\in\Omega, \quad (2.8)$$
$$y(x,t)=o, x\in\partial\Omega, t\in[o,T],$$

and y^* can be represented as

$$y^*(x,t,v,r)=\sum_{k=1}^{\infty}\frac{h_k}{\sqrt{\lambda_k}}\int_o^t \sin\sqrt{\lambda_k}(t-\tau)v''(\tau)d\tau\, z_k(x), \quad (2.9)$$

where

$$h_k=\int_\Omega r(x)z_k(x)dx,\ k=1,2,\ldots \quad (2.10)$$

Furthermore, there is a constant $c>o$ which is independent of r and v such that the following estimation holds

$$\int_\Omega\Big[y_t^*(x,t,v,r)^2+\sum_{i=1}^{n}y_{x_i}^*(x,t,v,r)^2+y^*(x,t,v,r)^2\Big]dx$$
$$\le c\int_\Omega r(x)^2dx\int_o^T v''(t)^2dt \text{ for all } t\in[o,T]. \quad (2.11)$$

Summarizing we have that for each $v\in X$ the unique generalized solution of (2.1),(2.2),(2.3) is given by

$$y(x,t,v)=\hat{r}(x)v(t)-y^*(x,t,v,\hat{r})+\hat{y}(x,t), \quad (2.12)$$

$x\in\bar{\Omega}, t\in[o,T]$, where $\hat{r}\in W_2^2(\Omega)$ is the solution of (2.5) and y^* and $\hat{y}$ are given by (2.9) and (2.7) respectively.

In order to formulate the control problem we assume that the control functions v in (2.3) are only allowed to vary in the subset

$$V=\{v\in X\,|\,|v''|\le 1 \text{ a.e.}\} \quad (2.13)$$

of X and then we are looking for optimal controls $\hat{v}\in V$ such that the vibration energy at the time T which is given by

$$E(y(\cdot,T,v))=\int_\Omega\Big[y_t(x,T,v)^2+\sum_{i=1}^{n}y_{x_i}(x,T,v)^2\Big]dx \quad (2.14)$$

attains its minimum value at $v=\hat{v}$, i.e.

$$E(y(\cdot,T,\hat{v}))\le E(y(\cdot,T,v)) \text{ for all } v\in V. \quad (2.15)$$

3. Existence of Optimal Controls

Let $r\in L_2(\Omega)$ be a fixed function. Then, for each $v\in X$, we define

$$S^r(v)(x)=\begin{pmatrix} y_t^*(x,T,v,r)-v'(T)\hat{r}(x) \\ \operatorname{grad}_x y^*(x,T,v,r)-v(T)\operatorname{grad}_x\hat{r}(x)\end{pmatrix}, \quad (3.1)$$

$x\in\Omega$, where y^* is the unique generalized solution (2.9) to (2.8), $\hat{r}\in W_2^2(\Omega)$ the solution to (2.5), and grad_x denotes the gradient vector with respect to x. From (2.11) we conclude that S^r defines a linear mapping from X into $Y=(L_2(\Omega))^{n+1}$. If we define

$$\hat{g}(x)=\begin{pmatrix}\hat{y}_t(x,T)\\ \mathrm{grad}_x\hat{y}(x,T)\end{pmatrix},\quad x\in\Omega, \tag{3.2}$$

where $\hat{y}$ is the solution (2.7) to (2.1),(2.2),(2.6) and denote the scalar product and the norm in Y by $\langle\cdot\,,\cdot\rangle_Y$ and $\|\cdot\|_Y$ respectively, we obtain, by virtue of (2.12),(2.14), and (3.1),

$$E(y(\cdot\,,T,v))=\langle\hat{g}-S^{\hat{r}}(v),\hat{g}-S^{\hat{r}}(v)\rangle_Y=\|\hat{g}-S^{\hat{r}}(v)\|_Y^2. \tag{3.3}$$

Therefore the control problem (2.15) can also be considered as a norm minimum problem in Y.

Using (2.11) it can be shown that for each pair $v_1,v_2\in X$ and each $r\in L_2(\Omega)$ we have

$$\|S^r(v_1)-S^r(v_2)\|_Y\leq K\|v_1-v_2\|_X \tag{3.4}$$

where K is given by

$$K=c^{1/2}\|r\|_{L_2(\Omega)}+(T\int_\Omega\hat{r}(x)^2dx+\frac{T^3}{3}\|\mathrm{grad}_x\hat{r}(x)\|^2_{L_2(\Omega)})^{1/2}.$$

Hence, $S^r:X\longrightarrow Y$ is a continuous mapping which implies that S^r is weakly continuous. Since the set V(2.13) of control functions is weakly compact in X the image $S^r(V)$ is weakly compact in Y and hence closed as it is convexe. By well known results on norm minimum problems in Hilbert space the existence of optimal controls $\hat{v}\in V$ with (2.15) is insured, by virtue of (3.3). Furthermore, there exists exactly one $S^{\hat{r}}(\hat{v})$ in $S^{\hat{r}}(V)$ such that

$$\|\hat{g}-S^{\hat{r}}(\hat{v})\|_Y\leq\|\hat{g}-S^{\hat{r}}(v)\|_Y \text{ for all } v\in V. \tag{3.5}$$

One can prove more than the closedness of the set $S^r(V)$ for each $r\in L_2(\Omega)$, namely the following

Theorem: For each $r\in L_2(\Omega)$ the set $S^r(V)$ with S^r given by (3.1) and V given by (2.13) is compact in Y. The proof of this theorem is the same as that of Satz 1 in [2] and shall be omitted.

More generally than stated above, for each $r\in L_2(\Omega)$, there is a unique element $S^r(\hat{v})\in S^r(V)$ such that (3.5) holds with $\hat{r}$ replaced by r and, furthermore, $S^r(v)$ is characterized by the maximum property

$$\langle \hat{g}-S^r(\hat{v}),S^r(\hat{v})\rangle_Y=\max_{v\in V}\langle \hat{g}-S^r(\hat{v}),S^r(v)\rangle_Y. \tag{3.6}$$

This again is a consequence of well known results on norm minimum problems in Hilbert space.

4. A Weak Bang-Bang-Principle

The maximum property (3.6) shall be further exploited to obtain more refined statements about v. For that purpose for each $N\in\mathbb{N}$ we define, for all $v\in X$,

$$S^{r_N}(v)(x)=\begin{pmatrix} y_t^*(x,T,v,r_N)-v'(T)\hat{r}(x) \\ \\ \mathrm{grad}_x y^*(x,T,v,r_N)-v(T)\mathrm{grad}_x\hat{r}(x)\end{pmatrix}, \tag{4.1}$$

$x\in\Omega$. Here y^* is the solution of (2.8) with

$$r_N(x)=\sum_{k=1}^{N}h_k z_k(x) \tag{4.2}$$

instead of r and the h_k's are defined by (2.10) for a given $r\in L_2(\Omega)$. y^* has the explicit form

$$y^*(x,t,v,r_N)=\sum_{k=1}^{N}\frac{h_k}{\sqrt{\lambda_k}}\int_0^t \sin\sqrt{\lambda_k}(t-\tau)v''(\tau)d\tau z_k(x). \tag{4.3}$$

Again $\hat{r}\in W_2^2(\Omega)$ is the solution of (2.5). Further, we have

$$\lim_{N\to\infty}\|r_N-r\|_{L_2(\Omega)}=0. \tag{4.4}$$

Because of

$$S^r(v)(x)-S^{r_N}(v)(x)=\begin{pmatrix} y_t^*(x,T,v,r-r_N) \\ \mathrm{grad}_x y^*(x,T,v,r-r_N)\end{pmatrix}$$

we obtain from (2.11) and (4.4)

$$\lim_{N\to\infty}\ \sup_{v\in V}\|S^r(v)-S^{r_N}(v)\|_Y=0. \tag{4.5}$$

This implies

$$\langle \hat{g}-S^r(\hat{v}),S^r(v)\rangle_Y=\lim_{N\to\infty}\langle \hat{g}-S^r(\hat{v}),S^{r_N}(v)\rangle_Y \tag{4.6}$$

for all $v\in V$. Using the explicit form (4.3) of y^* one calculates

$$\langle \hat{g}-S^r(\hat{v}),S^{r_N}(v)\rangle_Y=\int_0^T \hat{S}^N(t)v''(t)dt \tag{4.7}$$

where

$$\hat{S}^N(t)=\hat{a}_0+\hat{b}_0(T-t)+\sum_{k=1}^{N}\hat{a}_k\cos\sqrt{\lambda_k}(T-t)+\hat{b}_k\sin\sqrt{\lambda_k}(T-t) \tag{4.8}$$

and

$$\hat{a}_o = -\int_\Omega \hat{r}(x) y_t(x,T,\hat{v},r)dx,$$
$$\hat{b}_o = \sum_{i=1}^{n} \int_\Omega \hat{r}_{x_i}(x) y_{x_i}(x,T,\hat{v},r)dx,$$
$$\hat{a}_k = h_k \int_\Omega z_k(x) y_t(x,T,\hat{v},r)dx, \tag{4.9}$$
$$\hat{b}_k = \frac{h_k}{\sqrt{\lambda_k}} \sum_{i=1}^{n} \int_\Omega z_{kx_i}(x) y_{x_i}(x,T,\hat{v},r)dx,$$
$$k=1,\dots,N,$$

where

$$y(x,T,\hat{v},r) = \hat{r}(x)v(t) - y^*(x,T,\hat{v},r) + \hat{y}(x,T).$$

From this we obtain that the sequence (4.8) converges in the L_2-sense to $\hat{S} \in L_2[o,T]$ where

$$\hat{S}(t) = \hat{a}_o + \hat{b}_o(T-t) + \sum_{k=1}^{\infty} \hat{a}_k \cos\sqrt{\lambda_k}(T-t) + \hat{b}_k \sin\sqrt{\lambda_k}(T-t) \tag{4.10}$$

such that, by virtue of (4.6),(4.7), we have

$$\langle \hat{g} - S^r(\hat{v}), S^r(v) \rangle_Y = \int_o^T \hat{S}(t) v''(t) dt$$

for all $v \in V$. The maximum property (3.6) is therefore equivalent with

$$\int_o^T \hat{S}(t) \hat{v}''(t) dt = \max_{v \in V} \int_o^T \hat{S}(t) v''(t) dt = \int_o^T |\hat{S}(t)| dt. \tag{4.11}$$

Under the assumption

$$\langle \hat{g} - S^r(\hat{v}), S^r(\hat{v}) \rangle_Y \neq o \tag{4.12}$$

which implies $\hat{S} \neq \Theta_{L_2[o,T]}$ we therefore obtain as a necessary and sufficient condition for $\hat{v} \in V$ to be optimal

$$\hat{v}''(t) = \operatorname{sgn} \hat{S}(t) \text{ for all } t \in [o,T] \text{ such that } \hat{S}(t) \neq o. \tag{4.13}$$

This is called a weak bang-bang-principle for optimality.

If (4.12) is violated, by the maximum property (3.6) we conclude

$$\langle \hat{g} - S^r(\hat{v}), S^r(v) \rangle_Y = o \text{ for all } v \in V$$

and hence for all $v \in X$ which implies

$$\| \hat{g} - S^r(\hat{v}) \|_Y = \inf_{v \in X} \| \hat{g} - S^r(v) \|_Y.$$

References

[1] Butkovskiy,A.G.: Theory of optimal control of distributed parameter systems. Elsevier Publ.Comp.,New York-London-Amsterdam 1969

[2] Krabs,W.: Ein Kontroll-Approximationsproblem für die schwingende Saite. To appear in ISNM

[3] Roxin,E.: Optimierungsprobleme mit der Wellengleichung. Preprint Nr. 146 des Fachbereichs Mathematik der TH Darmstadt, Juli 1974

[4] Russell,D.L.: On boundary-value controllability of linear symmetric hyperbolic systems. "Proceedings of the Conference on the Mathematical Theory of Control", edited by A.V. Balakrishnan and L.W. Neustadt, Univ. of Southern Calif., Academic Press, New York 1967, pp.312-321

[5] Russell,D.L.: Nonharmonic Fourier series in the control of distributed parameter systems. J.Mathem.Anal.Appl.18 (1967), 542-560

[6] Russell,D.L.: Boundary value control of the higher-dimensional wave equation. SIAM J.Control 9 (1971),29-42

[7] Russell,D.L.: Boundary value control theory of the higher-dimensional wave equation, Part II. SIAM J.Control 9 (1971),4o1-419

[8] Triebel,H.: Höhere Analysis. VEB Deutscher Verlag der Wissenschaften, Berlin 1972

Address: W.Krabs, Fachbereich Mathematik der TH Darmstadt,
61 Darmstadt, Schloßgartenstr. 7

NONDIFFERENTIABLE OPTIMISATION

SUBGRADIENT AND ε - SUBGRADIENT METHODS

Claude LEMARECHAL IRIA - Laboria

Rocquencourt 78150 LE CHESNAY

We give some ideas which lead to descent methods for minimizing nondifferentiable functions . Such methods have been published in several papers and they all involve the same concept , namely the ε - subdifferential .
As an illustrative example we give a method for solving a real problem of inventory control . This nonlinear programming problem is interesting in that it has a lot of local optima whose cost is far higher than the global minimum . Despite it is not convex , we show that a satisfactory solution can be obtained by dual means .

1.- Let the problem be

(1) $\quad \text{Min} \{ f(x) \,/\, x \in F \}$

where f is a (proper) convex function defined on a Hilbert space H and $F \subset H$ is a closed convex set . We suppose x_o is given in F and we want to have a descent method starting from x_o , i.e. to construct a sequence $\{ x_n \} \subset F$ by the formula $x_{n+1} = x_n + \rho_n s_n$ where $\rho_n > 0$ is the stepsize and $s_n \in H$ is the descent direction satisfying :

(2) $\exists \bar{\rho} > 0$ such that $\rho \in]0,\bar{\rho}]$ implies $x_n + \rho s_n \in F$ and $f(x_n + \rho s_n) < f(x_n)$

2.- We assume only that f has bounded derivatives in the following sense : define $F_o = \{ x \in F \,/\, f(x) \leq f(x_o) \}$; then there exists a positive constant M such that

(3) $\forall x \in F_o$ such that f has a gradient $g(x)$ at x , $|g(x)| \leq M$

It must be enhanced that this does not imply any continuity on the gradients of f ; indeed grad $f(x)$ may not exist for every x in F_o. The following result gives a sufficient condition for boundedness :

THEOREM <u>Suppose f is bounded (from below) on F_o and suppose there exists a $\delta > 0$ such that f is bounded from above on the set $F_o + \delta B$ (where B is the unit ball in H) Then (3) holds</u> .

Proof- Let $x \in F_o$ such that the gradient $g(x)$ exists . Since f is convex we have :

$\forall y \in H \quad f(y) \geq f(x) + (g(x) , y - x)$

Let us choose $y = x + \delta \frac{g(x)}{|g(x)|}$. Then $y \in F_o + \delta B$ and $|g(x)| \leq \frac{f(y) - f(x)}{\delta}$ which is bounded when x and y describe F_o and $F_o + \delta B$ respectively ■

3.- It is known that f is differentiable on a dense subset of F_o [9,Thm 25.5] . Thus , given $x \in F_o$, we can construct sequences x_i converging (strongly) to x and such that the gradients $g(x_i)$ exist.The sequences $g(x_i)$ are bounded (from (3)) and , therefoer , have (weak) cluster points . Suppose we construct all the possible sequences in this framework , and consider the set G(x) of all the cluster points of all the possible sequences $g(x_i)$ thus obtained . Then G(x) is a bounded set and we can construct its convex hull . It can be shown that we then obtain [9,Thm 25.6] the so-called subdifferential of f at x , defined by

(4) $\quad \partial f(x) = \{ g \in H \ / \ \forall y \in H \quad f(y) \geq f(x) + (g,y-x) \}$

By construction $\partial f(x)$ is a closed bounded convex set which exactly describes the behaviour of grad f in the neighborhood of x .

4.- Suppose now that s is a descent direction at x (Cf (2)) . Then , for each subgradient $g \in \partial f(x)$ we have

$$f(x) > f(x + \rho s) \geq f(x) + (g, x + \rho s - x)$$

This implies that $(s,g) < 0 \quad \forall g \in \partial f(x)$ or , in other words

(5) $\qquad \max \{ (s,g) \ / \ g \in \partial f(x) \} < 0$

Applying the local definition of $\partial f(x)$, this means also that s has a negative scalar product with all the possible gradients of f in the neighborhood of x . It turns out that (5) is also a sufficient condition [9,Thm 23.2] . Therefore , if G(x) is not a singleton , knowing a descent direction at x implies to study f "all around" x , and our aim is now to show that such a study can be made by a finite process . In the sequel we suppose that , given $x \in F_o$, we can compute f(x) and one element "at random"in $\partial f(x)$ (in fact it will probably be <u>the</u> gradient since f is differentiable almost everywhere , but we do not insist on this) . This assumption must be compared to [1] and [2] where is required the knowledge of at least all $\partial f(x)$ for each x .

5.- Let $x \in F_0$ and $g^o \in \partial f(x)$. For simplicity assume first that $F = H$. We choose a direction $s^o = -g^o$, which is the simplest choice for having $(s^o, g^o) < 0$ (Cf (4)) . Is s^o a descent direction ? To check this , we tabulate $f(x + \rho s^o)$ for values of ρ going to 0 . Doing this , we assume that we encounter the worst situation , namely

$$f(x + \rho s^o) \geq f(x) \qquad \forall \rho > 0$$

But for any subgradient $g(\rho)$ computed at $x + \rho s^o$ we have

$$f(x + \rho s^o) \geq f(x) \geq f(x + \rho s^o) \; + \; (g(\rho) \, , x - (x + \rho s^o))$$

Hence $\quad (g(\rho), s^o) \geq 0 \quad \forall \rho > 0 \; , \; \forall g(\rho) \in \partial f(x + \rho s^o)$

Let g^1 be a cluster point of $g(\rho)$ when $\rho \to 0$. Then $(g^1, s^o) \geq 0$. Thus we have

$g^o \in \partial f(x)$, $g^1 \in \partial f(x)$ such that $(g^o, g^1) \leq 0$. Therefore

$$|g^o - g^1|^2 = |g^o|^2 - 2(g^o, g^1) + |g^1|^2 \geq |g^o|^2 + |g^1|^2$$

If $|g^o - g^1|$ is small , this implies that $|g^o|$ and $|g^1|$ are small , hence x is almost optimal . Roughly speaking :

Knowing $g^o \in \partial f(x)$ and being in the worst case (i.e. x is not almost optimal and the line-search does not diminish the objective) we are able to construct $g^1 \in \partial f(x)$ far from g^o (the construction is theoretical because it requires an infinite tabulation)

6.- Recursively suppose we know $g^o, g^1, \ldots, g^k$ all in $\partial f(x)$. We guess that , by a line-search along some direction s^k, either we obtain a decrease in the objective , or we obtain a new subgradient g^{k+1} such that $(g^{k+1}, s^k) \geq 0$.
Recall that we want to construct $\partial f(x)$. So g^{k+1} should be as far as possible from all the previous g^i's . To have this we feel intuitively that s^k should make the number $\max \{ (s^k, g^i) \, / \, i = 0,1, \ldots ,k \}$ as negative as possible (this is a minmax strategy) ; since s^k is a direction we have to normalize it and then , computing s^k is solving the problem

$$(6) \qquad \min_{|s|=1} \; \max_{i=0}^{k} \; (s, g^i)$$

It can be shown that solving this problem is equivalent to find the point of least norm in the convex hull of $\{ g^o, g^1, \ldots, g^k \}$ i.e. the set

$$\{ \textstyle\sum_{i=0}^{k} \lambda_i g_i \; / \; \lambda_i \geq 0 \; , \; \Sigma \lambda_i = 1 \}$$

(but this property is only true when s is normalized by the norm of H) . Thus solving (6) is a linear - quadratic problem and in [11] is given an excellent algorithm for solving it .

7.- It is now easy to imagine a descent method for solving (1) , each step of which would consist of a sequence of line - searches and projections . It can be proved that this sequence would be finite provided the minmax strategy (6) is adopted for s^k but however this algorithm would be computationally infeasible for two reasons :

- First each line - search should be infinite
- Second x_n would be likely to converge to a wrong point because we would have in fact an approximation of the steepest - descent algorithm , which is known to be nonconvergent in the nondifferentiable case [12] .

Hence the need for some tolerances which will eliminate these two drawbacks ; it turns out that all is well if we define the ε - subdifferential

$$(7) \qquad \partial_\varepsilon f(x) = \{ g \in H \; / \; \forall\, y \in H \quad f(y) \geq f(x) + (g, y-x) - \varepsilon \}$$

This set is an enlargement of $\partial f(x)$ and it is interesting for two reasons :

- First it consists of the convex hull of the gradients of f , but in a <u>fixed</u> neighborhood of x (we do not have to pass to the limit in §3)
- Second s is an ε - descent direction (i.e. $\exists\, \rho > 0$ such that $f(x + \rho s) < f(x) - \varepsilon$) if and only if $\max \{ (s,g) \; / \; g \in \partial_\varepsilon f(x) \} < 0$ (Cf (5)) .

Thus , we can construct the ε - subdifferential instead of the 0 - subdifferential by applying the same ideas as in §6 and now

- First the line-searches can be finite
- Second , when the construction is finished (and again this will happen if (6) is used for computing s^k) we have on hand an ε - descent direction and the minimization algorithm will obviously give an ε - optimal point after a finite number of steps .

8.- The ε - descentmethod has been published in [4] for the case without constraints. When F is a polyhedral set , defined by a finite number of linear inequality constraints , the key - idea is to minimize over H the new objective $[f + \delta_F](x)$ where δ_F is the indicator function of F , equal to 0 on F , $+\infty$ elsewhere . This new objective has unbounded subgradients but we have shown in [5] that all the calculations remain valid and the ε - descent method in this case gives an anti - zigzagging procedure which can be described basically as follows : $x_n \in F_0$ being given , we construct a sequence $\{ s^k \}$ of feasible directions <u>issued from x_n</u> . At step k , if a new constraint prevents a decrease by ε , then this constraint is penalized with a finite coefficient (which can be considered as a scaling factor between the gradients of f and of the constraint) and a new direction s^{k+1} is computed , issued from x_n

(and not from the new binded point) . This process is repeated until a decrease by ε can be obtained .

9.- The ε - descent method has thus been introduced from a local study of an unsmooth function and we are led quite naturally to compute the directions by projecting the origin onto the convex hull of some collection of gradients . Surprisingly we get the same conclusion if we start from the opposite viewpoint , namely the global study of a smooth function . Let us consider the conjugate gradient method applied to a quadratic objective . Zoutendijk [14] has observed that the relation of conjugation implies that all the gradients g_i's have the same projection onto the direction s_n . This suggests the optimality of the projection of the origin onto the <u>affine hull</u> of the g_i's . In [8] (see also [10]) is given the method based on this principle : at step n we have on hand the previous gradients $g_o, \dots, g_n$ and s_n is the unique solution of

$$(8) \qquad \begin{cases} \min \; |s|^2 \\ s = \sum_{i=o}^{n} \lambda_i g_i \\ \Sigma \lambda_i = 1 \end{cases}$$

But it is known also that if the objective is quadratic , then the gradients are orthogonal $(g_i, g_j) = 0$ if $i \neq j$; (8) is then very easy to solve and it turns out that the optimal λ_i's are all strictly positive . Hence in this case , s_n is not only the projection of the origin onto the affine hull but also onto the <u>convex hull</u> of $\{ g_o, \dots, g_n \}$.

Thus , as long as $g_o, \dots, g_n$ are enclosed in some <u>reasonable</u> ε - subdifferential it is natural to compute s_n by solving the problem

$$(9) \qquad \begin{cases} \min \; |s|^2 \\ s = \sum_{i=o}^{n} \lambda_i \, g_i \\ \Sigma \lambda_i = 1 \;, \;\; \lambda_i \geq 0 \end{cases}$$

Alternatively , if g_n is <u>too far</u> from $g_o, \dots, g_{n-1}$ then the sequence of projections is cut , we set $s_n = -g_n$ and $g_o, \dots, g_{n-1}$ are left away .

10.- According to the exact meaning chosen for the words "reasonable" and "too far" we obtain different algorithms . Let us choose a test $T(x_o, \dots, x_n; g_o, \dots, g_n)$

which means T is true $\Leftrightarrow \{$ there exists a reasonable ε - subdifferential gathering $x_o, \dots, x_n$ and $g_o, \dots, g_n$

then we can define the algorithm $A(T)$ as follows :

<u>Step 0</u> x_o is given in F_o. Compute g_o in $\partial f(x_o)$, set $n = 0$.

<u>Step 1</u> Solve (9) and obtain s_n.

<u>Step 2</u> Perform an (approximate) line - search along s_n and obtain ρ_n.
Set $x_{n+1} = x_n + \rho_n s_n$ and find $g_{n+1} \in \partial f(x_{n+1})$

<u>Step 3</u> If T then set $n = n + 1$ and go to 1
otherwise set $x_o = x_{n+1}$ and go to 0 ∎

Up to now we know three such algorithms :

- In [13] T is the most optimistic one, namely
$T \Leftrightarrow$ " $0 \notin \text{conv} \{ g_o, \dots, g_n \}$ "
In this case the projections are performed as long as (9) has a nonzero solution.

- The ε - descent method with very small ε corresponds to
$T \Leftrightarrow$ " $f(x_n) \geqslant f(x_o)$ "
and is therefore the most pessimistic one : we cut as soon as possible.

- Between these two extremes [7] is another example in which
$T \Leftrightarrow$ " $g_n \in \partial_{\varepsilon_n} f(x_o)$ where $\varepsilon_n = f(x_o) - f(x_n)$ " $\Leftrightarrow$ " $(g_n, x_n - x_o) \leq 0$ "

11.- As an illustrative example of §8 let us consider the following problem :

(10) $$\min \; J(x,y) = \sum_{i=1}^{n} \sum_{j=1}^{m} \left(\alpha_{ij}\, x_{ij} + \beta_{ij}\, y_{ij} + \gamma_{ij} \frac{x_{ij}}{y_{ij}}\right)$$

(11) $$\sum_{i=1}^{n} (a_{ij}\, x_{ij} + b_{ij}\, y_{ij}) \leq T_j \qquad j = 1, \dots, m$$

(12) $$\sum_{j=1}^{m} x_{ij} = D_i \qquad i = 1, \dots, n$$

(13) $$x_{ij} \geq 0 \quad y_{ij} \geq 0 \qquad i = 1, \dots, n \qquad j = 1, \dots, m$$

The (positive) coefficients $\alpha_{ij}\ \beta_{ij}\ \gamma_{ij}\ a_{ij}\ b_{ij}\ T_j\ D_i$ are listed in Table 1 for a real problem with $m = 7$, $n = 46$ (This problem has been given by Compagnie Saint-Gobain, France - Table 1 containes data only for allowed pairs (i,j). If a pair does not appear in Table 1 this means that the corresponding x_{ij} and y_{ij} must be held to 0 and this can be simulated by setting $\alpha_{ij} = \beta_{ij} = +\infty$).

For solving (10) - (13) we can adopt the following method : suppose we know a point x satisfying

(14) $$\sum_{i=1}^{n} a_{ij}\, x_{ij} < T_j \qquad j=1, \dots, m$$

(T(J),J=1,7) = 75. 75. 75. 87. 87. 87. 87.

I	J	ALFA(I,J)	BETA(I,J)	GAMA(I,J)	A(I,J)	B(I,J)	D(I)
1	1	2.695	670.5	0.10781	0.42428612E-03	4.5	4000.
1	4	2.706	704.6	0.10822	0.41322294E-03	4.5	
2	4	2.914	704.6	0.11655	0.41322294E-03	4.5	30000.
2	5	2.914	704.6	0.11655	0.41322294E-03	4.5	
3	1	2.635	670.5	0.10541	0.42428612E-03	4.5	65000.
3	4	2.646	704.6	0.10583	0.41322294E-03	4.5	
3	5	2.646	704.6	0.10583	0.41322294E-03	4.5	
4	7	3.386	582.8	0.13544	0.73346030E-03	4.0	8000.
5	3	2.526	681.3	0.10102	0.42428612E-03	4.5	11200.
6	7	4.500	582.8	0.17998	0.83465478E-03	4.0	1200.
7	2	2.622	669.7	0.10486	0.43200259E-03	4.5	2000.
7	7	3.307	582.8	0.13226	0.69156289E-03	4.0	
8	4	2.754	704.6	0.11017	0.41322294E-03	4.5	90000.
8	5	2.754	704.6	0.11017	0.41322294E-03	4.5	
9	4	2.555	704.6	0.10222	0.41322294E-03	4.5	56000.
10	4	2.690	704.6	0.10759	0.41322294E-03	4.5	24000.
10	5	2.690	704.6	0.10759	0.41322294E-03	4.5	
11	1	2.444	670.5	0.09775	0.42428612E-03	4.5	25000.
11	2	2.483	669.7	0.09930	0.43200259E-03	4.5	
12	2	2.712	669.7	0.10850	0.43200259E-03	4.5	5000.
12	4	2.684	704.6	0.10736	0.41322294E-03	4.5	
12	5	2.684	704.6	0.10736	0.41322294E-03	4.5	
13	2	2.542	669.7	0.10170	0.43200259E-03	4.5	8000.
14	1	2.428	670.5	0.09714	0.42428612E-03	4.5	25000.
14	2	2.467	669.7	0.09869	0.43200259E-03	4.5	
15	7	3.322	582.8	0.13288	0.68180263E-03	4.0	1000.
16	7	4.815	582.8	0.19261	0.89645898E-03	4.0	1520.
17	7	4.950	582.8	0.19799	0.89645898E-03	4.0	2000.
18	7	4.657	582.8	0.18627	0.86445361E-03	4.0	800.
19	7	3.883	582.8	0.15533	0.80677681E-03	4.0	3000.
20	7	4.894	582.8	0.19578	0.89645898E-03	4.0	4000.
21	7	4.593	582.8	0.18372	0.84925676E-03	4.0	1000.
22	7	4.857	582.8	0.19428	0.86445361E-03	4.0	800.
23	7	4.966	582.8	0.19862	0.86445361E-03	4.0	400.
24	7	4.301	582.8	0.17205	0.82047912E-03	4.0	9200.
25	4	2.645	704.6	0.10582	0.41242200E-03	4.5	48000.
25	5	2.645	704.6	0.10582	0.41242200E-03	4.5	
26	6	3.347	555.8	0.13386	0.69381786E-03	4.0	30000.
27	7	1.199	582.8	0.04797	0.67105074E-03	4.0	2000.
28	1	2.592	670.5	0.10369	0.42347750E-03	4.5	57000.
28	2	2.631	669.7	0.10524	0.43116440E-03	4.5	
29	1	2.436	655.6	0.09746	0.42347750E-03	4.5	30000.
29	2	2.475	669.7	0.09902	0.43116440E-03	4.5	
30	2	2.419	669.7	0.09677	0.43116440E-03	4.5	30000.
30	3	2.411	681.3	0.09646	0.42347750E-03	4.5	
31	6	3.107	555.8	0.12429	0.68390090E-03	4.0	5000.
31	7	3.277	582.8	0.13106	0.69022621E-03	4.0	
32	7	3.461	582.8	0.13843	0.74332859E-03	4.0	1600.
33	2	2.663	669.7	0.10652	0.43116440E-03	4.5	55000.
33	3	2.655	681.3	0.10621	0.42347750E-03	4.5	
34	2	2.784	669.7	0.11137	0.43116440E-03	4.5	26000.
34	3	2.777	681.3	0.11106	0.42347750E-03	4.5	
35	2	2.492	669.7	0.09968	0.43116440E-03	4.5	40000.
35	3	2.484	681.3	0.09937	0.42347750E-03	4.5	
36	1	2.553	670.5	0.10214	0.42347750E-03	4.5	27600.
36	2	2.592	669.7	0.10369	0.43116440E-03	4.5	
37	1	2.500	670.5	0.10002	0.42347750E-03	4.5	50000.
38	2	2.454	669.7	0.09817	0.43116440E-03	4.5	600.
39	1	2.586	670.5	0.10344	0.42347750E-03	4.5	2640.
39	2	2.625	669.7	0.10499	0.43116440E-03	4.5	
40	6	0.676	555.8	0.02703	0.17236921E-03	4.0	77840.
41	6	0.869	555.8	0.03475	0.18307303E-03	4.0	9000.
42	2	2.322	625.3	0.09289	0.42870594E-03	4.5	21120.
43	6	1.517	555.8	0.06067	0.36613923E-03	4.0	28000.
44	6	2.943	555.8	0.11772	0.73227868E-03	4.0	40000.
45	6	2.734	555.8	0.10938	0.37815748E-03	4.0	6000.
46	6	0.949	555.8	0.03796	0.19423515E-03	4.0	5000.

Table 1

Then we can minimize (10)-(11) with respect to y alone ; if (14) holds we obtain a well - defined function $M(x) < +\infty$. It can be seen that $M(x)$ is a barrier function forconstraints (14) and the problem (10)-(13) is equivalent to

$$
(15) \quad \begin{cases} \min \ M(x) \\ \sum_{j=1}^{m} x_{ij} = D_i \qquad i=1, \ldots, n \\ x_{ij} \geq 0 \end{cases}
$$

which should be solved by a feasible directions method starting from an initial point satisfying (14) . In [6] is shown that $M(x)$ can be computed numerically (by solving some nonlinear scalar equations by Newton method) and furthermore that $M(x)$ has continuous partial derivatives ; (15) is thus a standard nonlinear programming problem but there is a nontrivial difficulty : M is nonconvex and it turns out to have many local optima (there is a combinatorial aspect in (10)-(13)) .

For avoiding this , is needed a phase 1 procedure which not only finds a point x feasible in (14) but also a point (x,y) feasible in (11)-(13) such that $J(x,y)$ is "low". This enters the framework of the Generalized Lagrange Multiplier technique [3] . If we define the Lagrange function associated with (10)-(11)

$$
(16) \qquad L(x,y,\lambda) = J(x,y) + \sum_{j=1}^{m} \left[\sum_{i=1}^{n} (a_{ij} x_{ij} + b_{ij} y_{ij}) - T_j \right]
$$

then the GLM technique consists of maximizing the dual function $-h(\lambda)$ which is the minimum value of (16) with respect to x and y constrained by (12)-(13) . Here , this dual function is particularly easy to compute ; it is nondifferentiable and we can use the ε - descent method for maximizing it .

12.- For simplicity forget the constraints $\lambda \geq 0$. If $\bar{\lambda}$ solves the dual , $0 \in \partial h(\bar{\lambda})$. Besides $\bar{\lambda}$, the ε - descent method gives also $g^o, \ldots, g^k \in \partial h(\bar{\lambda})$ such that $0 \in \text{conv} \{ g^o, \ldots, g^k \}$ (this is shematic - we suppose an exact maximization) ; and it gives also the convex coefficients $u^o, \ldots, u^k$ such that $\sum_{i=o}^{k} u^i g^i = 0$. But these g^i's are here the values of the constraints (11) at specific points $(x,y)^o, \ldots, (x,y)^k$ which have been actually computed during the ε - descent method (in fact they are all minimizers of the Lagrange function (16) at $\bar{\lambda}$).

In summary , after running the ε - descent method , we have on hand

- $(x,y)^o, \ldots, (x,y)^k$ feasible in (12)-(13) and for which the constraints (11) have the values $g^o, \ldots, g^k$.
- $u^o, \ldots, u^k$ such that $\Sigma u^i g^i = 0$ (in fact , because of the constraints $\lambda \geq 0$ we have rather the Kuhn - Tucker conditions $\Sigma u^i g^i \leq 0$, $\bar{\lambda} \Sigma u^i g^i = 0$) .

Let us set $(\bar{x},\bar{y}) = \sum_{i=0}^{k} u^i (x,y)^i$

Then $(\bar{x},\bar{y})$ is feasible in (11)-(13) and it can be hoped that its cost $J(\bar{x},\bar{y})$ is not too high ; anyway $-h(\bar{\lambda})$ is a lower bound for the optimal cost and the quality of $(\bar{x},\bar{y})$ can be measured by $J(\bar{x},\bar{y}) + h(\bar{\lambda})$. Thus $\bar{x}$ can be used as an initial point in (15) .

It must be said that this is not a general technique for solving a nonconvex optimization problem : the structure of the problem must be suitable , and it is impossible to forecast the quality of the "dual quasi - optimum" . These §§ 11-12 have been developed in [6] .

REFERENCES

[1] D.P.Bertsekas,S.K.Mitter : A descent numerical method for optimization problems with nondifferentiable cost functionals . SIAM Journal on Control Vol 11 n°4 (1973) pp 637 - 652 .

[2] V.F.Demjanov : Algorithms for some minimax problems . Journal of Computer and Systems Science Vol 2 (1968) pp 342-380 .

[3] H.Everett III : Generalized Multipliers Method for solving problems of allocation of Resources . Operations Research Vol 11 (1963) pp 399-417 .

[4] C.Lemarechal : An Algorithm for Minimizing Convex Functions . Proceedings IFIP Congress (Stockholm , 1974) North Holland Publishing Company pp 552-556 .

[5] ———————— : Minimization of Nondifferentiable Functions with Constraints . Proceedings 12th Conference on Circuit Theory (Allerton , 1974) The University of Illinois at Urbana Champaign pp 16-24 .

[6] ———————— : Nondifferentiable Programming and GLM technique , Application to a Nonconvex Programming Problem . Working Paper IRIA-Laboria France .

[7] ———————— : An Extension of "Davidon" Methods to nondifferentiable problems . Mathematical Programming Study n°3 (to appear) North Holland Publishing Company .

[8] G.P.Mc Cormick,K.Ritter : Projection Method for Unconstrained Optimization . Journal of Optimization Theory and Applications Vol 10 n°2 (1972) pp 57-66 .

[9] R.T.Rockafellar : Convex Analysis . Princeton University Press , Princeton (1970)

[10] K.Ritter : A Method of Conjugate Directions for Linearly Constrained Nonlinear Programming Problems . SIAM Journal on Numerical Analysis Vol 12 n°3 (1975) pp 273-303 .

[11] P.Wolfe : Finding the Nearest Point in a Polytope . Mathematical Programming Study n°1 (to appear) North Holland Publishing Company .

[12] ———— : A Method of Conjugate Subgradients for Minimizing Nondifferentiable Functions . Proceedings , Allerton Conference (See[5]) .

[13] ———— : A Method of Conjugate Subgradients for Minimizing Nondifferentiable Functions . Nonlinear Programming Study n°3 (See [7]) .

[14] G.Zoutendijk : Methods of Feasible Directions . Elsevier , Amsterdam (1960) .

[illegible]

[illegible]

[illegible]

(1) [illegible]

(2) [illegible]

(3) [illegible]

(4) [illegible]

(5) [illegible]

(6) [illegible]

(7) [illegible]

(8) [illegible]

(9) [illegible]

(10) [illegible]

(11) [illegible]

(12) [illegible]

(13) [illegible]

APPROXIMATIONS TO STOCHASTIC OPTIMIZATION PROBLEMS

KURT MARTI

Institut für Operations Research
der Universität Zürich
Weinbergstrasse 59, CH-8006 Zürich

1.Introduction. A large class of problems in stochastic programming (see [7]), stochastic control (see [1,3]), estimation theory (see [5]) and optimal design of networks (see [6]) can be formulated within the following abstract framework:

Let X be a Banach space, Z a separable Banach space, $(\Omega,\mathbf{A},P)$ a probability space with elements ω, $T=T(\omega)$ a stochastic, linear operator from X to Z, describing the input-output behaviour $x \to T(\omega)x$ of some abstract stochastic linear control system and let $v=v(\omega)$ be a random variable in Z, playing the role of an abstract stochastic target which must be attained as good as possible by selecting a control variable x.

Let p be a continuous, sublinear, i.e. subadditive $(p(z_1+z_2)\leq p(z_1)+p(z_2),\ z_1,z_2 \in Z)$ and positively homogeneous $(p(tz)=tp(z),\ t\geq 0,\ z \in Z)$ functional on Z. It is assumed then that the violation of the constraint

$$T(\omega)x = v(\omega)$$

causes the loss

$$p(T(\omega)x - v(\omega)),$$

hence on an average we suffer the loss

$$R(x) = E_\omega p(T(\omega)x-v(\omega)),$$

where E_ω denotes the expectation operator with respect to ω. In order to minimize the average costs R(x) we have to solve the optimization problem

$$\text{minimize } R(x) \text{ subject to } x\in U \qquad (1)$$

where U is still a subset of X representing the constraints imposed on the decision or control variable x.

2.Properties of the optimization problem (1). We assume that $T=T(\omega)$ and $v=v(\omega)$ are normintegrable, i.e.

$$E_\omega\|T(\omega)\| <+\infty \text{ and } E_\omega\|v(\omega)\| <+\infty.$$

Because of the sublinearity of the lossfunction p, p is convex and has the properties $p(0)=0$, $-p(-z)\leq p(z)$ and

$$|p(z)| \leq \|p\|\ \|z\|,\ z\in Z,$$

$$|p(z_1)-p(z_2)| \leq \|p\|\,\|z_1-z_2\|,\ z_1,z_2\in Z,$$

where

$$\|p\| = \sup_{\|z\|=1}|p(z)|$$

is the norm of the sublinear functional p. Putting

$$N(x) = E_\omega \|T(\omega)x\|, \quad x \in X$$

we obtain the following proposition.

Theorem 2.1. The mean loss R=R(x) is a real valued, convex functional on X and it is

$$\begin{aligned} |R(x)| &\leq \|p\|(N(x)+E_\omega\|v(\omega)\|), \quad x \in X, \\ |R(x)-R(y)| &\leq \|p\| N(x-y), \quad x,y \in X. \end{aligned} \tag{2}$$

Since a convex functional defined on a Banach space is weakly lower semicontinuous ($R(x) \leq \liminf_{n\to\infty} R(x_n)$ for any sequence (x_n) converging weakly to x), the following theorem is a consequence (see e.g. [4]) from Theorem 2.1.

Theorem 2.2. If the set U of admissible controls in (1) is weakly sequentially compact (i.e. each sequence (x_n) in U contains a subsequence that converges weakly to a member of U), then (1) is solvable. If in addition X is reflexive, then (1) has a solution for any closed, convex and bounded U.

Note. If $R(x) \to +\infty$ for $\|x\| \to +\infty$, then we don't need the boundedness of U.

Let

$$\underline{p} = \inf_{\|z\|=1} p(z) \text{ and } \underline{N} = \inf_{\|x\|=1} N(x) = \inf_{\|x\|=1} E_\omega\|T(\omega)x\|. \tag{3}$$

If $\underline{p}>0$ -and therefore $p(z)\geq 0$, $z \in Z$- then

$$\begin{aligned} R(x) &\geq \underline{p}E_\omega\|T(\omega)x-v(\omega)\| \\ &\geq \underline{p}\,|E_\omega\|T(\omega)x\|-E_\omega\|v(\omega)\|\,| = \underline{p}\,|N(x)-E_\omega\|v(\omega)\|\,| \end{aligned}$$

and together with the inequalities (2) we find then that $R(x)\geq 0$ and

$$\underline{p}\,|N(x)-E_\omega\|v(\omega)\|\,| \leq R(x) \leq \|p\|(N(x)+E_\omega\|v(\omega)\|). \tag{4}$$

If $\bar{x}$ is now any solution of (1), then from (4) we obtain that

$$\frac{1}{\|p\|}\inf_{x\in U} R(x) - E_\omega\|v(\omega)\| \leq N(\bar{x}) \leq \frac{1}{\underline{p}}\inf_{x\in U} R(x) + E_\omega\|v(\omega)\|.$$

Provided that $\underline{N}>0$ we have finally the following norm-boundaries for $\bar{x}$:

$$\frac{1}{\|N\|}(-E_\omega\|v(\omega)\| + \frac{1}{\|p\|}\inf_{x\in U} R(x)) \leq \|\bar{x}\| \leq \frac{1}{\underline{N}}(E_\omega\|v(\omega)\| + \frac{1}{\underline{p}}\inf_{x\in U} R(x)). \tag{5}$$

Theorem 2.3. If $\underline{p}>0$ and $\underline{N}>0$, then $R(x)\to +\infty$ for $\|x\| \to +\infty$ and there is a $\rho>0$ such that $\{x \in U: \|x\| \leq \rho\} \neq \emptyset$ and

$$\inf_{x\in U} R(x) = \inf_{x\in U,\, \|x\|\leq\rho} R(x).$$

Proof. From the left hand side of (4) we obtain that

$$R(x) \geq \underline{p}(N(x)-E_\omega\|v(\omega)\|) \geq \underline{p}(\underline{N}\|x\|-E_\omega\|v(\omega)\|)$$

what yields the first part of the assertion. Let

$$\rho = \frac{1}{\underline{N}}(E_\omega\|v(\omega)\| + \frac{1}{\underline{p}}R(u_o)),$$

where u_o is an arbitrary, but fixed element of U. Then $\|u_o\|\leq\rho$ and $R(x)\geq R(u_o)$ for $\|x\|>\rho$. From this follows the second part of our theorem.

Notes. Obviously, $N(x)=E_\omega\|T(\omega)x\|$ is a sublinear and symmetric ($N(x)=N(-x)$, $x\in X$) non negative functional on X. Hence N is a norm on X if and only if $N(x)\neq 0$, $x\neq 0$, what is equivalent to the condition

$$P(\{\omega\in\Omega:\ T(\omega)x=0\}) < 1 \text{ for all } x\neq 0.$$

If N is a norm and $X=\mathbb{R}^n$, then N is equivalent to the norm $\|\cdot\|$ of X, hence there is a $\sigma>0$ such that $N(x)\geq\sigma\|x\|$, $x\in X$ and we see that $\underline{N}\geq\sigma>0$. Exists- in order to give an other example- the inverse of $\overline{T} = E_\omega T(\omega)$, then $\underline{N}\geq \|\overline{T}^{-1}\|^{-1}$.

3. Approximations to (1). Several procedures for obtaining approximative solutions of (1) have been studied, see e.g. [2,8,9]. We concentrate here on approximations to (1) which are constructed by means of

approximations to the lossfunction p.

As is seen later on, approximations of this type reduce (1) to problems with a finite dimensional output space Z, to problems where the approximative sublinear functional p_n satisfies $\underline{p}_n>0$ (see (3,4,5) and Theorem 2.3) and under an other version of this type of approximation we get comparatively simple expressions for the derivative of the functional R_n approximating R.

In addition to the properties of p already mentioned in section 2 we need some more information about p: Let Z* be the space of all continuous linear functionals on Z and

$$\mathcal{U}_p = \{f\in Z^*:\ fz\leq p(z),\ z\in Z\},$$

$$\mathcal{U}_p^z = \{f\in\mathcal{U}_p:\ fz=p(z)\},\ z\in Z.$$

By the theorem of Hahn-Banach it can be shown that $\mathcal{U}_p$ and $\mathcal{U}_p^z$ are not empty, furthermore they are convex and weakly sequentially compact (since Z is separable). Moreover it is (see e.g. [4])

$$p(z) = \max_{f\in\mathcal{U}_p} fz,\ z\in Z.$$

Taking a sequence $z_1,z_2,\ldots$ dense in the unit sphere $\|z\|=1$ of Z and selecting to each z_k a fixed element $f_k\in\mathcal{U}_p^{z_k}$ it holds also

$$p(z) = \sup_k f_k z,\ z\in Z, \qquad (6)$$

hence p can be represented as upper hull of a sequence of continuous, linear functionals on Z.

Our approximations of (1) are based on the following Lemma, which is easy to see.

Lemma 3.1. Let R_n, n=1,2,... be a sequence of functionals defined on X, such that $R_n(x) \to R(x)$, $n \to \infty$ uniformly for $x \in U$. Then

a) $\inf\{R_n(x): x \in U\} \to \inf\{R(x): x \in U\}$, $n \to \infty$;

b) If x_n is a solution to

$$\text{minimize } R_n(x) \text{ subject to } x \in U, \tag{1n}$$

then (x_n) is a minimizing sequence of (1), i.e.

$$R(x_n) \to \inf\{R(x): x \in U\}, \ n \to \infty$$

and each accumulation point $\bar{x} \in U$ of (x_n) is a solution of (1).

In the following we consider sequences R_n, n=1,2,... defined by

$$R_n(x) = E_\omega p_n(T(\omega)x - v(\omega)), \ x \in X,$$

where $p_1, p_2, \ldots$ is a sequence of "simpler" sublinear functionals on Z approximating p. Note that our theorems in section 2 are also valid if we replace p by p_n and R by R_n.

(i) Approximations of p by "finitely generated" sublinear functionals.

According to (6) we have that $p(z) = \sup_k f_k z$, $z \in Z$, where $f_1, f_2, \ldots$ is a given sequence of elements from Z^*. Therefore we set

$$p_n(z) = \max\{f_1 z, f_2 z, \ldots, f_n z\}, \ z \in Z,$$

n=1,2,... . Properties of (p_n) are shown in the following.

Lemma 3.2. It is $p_n(z) \uparrow p(z)$, $\|p_n\| \uparrow \|p\|$, letting $n \to \infty$. On relatively compact subsets of Z p_n converges uniformly to p.

Proof. Easy to see is the first part of the assertion. To show the second part, Let K be a relatively compact subset of Z, $\varepsilon > 0$ a non negative number, $z_1, z_2, \ldots, z_r$ an ε-net of K (to each $z \in K$ there is a z_i, $1 \leq i \leq r$, such that $\|z - z_i\| < \varepsilon$) and $n(\varepsilon)$ an integer such that $|p_n(z_i) - p(z_i)| < \varepsilon$, $n > n(\varepsilon)$ and all $1 \leq i \leq r$. Let z be an arbitrary element of K and z_i an element of the ε-net such that $\|z - z_i\| < \varepsilon$. For all $n > n(\varepsilon)$ we get then

$$|p_n(z) - p(z)| \leq |p_n(z) - p_n(z_i)| + |p_n(z_i) - p(z_i)| + |p(z_i) - p(z)|$$
$$< \|p_n\| \, \|z - z_i\| + \varepsilon + \|p\| \, \|z - z_i\| < \varepsilon(1 + 2\|p\|),$$

what shows the uniform convergence $p_n \to p$ on K.

Our first theorem of convergence is a consequence of Lemma 3.2.

Theorem 3.1. It is $R_n(x) \uparrow R(x)$, $n \to \infty$ for all $x \in X$ and the convergence is uniform on relatively compact subsets of X.

Proof. Since $p_n(T(\omega)x - v(\omega)) \uparrow p(T(\omega)x - v(\omega))$, $\omega \in \Omega$ and $|p_n(T(\omega)x - v(\omega))| \leq \|p\| (\|T(\omega)\| \, \|x\| + \|v(\omega)\|)$, the first part of our theorem follows from Lebesgue's convergence theorem. Because of $|R_n(x) - R_n(y)| \leq \|p\| \, \|x - y\| E_\omega \|T(\omega)\|$ and correspondingly $|R(x) - R(y)| \leq \|p\| \, \|x - y\| E_\omega \|T(\omega)\|$, $x, y \in X$ (see (2)), the remaining part of the assertion is proved as the corresponding part of Lemma 3.2.

Improvements of convergence.

(i.1) A first improvement of Theorem 3.1 is obtained if

$$T(\omega)x-v(\omega)\in K \text{ for all } x\in U \text{ and all } \omega\in\Omega,$$

where K is a relatively compact subset of K. In fact, let $z_1, z_2, \ldots$ be a sequence in K such that the first n elements form an ε_n-net of K and $\varepsilon_n > 0$, $\varepsilon_n \to 0$ for $n \to \infty$. Taking then again $f_k \in \mathcal{U}_p^{z_k}$ and defining p_n by $p_n(z) = \max_{1\leq k\leq n} f_k z$, $z\in Z$ it is $p_n(z_k) = p(z_k)$, $1\leq k\leq n$ and therefore

$$|p_n(z)-p(z)| \leq 2\|p\|\,\|z-z_k\| \leq 2\|p\|\varepsilon_n, \; z\in K,$$

where k is a given integer between 1 and n. Hence

$$|R_n(x)-R(x)| \leq 2\|p\|\varepsilon_n \text{ for all } x\in U$$

what shows the uniform convergence of R_n to R on U.

(i.2) An other improvement of the pointwise convergence $R_n \to R$ is obtained if $Z=\mathbb{R}^m$ and

$$p(z) = \inf\{\lambda>0: \lambda^{-1}z\in C\}, \; z\in Z$$

is the distance functional of a bounded, convex subset C of Z which contains the origin O of Z as an interior point. By [7],III, Theorem 2.3 there is a sequence of sublinear functionals $p_n(z) = \max_{1\leq k\leq r(n)} q_{nk}'z$, where for each $n=1,2,\ldots$ $q_{n1}, q_{n2}, \ldots, q_{nr(n)}$ are given m-vectors such that for each $n=1,2,\ldots$ and $z\in Z$

$$p_n(z) \geq p_{n+1}(z) \text{ and } 0 \leq p_n(z) - p(z) \leq \varepsilon_n p(z),$$

where $\varepsilon_n>0$ and $\varepsilon_n \to 0$, $n\to\infty$. Hence

$$|p_n(z) - p(z)| \leq \varepsilon_n\|p\|\,\|z\|, \; z\in Z.$$

This implies $R_n(x) \downarrow R(x)$ and

$$0\leq R_n(x)-R(x)\leq\varepsilon_n\|p\|(E_\omega\|T(\omega)\|\,\|x\| + E_\omega\|v(\omega)\|), \; x\in X,$$

what shows the uniform convergence of R_n to R on any bounded subset of X.

(i.3) Let p be non negative. In equation (6) representing p we may then assume that $f_1=0$. Hence

$$p_{n+\nu}(z) - p_n(z) = \max\{p_n(z), \max_{n+1\leq k\leq n+\nu} f_k z\} - p_n(z)$$

$$\leq \|z\| \max_{n+1\leq k\leq n+\nu} f_k$$

and therefore

$$0 \leq p(z) - p_n(z) \leq \|z\| \sup_{k\geq n+1} \|f_k\|,$$

letting $\nu\to\infty$. If

$$\sup_{k\geq n+1} \|f_k\| \to 0, \; n\to\infty$$

(the contribution of f_k to p vanishs with large index k), then $p_n \to p$ uniformly on bounded subsets of Z. As in (i.2) we obtain then that $R_n \to$ R uniformly on bounded subsets of X.

(ii) Approximation of a nonnegative p by sublinear functionals p_n such that $\underline{p}_n = \inf_{\|z\|=1} p_n(z) > 0$ (see (4,5) and Theorem 2.3).

We set

$$p_n(z) = \max\{p(z), \frac{1}{n}\|z\|\}, \quad z\in Z, \quad n=1,2,\dots .$$

Obviously, p_n is a sublinear functional such that $\underline{p}_n \geq \frac{1}{n} > 0$, $p_n(z) \geq p_{n+1}(z)$ and

$$0 \leq p_n(z) - p(z) \leq \frac{1}{n}\|z\|, \quad z\in Z.$$

Consequently we find that

$$0 \leq R_n(x) - R(x) \leq \frac{1}{n}(E_\omega \|T(\omega)\|\,\|x\| + E_\omega \|v(\omega)\|).$$

These relations show that $p_n \to p$ respectively $R_n \to R$, $n \to \infty$ uniformly on bounded subsets of Z respectively X.

(iii) Reduction of Z to finite dimensional output spaces Z_m.

Let $\beta_1, \beta_2, \dots$ be a basis of Z and Z_m the m-dimensional subspace of Z generated by the first m elements $\beta_1, \beta_2, \dots, \beta_m$ of the basis. Furthermore, let $\pi_m: Z \to Z_m$ be the continuous linear operator from Z to Z_m defined by

$$z = \sum_{i=1}^{\infty} z_i \beta_i \to \pi_m z = \sum_{i=1}^{m} z_i \beta_i .$$

We know that $\|\pi_m\| \leq \gamma$, where $\gamma > 0$ is a fixed positive number for all m=1, 2,..., hence $\pi_m z \to z$ and $p(\pi_m z) \to p(z)$ for $m \to \infty$ uniformly on relatively compact subsets of Z. Since $z \to p(\pi_m z)$, $z\in Z$ is for each fixed m a sublinear functional with norm $< \gamma \|p\|$ we can show as in Theorem 3.1 that

$$R_{[m]}(x) = E_\omega p(\pi_m(T(\omega)x - v(\omega))) \to R(x), \quad m \to \infty$$

uniformly on relatively compact substes of X. Since

$$p(\pi_m z) = \sup_{f\in \mathcal{U}_p} f\pi_m z = \sup_{f\in \mathcal{U}_p} (f\beta_1, f\beta_2, \dots, f\beta_m) \begin{pmatrix} z_1 \\ z_2 \\ \vdots \\ z_m \end{pmatrix}$$

we may consider $R_{[m]}$ as average loss in a decision problem as described in the introduction where now the outputspace Z_m is finite dimensional.

Using again equation (6), defining $p_n(z) = \max_{1\leq k\leq n} f_k z$ and

$$R_{[m]n}(x) = E_\omega p_n(\pi_m(T(\omega)x - v(\omega))) = E_\omega \max_{1\leq k\leq n} (f_k\beta_1, \dots, f_k\beta_m) \begin{pmatrix} (T(\omega)x - v(\omega))_1 \\ \vdots \\ (T(\omega)x - v(\omega))_m \end{pmatrix}$$

we obtain the following proposition.

Theorem 3.2. It is $R_{[m]n} \to R$ for $m,n \to \infty$ uniformly on relatively compact subsets of X.

Proof. The pointwise convergence $R_{[m]n}(x) \to R(x)$, $m,n \to \infty$, $x\in X$ follows from $\|p_n\| \leq \|p\|$,

$$\begin{aligned}|R_{[m]n}(x) - R(x)| &\leq E_\omega|p_n(\pi_m(T(\omega)x-v(\omega))) - p(T(\omega)x-v(\omega))|\\ &\leq E_\omega|p_n(\pi_m(T(\omega)x-v(\omega))) - p_n(T(\omega)x-v(\omega))|\\ &\quad + E_\omega|p_n(T(\omega)x-v(\omega)) - p(T(\omega)x-v(\omega))|\\ &\leq \|p_n\| E_\omega\|\pi_m(T(\omega)x-v(\omega)) - (T(\omega)x-v(\omega)\|\\ &\quad + E_\omega|p_n(T(\omega)x-v(\omega)) - p(T(\omega)x-v(\omega)|\end{aligned}$$

and the observation that the last two terms converge to 0 by Lebesgue's convergence theorem. Since

$$|R_{[m]n}(x) - R_{[m]n}(y)| \leq \gamma\|p\| E_\omega\|T(\omega)\| \|x\|$$

we prove the asserted uniform convergence on relatively compact subsets of X as the corresponding part of Lemma 3.2.

In order to illustrate this type of approximation of R by the double sequence $(R_{[m]n})$ we consider the following example: Let X=Z=c be the separable Banach space c of all convergent sequences of real numbers, $x=(u_0,u_1,\dots,u_j,\dots)'$,

$$T(\omega) = (K_{ij}(\omega)) \text{ and } v(\omega) = (v_i(\omega)),\ i=1,2,\dots,\ j=0,1,2,\dots,$$

where

$$K_{ij}(\omega) = \begin{cases} h(\omega)f(\omega)^{i-1-j}g(\omega) & \text{for } 0\leq j<i,\ i=1,2,\dots \\ 0 & \text{otherwise}\end{cases}$$

and

$$v_i(\omega) = z_i^0(\omega) - s_0(\omega)h(\omega)f(\omega)^i.$$

In these equations $s_0=s_0(\omega)$, $f=f(\omega)$ with $-1<f(\omega)<1$, $g=g(\omega)$ and $h=h(\omega)$ are real random variables with a given joint distribution. Furthermore $z^0=z^0(\omega)=(z_i^0(\omega))$ is a random vector in Z=c. It can be seen (see [8]) that $v=v(\omega)$ is a random vector in c and $T=T(\omega)$ is a stochastic linear operator from c into c. It should be clear that we are considering here a discrete time stochastic linear control process (with infinite many stages), given by the difference equations

$$s_{j+1} = f(\omega)s_j + g(\omega)u_j,\ j=0,1,2,\dots \tag{7a}$$

$$z_i = h(\omega)s_i, I=1,2,\dots \tag{7b}$$

where (7a) is the input-state-, (7b) the state-output-equation. The problem is then to reach the target $z^0(\omega)=(z_i^0(\omega))$ as good as possible; the divergence between the output $z=z(\omega)$ and the desired output $z^0(\omega)$ is measured by a sublinear functional p on Z=c. To simplify our consideration we assume that the costs arise only from the deviation between output and target (measured by p).

To apply now the procedure described above we take the basis

$$\beta_0 = (1,1,1,\dots)',\ \beta_1 = (1,0,0,\dots)',\ \beta_2 = (0,1,0,\dots)',\dots .$$

For the elements $z = (z_i)$ of c we get then

$$z = z_\infty \beta_0 + \sum_{i=1}^{\infty} (z_i - z_\infty)\beta_i,$$

where $z_\infty = \lim_{i\to\infty} z_i$ (note that $z \to z_\infty$ is an element of c*) and

$$\pi_m z = z_\infty \beta_0 + \sum_{i=1}^{m} (z_i - z_\infty)\beta_i = z_\infty \beta_{om} + \sum_{i=1}^{m} z_i \beta_i,$$

where $\beta_{om} = \beta_0 - \sum_{i=1}^{m} \beta_i$. Therefore

$$\pi_m(T(\omega)x - v(\omega)) = \left(\frac{g(\omega)h(\omega)}{1-f(\omega)} x_\infty - v_\infty(\omega)\right)\beta_{om} + \sum_{i=1}^{m} ((T(\omega)x)_i - v_i(\omega))\beta_i$$

and

$$f_k\pi_m(T(\omega)x - v(\omega)) = (f_k\beta_1,\dots,f_k\beta_m, f_k\beta_{om})(T_{[m]}(\omega)x_{[m]} - v_{[m]}(\omega))$$

if

$$T_{[m]} = \begin{pmatrix} 1 & 0 & 0 & \dots & 0 \\ f & 1 & 0 & \dots & 0 \\ f^2 & f & 1 & \dots & 0 \\ \vdots & \vdots & \vdots & & \vdots \\ f^{m-1} & f^{m-2} & f^{m-3} & \dots & 1 \\ 0 & 0 & 0 & \dots & (1-f)^{-1} \end{pmatrix}, \qquad x_{[m]} = \begin{pmatrix} u_0 \\ u_1 \\ u_2 \\ \vdots \\ u_{m-1} \\ u_\infty \end{pmatrix}$$

and

$$v_{[m]} = \begin{pmatrix} v_1 \\ v_2 \\ \vdots \\ v_m \\ v_\infty \end{pmatrix} = \begin{pmatrix} z_1^o - s_o hf \\ z_2^o - s_o hf^2 \\ \vdots \\ z_m^o - s_o hf^m \\ z_\infty^o \end{pmatrix} .$$

Though our general procedure described in (iii) may always be applied if Z has a basis, we don't get in each case approximative problems with an easy structure as in the above example. A possibility how to proceed the is shown in the following example: Let $X=Z=C[0,1]$ be the space of all real valued continuous functions on the bounded interval $0 \leq t \leq 1$. By (6) it is

$$p(z) = \sup_k f_k z = \sup_k \int_0^1 z(t)\,d\bar{f}_k(t),$$

where $\bar{f}_k$ is the function of bounded variation on the interval $0 \leq t \leq 1$ defined by the continuous, linear functional f_k on $C[0,1]$. We consider the the sublinear functionals $p_{[m]n}$, defined by

$$p_{[m]n}(z) = \max_{1\leq k\leq n} \sum_{i=1}^{m} z\left(\frac{i}{m}\right)\left(\bar{f}_k\left(\frac{i}{m}\right) - \bar{f}_k\left(\frac{i-1}{m}\right)\right),\quad z \in Z.$$

It holds $\|p_{[m]n}\| \leq \|p\|$ and $p_{[m]n}(z) = \tilde{p}_{[m]n}(z_{[m]})$, where $z_{[m]} = (z(\frac{1}{m}), z(\frac{2}{m}),\dots,z(1))'$ and $\tilde{p}_{[m]n}$ is the sublinear functional on $\mathbb{R}^m$, given by

$$\tilde{p}_{[m]n}((z_1,\dots,z_m)') = \max_{1\le k\le n} \sum_{i=1}^{m} z_i(\bar{f}_k(\tfrac{i}{m})-\bar{f}_k(\tfrac{i-1}{m})).$$

As above we set $p_n(z) = \max_{1\le k\le n} f_k z$. With help of Stieltjes'integral we have for each $z\in Z$ and for an arbitrary $\varepsilon>0$ an integer $m(z,\varepsilon)$ such that

$$|p_{[m]n}(z)-p(z)| \le 2\varepsilon\|p\| + |p_n(z)-p(z)|$$

for $m\ge m(z,\varepsilon)$ and all $n=1,2,\dots$. Therefore $p_{[m]n}$ converges pointwise to p, letting $m,n\to\infty$. This yields -again as in Lemma 3.2- the uniform convergence on relatively compact subsets of Z. Since

$$|p_{[m]n}(T(\omega)x-v(\omega))| \le \|p\|\,\|T(\omega)x-v(\omega)\|$$

we get

$$R_{[m]n}(x) = E_\omega p_{[m]n}(T(\omega)x-v(\omega)) \to R(x),\quad m,n\to\infty$$

from Lebesgue's theorem. We observe that also this convergence is uniform on relatively compact subsets of X. In the case that

$$(T(\omega)x-v(\omega))(t) = \int_0^t K(t,\tau)x(\tau)d\tau - v(\omega)(t),$$

where $K(t,\tau) = a(t)b(\tau)$ it is

$$(T(\omega)x-v(\omega))_{[m]} = T_{[m]}x_{[m]} - v_{[m]},$$

if

$$T_{[m]} = \begin{pmatrix} a(\frac{1}{m}) & 0 & \dots & 0 \\ a(\frac{2}{m}) & a(\frac{2}{m}) & \dots & 0 \\ \vdots & \vdots & \vdots & \vdots \\ a(1) & a(1) & \dots & a(1) \end{pmatrix}, \quad x_{[m]} = \begin{pmatrix} \int_0^{1/m} b(\tau)x(\tau)d\tau \\ \int_{1/m}^{2/m} b(\tau)x(\tau)d\tau \\ \vdots \\ \int_{\frac{m-1}{m}}^{1} b(\tau)x(\tau)d\tau \end{pmatrix}$$

and

$$v_{[m]} = \begin{pmatrix} v(\frac{1}{m}) \\ v(\frac{2}{m}) \\ \vdots \\ v(1) \end{pmatrix}. \tag{8}$$

Let us consider now for the functions $\bar{f}_k$ of bounded variation the two examples

(a) $\bar{f}_k(t) = \hat{f}_k t,\ 0\le t\le 1,\ k=1,2,\dots$

(b) $\bar{f}_k(t) = \begin{cases} 0 & \text{for } 0<t\le 1 \\ \hat{f}_k & \text{for } t=1 \end{cases},\ k=1,2,\dots,$

where $\hat{f}_k$, $k=1,2,\dots$ is a given sequence of real numbers such that $\hat{f}_k<0$ and $\hat{f}_\kappa>0$ for at least one integer k respectively κ. We find then that

$$\tilde{p}_{[m]n}(z_1,z_2,\dots,z_m) = \max_{1\le k\le n} \sum_{i=1}^{m} z_i(\bar{f}_k(\tfrac{i}{m})-\bar{f}_k(\tfrac{i-1}{m})) = \begin{cases} \max\limits_{1\le k\le n} \sum_{i=1}^{m} z_i(\frac{1}{m}f_k) & \text{for(a)} \\ \max\limits_{1\le k\le n} z_m f_k & \text{for(b)} \end{cases}$$

Setting

$$c_{n1} = \min_{1\leq k\leq n} \hat{f}_k \text{ and } c_{n2} = \max_{1\leq k\leq n} \hat{f}_k$$

as well as

$$\tilde{p}_n(t) = \max_{1\leq k\leq n} \hat{f}_k t = \begin{cases} t \min\limits_{1\leq k\leq n} \hat{f}_k & \text{for } t\leq 0 \\ t \max\limits_{1\leq k\leq n} \hat{f}_k & \text{for } t\geq 0 \end{cases} = \begin{cases} tc_{n1} & \text{for } t\leq 0 \\ tc_{n2} & \text{for } t\geq 0 \end{cases}$$

we obtain that

$$p_{[m]n}(z) = \tilde{p}_{[m]n}(z_{[m]}) = \begin{cases} \tilde{p}_n(\frac{1}{m}\sum_{i=1}^{m} z(\frac{i}{m}) & \text{for (a)} \\ \tilde{p}_n(z(1)) & \text{for (b)} \end{cases}.$$

We observe that $p_{[m]n}(z)$ depends essentially only on the onedimensional variable $\frac{1}{m}\sum_{i=1}^{m} z(\frac{i}{m})$ respectively $z(1)$. From this representation of $p_{[m]n}(z)$ follows -using the notation introduced in (8)-

$$R_{[m]n}(x) = E_\omega p_{[m]n}(T(\omega)x-v(\omega)) = R_{[m]n}(x_{[m]}) = \tag{9}$$

$$=E_\omega \tilde{p}_{[m]n}(T_{[m]}x_{[m]}-v(\omega)_{[m]}) = \begin{cases} E_\omega \tilde{p}_n(\frac{1}{m}\sum_{i=1}^{m} T_{[m]i}x_{[m]} - \frac{1}{m}\sum_{i=1}^{m} v(\omega)(\frac{i}{m})) & \text{for (a)} \\ E_\omega \tilde{p}_n(T_{[m]m}x_{[m]}-v(\omega)(1)) & \text{for (b)} \end{cases}$$

where $T_{[m]i}$ denotes the ith row of the (m,m)-matrix $T_{[m]}$. We consider this example again in the next section 4 after having presented some properties of differentiability of R and its approximatives R_n and $R_{[m]n}$.

<u>4. Differentiability of R, R_n and $R_{[m]n}$.</u> Let $Dp(z)$, $DR(x)$,... denote the Frèchet(F-)-derivative of p, R,... at z respectively at x; Note that by definition of these derivatives $Dp(z)\in Z^*$, $DR(x)\in X^*$, the evaluation of these functionals at $y\in Z$ respectively at $h\in X$ is denoted by $Dp(z)y$ respectively $DR(x)h$. Furthermore $\Delta_F p$ denotes the set of elements $z\in Z$ at which p is F-differentiable.

<u>Theorem 4.1.</u> Let μ be the distribution of $T(\omega)$ and Φ_T the conditional distribution of $v(\omega)$ given $T(\omega)=T$. If

$$P(\{\omega\in\Omega:\ T(\omega)x-v(\omega)\in\Delta_F p\}) = 1, \tag{10}$$

then R is F-differentiable at x and it is

$$DR(x)h = \int(\int_{v\in Tx-\Delta_F p} Dp(Tx-v)Th\Phi_T(dv))\mu(dT),\ h\in X.$$

Proof: see [7] and [8].

Note that this theorem is also valid if we replace p,R by p_n,R_n or by $p_{[m]n}$,$R_{[m]n}$.

We put now again as in section 3 (see (6)) $p(z)=\sup_k f_k z$ and $p_n(z)=\max_{1\leq k\leq n} f_k z$. Furthermore let C_k, k=1,2,...,n be defined by

$$C_k = \{z\in Z:\ p(z)=f_k z\} = \{z\in Z:\ f_i z\leq f_k z,\ i=1,2,...,n\}.$$

We see that C_k, $k=1,2,\ldots,n$ are closed convex cones in Z such that $\bigcup_{k=1}^{n} C_k = Z$ and the interior int C_k of C_k is given by

$$\text{int } C_k = \{z \in Z:\ f_i z < f_k z,\ i=1,2,\ldots,n,\ i \neq k\}$$

Without any loss of generality we may suppose that int $C_k \neq \emptyset$ for all $k=1,2,\ldots,n$. Since $p(z)=f_k z$ for $z \in C_k$ we have immediately the following proposition.

<u>Theorem 4.2.</u> If there is an integer k between 1 and n and an element $x \in X$ such that

$$T(\omega)x - v(\omega) \in C_k \text{ almost sure,}$$

then

$$R(x) = f_k(E_\omega T(\omega)x - E_\omega v(\omega)).$$

Note. There are elements k,x with the properties required above e.g. if $\omega \to \|T(\omega)\|$ and $\omega \to \|v(\omega)\|$ are bounded random variables.

<u>Lemma 4.1.</u> $\Delta_F p = \bigcup_{k=1}^{n} \text{int } C_k$.

If $Z=\mathbb{R}^m$ and the conditional distributions Φ_T of $v(\omega)$ given $T(\omega)=T$ are absolutely continuous with respect to the Lebesgue measure on $\mathbb{R}^m$, then equation (10) is satisfied since the complement of $\bigcup_{k=1}^{n} \text{int } C_k$ is contained in the union of finitely many hyperplanes of $\mathbb{R}^m$.

Combining Lemma 4.1 with Theorem 4.1 we find our next result.

<u>Theorem 4.3.</u> If for a particular $x_0 \in X$

$$P(\{\omega \in \Omega:\ T(\omega)x_0 - v(\omega) \in \bigcup_{k=1}^{n} \text{int } C_k\}) = 1,$$

then $R_n(x)=E_\omega p_n(T(\omega)x-v(\omega))$ is F-differentiable at x_0 and it holds

$$DR_n(x)h = \sum_{k=1}^{n} \int \Phi_T(Tx-C_k) f_k Th\, \mu(dT).$$

Denoting the adjoint operator of T by T^* we may also write

$$DR_n(x) = \sum_{k=1}^{n} \int \Phi_T(Tx-C_k) T^* f_k\, \mu(dT). \tag{11}$$

Notes. If "$\leq_k$" denotes the partial order

$$z_1 \leq_k z_2 \iff z_2 - z_1 \in C_k,\ z_1, z_2 \in Z$$

induced on Z by the cone C_k, then

$$\Phi_T(z-C_k) = P(v(\omega) \leq_k z \mid T(\omega)=T),$$

hence $z \to \Phi_T(z-C_k)$ may be considered (for each fixed pair (k,T)) as a generalized distribution function. - If $U=X$ and equation (11) holds for all $x \in X$, then each solution $\bar{x}$ of the approximative convex optimization problem (1n) is characterized by the equation

$$0 = DR_n(\bar{x}) = \sum_{k=1}^{n} \int \Phi_T(T\bar{x}-C_k) T^* f_k\, \mu(dT). \tag{12}$$

If $\sum_{k=1}^{n} E_\omega \|T(\omega)^* f_k - E_\omega T(\omega)^* f_k\|$ is small -especially if $T(\omega)$ is constant almost sure-, then we may replace (12) by

$$0 = \sum_{k=1}^{n} \int \Phi_T(Tx-C_k)\mu(dT) \int T^* f_k \mu(dT) = \sum_{k=1}^{n} P(v(\omega) \leq_k T(\omega)x) E_\omega T(\omega)^* f_k$$

what is a system of chance-constraints.

We conclude our paper by applying the results of this section to the example already considered at the end of section 3:

Let the stochastic process $v=v(\omega)(t)$ be such that for all $z\in Z$

$$P(\{\omega\in\Omega:\ \sum_{i=1}^{m} v(\omega)(\tfrac{i}{m})=z\}) = 0,\ m=1,2,\dots \quad \text{for (a)}$$

and

$$P(\{\omega\in\Omega:\ v(\omega)(1)=z\}) = 0 \qquad \text{for (b)}.$$

Applying Theorem 4.3 to $R_{[m]n}$, given by equation (9), yields then

$$\frac{\partial}{\partial x_{[m]}}R_{[m]n}(x)=\begin{cases} c_{n1}+(c_{n2}-c_{n1})F_{\frac{1}{m}\sum_{i=1}^{m}v(\frac{i}{m})}\left(\frac{1}{m}\sum_{i=1}^{m}T_{[m]i}x_{[m]}\right)\begin{pmatrix}\sum_{i=1}^{m}a(\frac{i}{m})\\ \sum_{i=2}^{m}a(\frac{i}{m})\\ \vdots\\ a(1)\end{pmatrix} & \text{for (a)}\\ c_{n1}+(c_{n2}-c_{n1})f_{v(1)}(T_{[m]m}x_{[m]})\begin{pmatrix}a(1)\\ a(1)\\ \vdots\\ a(1)\end{pmatrix} & \text{for (b)},\end{cases} \tag{14}$$

where $F_{\frac{1}{m}\sum_{i=1}^{m}v(\frac{i}{m})}$ respectively $F_{v(1)}$ denotes the distribution function of the real random variable $\omega\to\frac{1}{m}\sum_{i=1}^{m}v(\omega)(\frac{i}{m})$ respectively $\omega\to v(\omega)(1)$ and $\frac{\partial}{\partial x_{[m]}}R_{[m]n}(x)$ denotes the F-derivative of $R_{[m]n}(x)$ -depending actually on $x_{[m]}$- with respect to $x_{[m]}$. Let $a(1)\neq 0$. By (14) we have then that $DR_{[m]n}(x)=0$ holds if and only if

$$F_{\frac{1}{m}\sum_{i=1}^{m}v(\frac{i}{m})}\left(\frac{1}{m}\sum_{i=1}^{m}T_{[m]i}x_{[m]}\right) = \alpha_n \equiv \frac{-c_{n1}}{c_{n2}-c_{n1}} \quad \text{for (a)} \tag{15a}$$

respectively

$$F_{v(1)}(T_{[m]m}x_{[m]}) = \alpha_n \qquad \text{for (b)} \tag{15b}$$

Since $c_{n1}<0$ and $c_{n2}>0$ (see the assumptions concerning $\bar{f}_k$ in section 3) for sufficiently large n it is $0\leq\alpha_n\leq 1$ for large integers n.

Now let $v=v(\omega)(t)$ be a normal process with

$$\overline{v(t)} = E_\omega v(\omega)(t) = at$$

and

$$r(s,t) = E_\omega(v(\omega)(t)-\overline{v(t)})(v(\omega)(s)-\overline{v(s)}) = \sigma^2\min(s,t).$$

Clearly $v(\omega)(1)$ has under these assumptions a normal distribution with mean a and variance σ^2, whereas $\frac{1}{m}\sum_{i=1}^{m}v(\omega)(\frac{i}{m})$ is normal with

$$\text{mean} = \frac{1}{m}a\left(\frac{1}{m}+\frac{2}{m}+\dots+\frac{m}{m}\right) = \frac{a}{m^2}\frac{m(m+1)}{2} = \frac{a}{2}\frac{m+1}{m}$$

and

$$\text{variance}=E_\omega\left(\frac{1}{m}\sum_{i=1}^{m}v(\omega)(\tfrac{i}{m})-\frac{1}{m}\sum_{i=1}^{m}E_\omega v(\omega)(\tfrac{i}{m})\right)^2 = E_\omega\frac{1}{m^2}\left(\sum_{i=1}^{m}(v(\omega)(\tfrac{i}{m})-E_\omega v(\omega)(\tfrac{i}{m}))\right)^2$$

$$=\frac{1}{m^2}\sum_{i,j=1}^{m}E_\omega(v(\omega)(\tfrac{i}{m})-E_\omega v(\omega)(\tfrac{i}{m}))(v(\omega)(\tfrac{j}{m})-E_\omega v(\omega)(\tfrac{j}{m}))$$

$$=\frac{1}{m^2}\sum_{i,j=1}^{m}r(\tfrac{i}{m},\tfrac{j}{m}) = \frac{\sigma^2}{m^2}\left(\frac{m+1}{2}+\frac{m^2-1}{3}\right) = \sigma^2\left(\frac{m+1}{2m^2}+\frac{1}{3}-\frac{1}{3m^2}\right).$$

Denoting the distribution function of the normal distribution with mean zero and variance 1 by Ψ we may write (15a) in the form

$$\frac{1}{m}\sum_{i=1}^{m} {}^{T}[m]\,i^{\times}[m] = F^{-1}_{\frac{1}{m}\sum_{i=1}^{m} v(\frac{i}{m})}(\alpha_n) = \frac{a}{2}\,\frac{m+1}{m} + \Psi^{-1}(\alpha_n)\,\sigma\left(\frac{m+1}{2m^2}+\frac{1}{3}-\frac{1}{3m^2}\right)^{1/2}$$

or

$$\sum_{i=1}^{m} {}^{T}[m]\,i^{\times}[m] = \frac{a}{2}(m+1) + \Psi^{-1}(\alpha_n)\,\sigma\left(\frac{m+1}{2}+\frac{m^2-1}{3}\right)^{1/2}$$

and (15b) has the simpler form

$${}^{T}[m]\,m^{\times}[m] = F^{-1}_{v(1)}(\alpha_n) = a + \Psi^{-1}(\alpha_n)\,\sigma.$$

<u>References:</u>

1. Aoki, M.: Optimization of Stochastic Systems (Topics in Discrete-Time Systems). New York-London: Academic Press 1967

2. Berkovich, E.M.: The Approximation of Two-Stage Stochastic Extremal Problems. USSR Computational Mathematics and Physics 11, Nr. 5, 1150-1165(1971)

3. Berkovich, E.M.: A two-stage stochastic optimum control problem. Moscow Univ. Math. Bull. 25, No.3-4, English p. 49-55(1970)

4. Demyanov, V.F., Rubinov, A.M.: Approximative Methods in Optimization Problems. New York: American Elsevier 1970

5. De Groot, M.H., Rao, M.M.: Bayes estimation with convex loss. Ann. Math. Statist. 34, 839-846(1963)

6. Director, S.W.: Survey of Circuit-Oriented Optimization Techniques. IEEE Transactions on Circuit Theory, Vol.CT-18,No.1,3-10(1971)

7. Marti, K.: Entscheidungsprobleme mit linearem Aktionen-und Ergebnisraum. Z.Wahrscheinlichkeitstheorie verw.Geb.23,133-147(1972)

8. Marti, K.: Approximation der Entscheidungsprobleme mit linearer Ergebnisfunktion und positiv homogener, subadditiver Verlustfunktion. Z.Wahrscheinlichkeitstheorie verw.Geb.31,203-233(1975)

9. Marti, K.: Convex Approximation of Stochastic Optimization Problems. Methods of Operations Research 20, 66-76(1975)

MODELS IN RESOURCE ALLOCATION — TREES OF QUEUES

Bernd Meister

IBM Zurich Research Laboratory, 8803 Rüschlikon, Switzerland

Summary: Queueing models play an important role in resource-allocation problems. In this paper a tree consisting of terminals (sources of requests), queues and synchronous servers is investigated. The generating function and the first moment of the holding time of a request generated by an arbitrary terminal are calculated.

1. Description of Model

First, we define a class of input processes, i.e., the streams of requests entering the queueing system.

Definition 1:

A process $\mathcal{X} = \{X_j\}$, $j = 1, 2, 3, \ldots$ is called a process with stationary independent increments (s.i.i. process) if the random variables X_j are non-negative, integer-valued and independent with a common distribution

$$\Pr \{X_j = k\} = p_k, \quad k = 0, 1, 2, \ldots \quad . \tag{1.1}$$

The subscript j will be interpreted as time and X_j as the number of requests generated by a terminal in the time interval $[j - 1, j)$. The time required to serve such a request is assumed to be a positive, integer-valued random variable s, independent of the X_j with a distribution

$$\Pr \{s = k\} = q_k, \quad k = 1, 2, 3, \ldots \quad . \tag{1.2}$$

The service time for all requests of an s.i.i. process is given by the same random variable s.

Definition 2:

Two s.i.i. processes $\mathcal{X}_1, \mathcal{X}_2$ are called independent if the processes $\mathcal{X}_1, \mathcal{X}_2$ are independent and their respective service times s_1, s_2 are independent random variables.

The tree of queues which will be investigated is inductively defined. First we define a 1-tree.

Definition 3:

A 1-tree consists of a finite number of input terminals $T_1, T_2, \ldots, T_N$, ordered according to decreasing priorities, and a single server S. The terminals have waiting rooms of infinite capacity (Fig. 1).

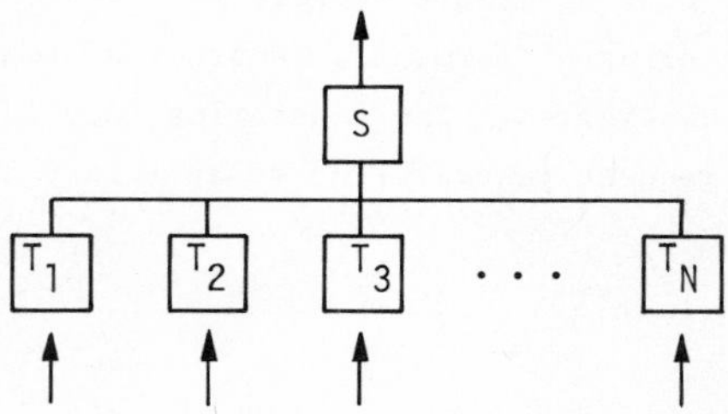

Fig. 1. 1-tree.

The requests which enter the 1-tree through the terminals T_i and wait there for service are given by independent s.i.i. processes

$$\mathcal{X}_i = \{X_{ij}\} \qquad i = 1, 2, \ldots, N, \quad j = 1, 2, 3, \ldots \quad . \tag{1.3}$$

The corresponding service times are called s_i. The service discipline can be described as follows. Service starts periodically at the discrete times $j = 1, 2, 3, \ldots$ and the duration of one service period is of length 1. Let $I_j(T_i)$ be the number of service periods needed to serve all requests waiting in the queue at T_i if this service starts at time j and no interruptions or arrivals occur. Then the following equations are assumed to hold:

$$I_{j+1}(T_i) = \left(I_j(T_i) - \chi_{\sum_{k=1}^{i-1} I_j(T_k) = 0}\right)^+ + X_{ij}$$

$$i = 1, 2, \ldots, N; \quad j = 1, 2, 3, \ldots \tag{1.4}$$

where $a^+ = \max(a,0)$, $\chi_A = 1$ if the event A occurs and $\chi_A = 0$ otherwise. In a 1-tree the queues are scanned in the order of decreasing priorities and the first non-empty queue receives service for one unit of time. Within each queue, service in the order of arrivals is assumed (first come - first served). Therefore, a 1-tree is a time discrete queueing system with N priority classes and preemptive priorities investigated in [1], [2]. It is assumed in equations (1.4) that requests arriving exactly at the service times $j = 1, 2, 3, \ldots$ have to wait for at least one unit of time until they are served. The stream of requests which leaves the server after having completed service is called the output process of the 1-tree.

To construct more general trees we need the notion of a tree element.

Defintion 4:

A tree element consists of a finite number of infinite capacity buffers B, input terminals T ordered according to decreasing priorities and a single server S (Fig. 2). The service discipline for the requests waiting at the terminals T and in the buffers B is the same as in a 1-tree.

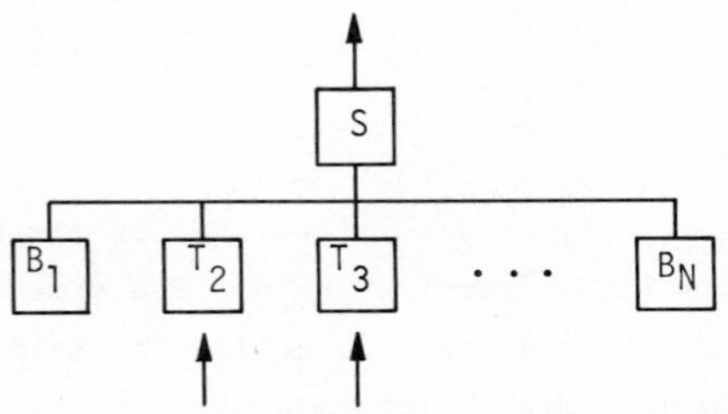

Fig. 2. Tree element.

Both terminals and buffers have infinite waiting rooms; the difference is that input terminals are sources of requests, whereas the buffers are passive and have to be connected to other sources. In our application the input stream into a buffer will be the output stream of a certain server.

Now we are prepared to define a k-tree ($2 \leq k < \infty$).

Definition 5:

A k-tree is a tree element all buffers of which are connected to k'-trees, $1 \leq k' \leq k - 1$, such that the input processes into these buffers are the output processes of the respective k'-trees. At least one buffer of the tree element is connected to a (k - 1)-tree (Fig. 3).

All servers in a k-tree work synchronously, i.e., their service operation starts at times $j = 1, 2, 3, \ldots$ and the duration of a service period is 1.

A k-tree has a unique server, the output of which does not enter another buffer but leaves the tree. This server is called the output server S_0. In a k-tree, a unique, finite, directed chain of servers S and buffers B connects each terminal T with the output server S_0: $T \rightarrow S \rightarrow B \rightarrow S \rightarrow B \rightarrow \ldots \rightarrow B \rightarrow S_0$. This chain is called

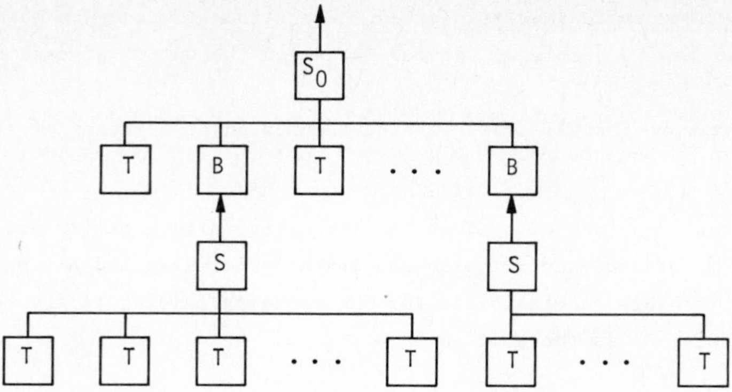

Fig. 3. 2-tree.

the service chain of terminal T. The service time for a request generated at T is the same for all servers in its service chain. The time which is needed for such a request to pass the whole service chain is called its holding time, i.e., the holding time includes all waiting times at intermediate buffers and service times by the different servers. The number of intermediate buffers B in the service chain is called the rank of a terminal. The ranks of the terminals in a k-tree lie between 0 and k - 1; at least one terminal has rank k - 1.

Now we investigate the priorities of the terminals in a k-tree. We consider two arbitrary terminals T and T'.

Case 1: Both terminals are served by a common server S. Then we accord higher priority to T in the tree if T has higher priority than T' for server S.

Case 2: Both terminals are served by different servers. We then consider the service chains of T and T'. There is one first unique server S' on both chains which serves requests from both terminals such that the subchains $S' \rightarrow \ldots \rightarrow S_0$ become identical for both terminals. Then terminal T has a higher priority than T' if the requests generated at T have a higher priority than the requests generated at T' for server S'.

In this way we obtain a complete ordering of all terminals in a k-tree which will be used in the following.

2. Calculation of Holding Time

We now consider a k-tree $\mathcal{T}_k$ with M terminals $T_1, T_2, \ldots, T_M$ which are ordered according to decreasing priorities in the tree. In addition, this tree may have a finite number of intermediate buffers and servers. Let the input processes of these terminals be $\mathcal{X}_1, \mathcal{X}_2, \ldots, \mathcal{X}_M$ with generating functions

$$E(z^{X_{ij}}) = P_i(z), \; i = 1, 2, \ldots, M \quad , \tag{2.1}$$

where E denotes expectation.

The service times are denoted by $s_1, s_2, \ldots, s_M$ with generating functions

$$E(z^{s_i}) = Q_i(z), \quad i = 1, 2, \ldots, M \quad . \tag{2.2}$$

First, we consider the case $k = 1$ which was analyzed in [2]. The results of [2] will be briefly summarized and in the remainder of this paper it will be shown that the case $k > 1$ can be reduced to $k = 1$.

Consider a terminal T_ℓ, ℓ arbitrary but fixed, $1 \leq \ell \leq M$, of a 1-tree $\mathcal{T}_1$.

Let

$$W_{\ell j} = \sum_{i=1}^{\ell} I_j(T_i) \tag{2.3}$$

be the number of service periods required to serve all requests waiting at terminals $T_1, T_2, \ldots, T_\ell$ if the service starts at time j, no interruptions occur and no additional requests arrive.

Under the assumption

$$\sum_{i=1}^{\ell} E(X_{ij})\, E(s_i) < 1 \tag{2.4}$$

the $W_{\ell j}$ have a stationary distribution as $j \to \infty$ with generating function

$$H_\ell^*(z) = \left[1 - \sum_{i=1}^{\ell} E(X_{ij})\, E(s_i)\right] \frac{(z-1) \prod_{i=1}^{\ell} R_i(z)}{z - \prod_{i=1}^{\ell} R_i(z)} \tag{2.5}$$

and

$$R_i(z) = P_i\,[Q_i(z)] \quad . \tag{2.6}$$

We now assume the system to be in the steady state and insert a test or virtual request at the end of the queue at T_ℓ. The service time of this test request is a random variable s, independent of the X_{ij} and the s_i with generating function

$$E(z^s) = S(z) \qquad . \tag{2.7}$$

We assume here that the test request arrives at a time $j - 0$, i.e., immediately before the beginning of a service period. Other cases can be treated in a similar way [2].

The time required to complete service of this test message is called the (virtual) holding time at terminal T_ℓ and denoted by $d_\ell(s)$.

Its generating function is given by [3], [2]:

$$E\left[z^{d_\ell(s)}\right] = H_\ell^*[\theta_\ell(z)]\ S[\theta_\ell(z)] \tag{2.8}$$

with $\theta_\ell(z)$ being the solution of

$$\theta_\ell(z) = z \prod_{i=1}^{\ell-1} R_i[\theta_i(z)] \tag{2.9}$$

for $\ell < 1$ and $\theta_1(z) = z$. Let $\mathscr{Y}_i$ be the process with generating function $R_i(z)$. Then the expectation is given by

$$E[d_\ell(s)] = \frac{1}{2}\ \frac{\dfrac{\sum_{i=1}^{\ell} \operatorname{Var}(\mathscr{Y}_i)}{1 - \sum_{i=1}^{\ell} E(\mathscr{Y}_i)} + \sum_{i=1}^{\ell} E(\mathscr{Y}_i) + 2\,E(s)}{1 - \sum_{i=1}^{\ell-1} E(\mathscr{Y}_i)} \quad , \tag{2.10}$$

where Var denotes variance. This expectation is finite if, in addition to (2.4) we have $\operatorname{Var}(\mathscr{X}_i) < \infty$, $\operatorname{Var}(s_i) < \infty$, $1 \le i \le M$. The second moment of $d_\ell(s)$ can be found in [2].

We now turn to the general case and consider a k-tree $\mathscr{T}_k$, $k > 1$ and again a terminal T_ℓ, $1 \le \ell \le M$, ℓ arbitrary but fixed. The input processes and service times are again given by (2.1) and (2.2). We are interested in the holding time $d_\ell(s)$ of a virtual request having service time s which arrives at the end of the queue at T_ℓ just before the beginning of a service period when the system is in the steady state. First, we replace all input processes $\mathscr{X}_i$ by processes $\mathscr{Y}_i$ with generating functions

$$E(z^{Y_{ij}}) = P_i[Q_i(z)] = R_i(z) \tag{2.11}$$

and we set all service times

$$s_i = 1, \quad 1 \leq i \leq M \quad . \tag{2.12}$$

The processes $\mathscr{Y}_i$ are again s.i.i. processes and we call the resulting k-tree $\mathscr{T}_k'$. Then the following equivalence between $\mathscr{T}_k$ and $\mathscr{T}_k'$ holds:

Let $W'_{\ell j}$ be the total number of requests waiting at all terminals and buffers of $\mathscr{T}_k'$ which have priority higher than or equal to T_ℓ, the terminal under consideration, just before the start of the j-th service period, i.e., at time $j - 0$, and let $W_{\ell j}$ be the number of service periods required by all intermediate servers and the output server of $\mathscr{T}_k$ to serve all requests waiting at buffers or terminals with a priority higher than or equal to the priority of terminal T_ℓ. There is assumed that this service starts at time j and no interruptions or arrivals take place. Here a buffer B has a higher priority than T_ℓ if the requests waiting in B have been generated at terminals T_k with a higher priority than T_ℓ. Then we have the following

LEMMA

$$W'_{\ell j} = W_{\ell j} \tag{2.13}$$

provided

$$W'_{\ell 0} = W_{\ell 0} \tag{2.14}$$

holds for the corresponding initial values.

This equivalence allows to calculate the holding time at a terminal by just counting the number of requests which will be served before the test request in $\mathscr{T}_k'$. The proof is based on the fact that the time required to serve one request with service time s is the same as the time required to serve s requests with service time one each [2].

We now consider the tree $\mathscr{T}_k'$. It follows from [4] that this tree is equivalent to a k-loop with capacity $C = 1$ for each server. In [4] an equivalence theorem for k-loops has been proved which we now express in terms of k-trees.

Let the ranks of the terminals in $\mathscr{T}_k'$ be r_i, $i = 1, 2, \ldots, M$. These ranks are the same as in $\mathscr{T}_k$.

$$\text{Set} \quad \mathcal{Y}_i' = \{Y_{ij}'\} = \begin{cases} 0 & \text{for} \quad 1 \le j \le r_i \\ Y_{ij-r_i} & \text{for} \quad j > r_i \end{cases}$$

and let $\mathcal{T}_1'$ be the 1-tree with the ordered set of input terminals $T_1, T_2, \ldots, T_M$ and corresponding input processes $\mathcal{Y}_1', \mathcal{Y}_2', \ldots, \mathcal{Y}_M'$.

Define $W_{\ell j}^{(1)}$ as the total number of requests wating of the terminals $T_1, T_2, \ldots, T_\ell$ of $\mathcal{T}_1'$ at time $j - 0$.

Then we have [4]:

THEOREM

$$W_{\ell j}' = W_{\ell j}^{(1)}$$

provided $W_{\ell 0}' = W_{\ell 0}^{(1)}$ holds for the corresponding initial values.
In addition, the output streams leaving the output servers of $\mathcal{T}_k'$ and $\mathcal{T}_1'$ are identical in time.

Since the difference between the processes $\mathcal{Y}_i$ and $\mathcal{Y}_i'$ does not influence the steady state we conclude from the lemma and the theorem that under the assumption (2.4) $W_{\ell j}$, $W_{\ell j}'$ and $W_{\ell j}^{(1)}$ prossess the same stationary distribution with a generating function given by (2.5) and (2.6).

The generating function of the virtual holding time $d_\ell(s)$ defined as before is the same for trees $\mathcal{T}_k$, $\mathcal{T}_k'$ and $\mathcal{T}_1'$ and is given by (2.8) and (2.9). The expectation of $d_\ell(s)$ is again given by (2.10). The virtual holding time for a test request which arrives just after the beginning of a service period, i.e., at a time $j + 0$, can be calculated as in [2]. In [2] also bounds for the holding time of a test request arriving at an arbitrary instant of time and explicit expressions for second moments are given which can be directly applied to k-trees.

Finally, an invariance property of k-trees will be mentioned. The holding time $d_\ell(s)$ at terminal T_ℓ in the 1-tree $\mathcal{T}_1'$ is invariant against permutations of terminals $T_1, T_2, \ldots T_{\ell-1}$, since it is influenced only by the sum of the corresponding input processes. We now consider the service chain of T_ℓ in the equivalent k-tree $\mathcal{T}_k$:

$$T_\ell \to S_r \to B \to S_{r-1} \to \ldots \to B \to S_0.$$

All requests generated according to processes $\mathcal{X}_1, \mathcal{X}_2, \ldots, \mathcal{X}_{\ell-1}$ have somewhere along this chain a higher priority than requests generated at T_ℓ. Let $\mathcal{X}^{(\rho)}$, $\rho = 0, 1, \ldots, r$ be the sum of those of the $\mathcal{X}_i$, $i = 1, 2, \ldots, \ell - 1$, which have for server S_ρ a priority higher than that of $\mathcal{X}_\ell$. Here we mean by priority of a process, the priority of the corresponding terminal.

We now write the service chain in the form:

$$T_\ell \to \left(S_r, \mathcal{X}^{(r)}\right) \to B \to \left(S_{r-1}, \mathcal{X}^{(r-1)}\right) \to B \ldots \to B \to \left(S_0, \mathcal{X}^{(0)}\right) .$$

Then the holding time at T_ℓ is invariant against permutations of the different pairs $\left(S_\rho, \mathcal{X}^{(\rho)}\right)$ in this chain. This means that the requests of higher priority may enter at any of the servers in the service chain.

Especially, the holding time does not change, if the flow of requests in a k-tree is reversed, i.e., all requests enter at the output s′rver and then proceed to the different terminals.

3. References

[1] A. G. KONHEIM and B. MEISTER, Service in a loop system, *J. Assoc. Comput. Mach.* 19 (1972), 92-108.

[2] B. MEISTER, Ein zeitdiskretes Wartesystem mit unterbrechenden Prioritäten, Proceedings of the GI-Jahrestagung 1975, Lecture Notes in Computer Science, Springer-Verlag, Berlin, 1975.

[3] A. G. KONHEIM and B. MEISTER, Waiting lines and times in a system with polling, *J. Assoc. Comput. Mach.* 21 (1974), 470-490.

[4] A. G. KONHEIM and B. MEISTER, Waiting lines in multiple loop systems, *J. Math. Anal. Appl.* 39 (1972), 527-540.

DUAL METHODS IN CONVEX CONTROL PROBLEMS

C. P. Ortlieb

Institut für Angewandte Mathematik der Universität Hamburg
2 Hamburg 13, Bundesstraße 55 / Germany

The concept of duality is playing a central role in theory and numerical treatment of convex optimization problems. Particularly this is true, if we consider problems with nondifferentiable cost criteria and complicated constraints. In that case it may happen that the primal problem is far from being as simple as the dual. Following LUENBERGER (69), p.299 a dual method for a convex optimization problem means treating the dual problem by any (numerical) method. LUENBERGER (69) suggests to determine an optimal solution of the dual in this way, and then to solve the primal by "minimizing the corresponding Lagrangian". However, determining an optimal solution of the dual will not be possible in a finite number of steps but only by an infinite process, in general. Therefore, we have to construct approximation solutions of the primal problem already before a dual optimal solution is known.

LUENBERGER (73), p. 312 - 316 reflects on questions of that kind concerning a class of differentiable programs, which have to satisfy merely local convexity assumptions. In the following we shall deal with a class of global convex but not neccessarily differentiable programs given in the form of convex control problems, which, nevertheless, is as general as the concept of "generalized convex programs" (ROCKAFELLAR (69 b)). We shall show that treating the dual problem by a combination of Ritz' method with the algorithm of WOLFE (74 a), (74 b) (or any algorithm having similar properties, see 4.1) yields in fact primal approximation solutions.

1. Control Problem and Dual Control Problem

We consider a convex control problem in the following general form :

(S) Minimize $I(u) := f(Su,u)$ subject to $u \in U$

where U and X are real topological linear spaces, $S : U \longrightarrow X$ is a linear and continuous operator, and $f : X \times U \longrightarrow R \cup \{\infty\}$ is a proper convex functional (ROCKAFELLAR (69 a), p.24). Let $\text{dom } f := \{(x,u) \in X \times U : f(x,u) < \infty\}$. Then the constraints of (S) are given by :

$$(Su,u) \in \text{dom } f \quad .$$

The connection between problem (S) and ordinary problems of optimal control will become clear, if we interpret U as a space of "controls", X as a space of "states", and S as a "control operator" that assigns to every control u a uniquely determined state $x = Su$. In ordinary control problems U and X will be function spaces (see 5.).

We consider a second convex control problem :

(D) Minimize $J(y) := f^{*}(y,-S^{*}y)$ subject to $y \in Y$

where V and Y are the dual spaces of U and X respectively, S^{*} is the adjoint operator of S, and

$$f^{*}(y,v) := \sup_{(x,u)\in X\times U} [\langle x,y\rangle + \langle u,v\rangle - f(x,u)] \quad ((y,v)\in Y\times V) .$$

f^{*} is called the conjugate convex functional of f (see BRØNDSTED (64)). Problem (D) is called the dual control problem of problem (S). If $u \in U$, $y \in Y$, we have

$$I(u) + J(y) = f(Su,u) + f^{*}(y,-S^{*}y) \geq \langle Su,y\rangle + \langle u,-S^{*}y\rangle = 0 .$$

Hence we get the following theorem about upper and lower bounds for

$$\inf (S) := \inf \{ I(u) : u \in U \} :$$

1.1 T h e o r e m : If $u \in U$ and $y \in Y$, then the following inequality holds :

$$- J(y) \leq - \inf (D) \leq \inf (S) \leq I(u) .$$

In the next paragraph we shall give sufficient conditions such that the equality $- \inf (D) = \inf (S)$ is valid.

2. Duality Theorems

2.1 T h e o r e m : Let the origin of X be an interior point of the set

$$A_{\bar{b}} := \{ x - Su : (x,u)\in X\times U,\ f(x,u) \leq \bar{b} \}$$

for some real $\bar{b}$. Then the equality $\inf (S) = - \min (D)$ is valid, and the minimum of problem (D) is achieved by some $y \in Y$.

Proof :

Let $B := \{(b,z)\in R\times X : z = x - Su,\ f(x,u) \leq b$ for some $(x,u)\in X\times U \}$. By definition $m := \inf (S) = \inf \{b : (b,0)\in B\}$. If $m = -\infty$, then $J(y) = \infty$ for all y, and the theorem is proved.

Let m be finite. Then $(\bar{b}+1,0)$ is an interior point of the convex set B, and $(m,0)$ is not. Therefore, by the separation theorem of Eidelheit (KÖTHE (66), p.191), there is a nonzero continuous linear functional $(a,y)\in R\times Y$ separating $(m,0)$ from B :

$$a\, m \leq a\, b - \langle z,y\rangle \quad \text{for all } (a,z)\in B . \tag{2.1}$$

In (2.1) b may be arbitrarily large, hence $a \geq 0$. If $a = 0$, then y would separate the origin of X from $A_{\bar{b}}$, which is not possible. Therefore, $a > 0$ and, without loss of generality, we may assume $a = 1$. Then it follows from (2.1) that

$$m \leq f(x,u) - \langle x - Su,y\rangle = f(x,u) - \langle x,y\rangle - \langle u,-S^{*}y\rangle \quad \text{for all } (x,u)\in \text{dom } f . \tag{2.2}$$

Since the infimum of the right hand side of (2.2) is $- f^{*}(y,-S^{*}y)$, we get

$$\inf (S) = m \leqslant - f^*(y,-S^*y) \leqslant - \inf (D)$$

and the assertion follows from 1.1 .

The proof of 2.1 is standard for duality theorems of this kind (see e.g. DIETER (66), ROCKAFELLAR (67), LUENBERGER (69), p.201, p.217, LEMPIO (71)). A similar, but more sophisticated argument using the interior mapping principle (DUNFORD/SCHWARTZ (57), p.55) yields :

2.2 Theorem : Let U and X be Banach spaces, let f be lower semicontinuous, let dom f contain relative interior points, and let f be continuous throughout its relative interior. Assume that an $\bar{u} \in U$ exists such that the co-dimension of

$$\{x \in X : (x,\bar{u}) \in \text{aff (dom f)}\}$$

is finite (aff (dom f) denotes the affine hull of dom f). Suppose further that (Su,u) is a relative interior point of dom f for some $u \in U$. Then the assertion of 2.1 holds.

Dually, min (S) = - inf (D), if V, Y, f^*, $-S^*$ satisfy the assumptions of 2.1 :

2.3 Theorem : Let U and X be dual spaces of the normed spaces V and Y respectively. Let $S : U \longrightarrow X$ be linear and continuous with respect to the weak V-topology of U and Y-topology of X (DUNFORD/SCHWARTZ (57), p.419), and let $f : X \times U \longrightarrow R \cup \{\infty\}$ be proper convex and lower semicontinuous with respect to these topologies.

2.3.1 : For $y \in Y$ and $v \in V$ the following statements are equivalent :

(a) The sets $\{(x,u) \in X \times U : b \leqslant \langle x,y \rangle + \langle u,v \rangle - f(x,u)\}$ are bounded with respect to the dual norm on $X \times U$ for all $b \in R$.

(b) f^* is finite and continuous on some neighbourhood of (y,v).

2.3.2 : If the statements (a), (b) of 2.3.1 are true for some $y \in Y$ and $v = -S^*y$, then min (S) = - inf (D), and the minimum of (S) is achieved by some $u \in U$.

2.3.3 : If dom f is bounded with respect to the dual norm on $X \times U$, then dom $f^* = Y \times V$, and the assertion of 2.3.2 holds.

Proof :

2.3.1 : The equivalence of (a) and (b) follows from ROCKAFELLAR (66), p.60 .

2.3.2 : Let $(\bar{y},-S^* \bar{y}) \in Y \times V$ satisfy condition (b). Then $f^*(\bar{y},\cdot)$ is bounded above by some real $\bar{b}$ on a neighbourhood of $-S^* \bar{y}$. Therefore, the set $\{ v + S^* y : (y,v) \in Y \times V, f^*(y,v) \leqslant \bar{b} \}$ contains the origin of Y as an interior point. Hence V, Y, f^* , $-S^*$ satisfy the assumptions of 2.1 .

2.3.3 : If dom f is bounded, then condition (a) is true for all $(y,v) \in Y \times V$.

Remark : The assumption that dom f is bounded does not seem too restrictive in practical problems, since quantities appearing in reality do not become arbitrarily large, in general. Therefore, often it will be possible to add artificial constraints without changing the optimal solutions of (S) but making dom f bounded.

3. $\mathcal{E}$ - Subgradients

3.1 D e f i n i t i o n (ROCKAFELLAR (69 a), p.219) : Let U be a linear topological space and V its dual space. Let $I : U \longrightarrow R \cup \{\infty\}$ be proper convex. Let $\bar{u} \in U$ and $\mathcal{E} \geqslant 0$. $v \in V$ is called "$\mathcal{E}$ - subgradient" of I at $\bar{u}$, if

$$I(\bar{u}) + \langle u - \bar{u}, v \rangle \leqslant I(u) + \mathcal{E} \quad \text{for all} \quad u \in U .$$

The set of all $\mathcal{E}$ - subgradients of I at $\bar{u}$ is denoted by $\partial_{\mathcal{E}} I(\bar{u})$. If $\mathcal{E} = 0$, we say "subgradient" instead of 0 - subgradient, and the set of all subgradients of I at $\bar{u}$ is denoted by $\partial I(\bar{u})$.

The concept of $\mathcal{E}$ - subgradients gets more and more important in nondifferentiable convex minimization (see PSHENICHNIY (65 a), (65 b), LEVITIN (69), BERTSEKAS/MITTER (73), LEMARECHAL (74 a), (74 b), WOLFE (74 a), (74 b)). For later use we will characterize the $\mathcal{E}$ - subgradients of functionals I, J as considered in 1. and 2. and related functionals.

Let I be defined as in 1., let $\bar{u} \in U$ and $\mathcal{E} \geqslant 0$. By definition $w \in \partial_{\mathcal{E}} I(\bar{u})$ if and only if

$$f(S\bar{u},\bar{u}) - \langle \bar{u}, w \rangle \leqslant f(Su,u) - \langle u, w \rangle + \mathcal{E} \quad \text{for all } u \in U .$$

Consider the following optimization problems :

(S_w) Minimize $f(Su,u) - \langle u,w \rangle$ subject to $u \in U$

(D_w) Minimize $f^*(y, w - S^*y)$ subject to $y \in Y$.

(D_w) is the dual problem of (S_w). If the assumptions of 2.1 resp. 2.2 are satisfied, then $\inf (S_w) = - \min (D_w)$. Therefore, $w \in \partial_{\mathcal{E}} I(\bar{u})$ if and only if

$$f(S\bar{u},\bar{u}) - \langle \bar{u},w \rangle \leqslant \inf (S_w) + \mathcal{E} = - \min (D_w) + \mathcal{E} .$$

However, this inequality is valid if and only if there is a y such that

$$f(S\bar{u},\bar{u}) - \langle \bar{u},w \rangle \leqslant - f^*(y,w - S^*y) + \mathcal{E} .$$

Substituting $v = w - S^*y$ we get the equivalent statement

$$f(S\bar{u},\bar{u}) + f^*(y,v) \leqslant \langle \bar{u}, v + S^*y \rangle + \mathcal{E} = \langle S\bar{u}, y \rangle + \langle \bar{u}, v \rangle + \mathcal{E} .$$

By definition of $\partial_{\mathcal{E}} f(S\bar{u},\bar{u})$ the last statement is equivalent to $(y,v) \in \partial_{\mathcal{E}} f(S\bar{u},\bar{u})$. Therefore, we have proved :

3.2 T h e o r e m : Let the assumptions of 2.1 resp. 2.2 be satisfied. Let $u \in U$ and $\mathcal{E} \geqslant 0$. Then $w \in \partial_{\mathcal{E}} I(u)$ if and only if there is a pair (y,v) such that $(y,v) \in \partial_{\mathcal{E}} f(Su,u)$, and $w = v + S^*y$.

R e m a r k : The representation of $\mathcal{E}$ - subgradients of I, given in 3.2, holds whenever $\inf (S_w) = - \min (D_w)$ for all $w \in V$.

We will give two corollaries of 3.2 concerning the functional J and some related functional that occurs in problem (D) when treated by Ritz' method :

3.3 Corollary : Let the assumptions of 2.3.2 be satisfied. Let $y \in Y$ and $\varepsilon \geq 0$. Then $z \in \partial_\varepsilon J(y)$ if and only if there is a pair (x,u) such that $(x,u) \in \partial_\varepsilon f^*(y,-S^*y)$, and $z = x - Su$.
Particularly $z \in \partial J(y)$ if and only if $z = \bar{x} - S\bar{u}$ where the functional $f(x,u) - \langle x,y\rangle + \langle u,S^*y\rangle$ achieves its minimum in $(\bar{x},\bar{u})$.

Remark : If J is finite and continuous in y, then $\partial J(y) \neq \emptyset$ (see ROCKAFELLAR (69 a),p.217).

3.4 Corollary : Let the assumptions of 2.3.2 be satisfied. Let M be a finite dimensional linear subspace of Y and $\{y_1, \dots ,y_n\}$ a basis of M. Assume that M contains an interior point of dom J. For all real n-vectors $a = (a_1, \dots ,a_n)^T$ let

$$g(a) := J\left(\sum_{i=1}^{n} a_i y_i \right) .$$

Let $a \in R^n$ and $\varepsilon \geq 0$. Then $b \in \partial_\varepsilon g(a)$ if and only if there is an ε- subgradient z of J at $\sum_{i=1}^{n} a_i y_i$ such that $b_i = \langle z,y_i\rangle$ $(i = 1, \dots ,n)$, $b = (b_1, \dots ,b_n)^T$.

4. Dual Methods

If the assumptions of 2.3.2 are satisfied, problem (D) may be easier to be treated than problem (S). Especially, if the assumptions of 2.3.3 are satisfied, the constraints of (S) may be very complicated, while there are no constraints in problem (D). Hence we will apply some minimization methods to problem (D). Nevertheless, we are interested in solving problem (S). That is why we have to look for optimization methods that yield approximations to solutions of problem (S), too :
We will apply Ritz' method to problem (D). Any finite dimensional auxiliary problem, which will occur, we will treat by some known minimization procedure doing the following :

4.1 Algorithm : Let $g : R^n \longrightarrow R$ be convex and bounded below. Assume that, given $a \in R^n$, we have a finite process which will find one subgradient of g at a. Suppose further that there is a constant C such that $|b| \leq C$ for all subgradients b of g ($|\ |$ denotes the Euclidean norm in R^n). Then, given $\gamma > 0$, $\varepsilon > 0$ and $a^o \in R^n$, the algorithm terminates after a finite number of steps having constructed $a \in R^n$ and $b \in \partial_\varepsilon g(a)$ such that $g(a) \leq g(a^o)$, and $|b| \leq \gamma$.

Remark : The "Method of Conjugate Subgradients" (WOLFE (74 a), (74 b)) is an algorithm which has the desired properties. If g satisfies some additional conditions, then other known algorithms are applicable, too (see e.g. LEVITIN (69), LEMARE-chal (74 a), (74 b)).

Algorithm 4.1 will be applied to functions occuring in Ritz' method :

4.2 L e m m a : Let the assumptions of 2.3.3 be satisfied. Let $I(u) < \infty$ for some $u \in U$. Let M be a finite dimensional linear subspace of Y and $\{y_1, \dots, y_n\}$ a basis of M. For all real n-vectors $a = (a_1, \dots, a_n)^T$ let

$$g(a) := J\Big(\sum_{i=1}^{n} a_i y_i\Big) \quad .$$

Assume that, given $y \in M$, we have a finite process which will find $(\bar{x}, \bar{u})$ minimizing the functional $f(x,u) - \langle x,y \rangle + \langle u, S^* y \rangle$. Then 4.1 is applicable to g and yields, given $\gamma > 0$, $\varepsilon > 0$ and $y^o \in M$, elements $y \in M$, $u \in U$, $z \in X$ such that

$$J(y) \leqslant J(y^o) \tag{4.1}$$

$$f(z + Su, u) + f^*(y, -S^* y) \leqslant \langle z,y \rangle + \varepsilon \tag{4.2}$$

$$\| z \|_M \leqslant \gamma \tag{4.3}$$

where $\| z \|_M := \sup \{ | \langle z,y \rangle | : y \in M, \| y \| \leqslant 1 \}$.

Proof :
Consider the representation of subgradients and ε- subgradients of g, given in 3.3, 3.4 . The applicability of 4.1 to g follows immediately ($-\min(S) > -\infty$ is a lower bound for g). Therefore, 4.1 yields $a \in R^n$, $y = \sum_{i=1}^{n} a_i y_i$, $(x,u) \in \partial_\varepsilon f^*(y, -S^* y)$,

$z = x - Su$, $b_i = \langle z, y_i \rangle$ such that $| b | = \Big(\sum_{i=1}^{n} b_i^{\,2} \Big)^{1/2} \leqslant \bar{\gamma}$, and $J(y) \leqslant J(y^o)$.

$(x,u) \in \partial_\varepsilon f^*(y, -S^* y)$ implies (4.2). If b_i tends to zero ($i = 1, \dots, n$), so does $\| z \|_M$. Therefore, choosing $\bar{\gamma}$ sufficiently small, we get 4.3 .

R e m a r k : If we use Wolfe's method in 4.1, then (x,u) can be effectively constructed as convex combination of subgradients of f^*.

As a result of 4.2 we are able to treat problem (D) by the following procedure :

4.3 D u a l R i t z M e t h o d : Let the assumptions of 2.3.3 be satisfied. Let $I(u) < \infty$ for some $u \in U$. Let Y be separable. - Choose a sequence $\{ M_n : n \in N \}$ of finite dimensional linear subspaces of Y such that $M_n \subset M_{n+1}$ for all n, $\bigcup_{n \in N} M_n$ is dense in Y, and, given $y \in M_n$ ($n \in N$), there is a finite process which will find $(\bar{x}, \bar{u})$ minimizing the functional (the "corresponding Lagrangian", see introduction) $f(x,u) - \langle x,y \rangle + \langle u, S^* y \rangle$. Choose sequences of positive reals γ_n, ε_n converging to zero. Choose $y_o \in M_1$. - For every n by applying 4.1 find $y_n \in M_n$, $u_n \in U$, $z_n \in X$ such that

$$J(y_n) \leqslant J(y_{n-1}) \tag{4.4}$$

$$f(z_n + Su_n, u_n) + f^*(y_n, -S^*y_n) \leqslant \langle z_n, y_n \rangle + \varepsilon_n \tag{4.5}$$

$$\|z_n\|_{M_n} \leqslant \gamma_n \quad . \tag{4.6}$$

R e m a r k : Whether it is possible or not to minimize the functionals $f(x,u) - \langle x,y \rangle + \langle u,S^*y \rangle$ depends on the concrete problem (S) to be solved. However, minimizing that functionals appears to be much more easier than solving problem (S), because the constraint $x = Su$ needs not to be satisfied.

If it would be possible to get y_n, u_n, z_n satisfying (4.5), (4.6) such that $M_n = Y$ and $\gamma_n = \varepsilon_n = 0$, then from (4.5), (4.6) it would follow that $z_n = 0$, $f(Su_n,u_n) + f^*(y_n,-S^*y_n) \leqslant 0$, and, therefore, by 1.1 (S) and (D) would achieve their minima in u_n and y_n respectively, However, we are getting this result only approximately, in general :

4.4 T h e o r e m : Let the assumptions of 4.3 be satisfied. Assume that the sequence $\{y_n\}$ generated in 4.3 is bounded. Then $\{y_n\}$ is a minimizing sequence of problem (D), i.e. $J(y_n)$ converges to inf (D) . $\{u_n\}$ contains a weakly convergent subsequence, and every weakly convergent subsequence of $\{u_n\}$ converges weakly to some $u \in U$ in which (S) achieves its minimum.

Proof :

a) We will prove $\{z_n\}$ converging weakly to 0 : Let $y \in Y$, $\varepsilon > 0$. Since dom f is bounded, so is $\{z_n\}$. Let $K > 0$ such that $\|z_n\| \leqslant K$ $(n \in N)$. Choose n_o, $\bar{y} \in M_{n_o}$ such that

$$\|y - \bar{y}\| \leqslant \varepsilon/(2K) \quad .$$

Choose $n_1 \geqslant n_o$ large enough to have

$$\gamma_n \leqslant \varepsilon/(2\|\bar{y}\|) \quad \text{whenever} \quad n \geqslant n_1 \ .$$

Therefore, if $n \geqslant n_1$,

$$|\langle z_n,y \rangle| \leqslant |\langle z_n, y - \bar{y} \rangle| + |\langle z_n,\bar{y} \rangle| \leqslant \|z_n\| \cdot \|y - \bar{y}\| + \gamma_n\|\bar{y}\| \leqslant \varepsilon/2 + \varepsilon/2 = \varepsilon \ .$$

Hence $\langle z_n,y \rangle \longrightarrow 0$ for all y, that is, z_n converges weakly to 0 .

b) From (4.5) it follows that

$$f(z_n + Su_n,u_n) + f^*(y_n,-S^*y_n) \leqslant \gamma_n\|y_n\| + \varepsilon_n \quad . \tag{4.7}$$

Since y_n is bounded, the right hand side of (4.7) converges to zero. Let

$$f_o^* := \lim_{n \to \infty} f^*(y_n,-S^*y_n) \quad .$$

Since $\{u_n\}$ is bounded, $\{u_n\}$ contains a weakly convergent subsequence. Let $\{u_{n_j}\}$ be a subsequence of $\{u_n\}$ such that $u_{n_j} \to \bar{u}$ weakly. From a) we get :

$$x_{n_j} = z_{n_j} + Su_{n_j} \to S\bar{u} \text{ weakly} .$$

b$_1$) Let $f(S\bar{u},\bar{u}) = \infty$. Since f is weakly lower semicontinuous, the sets

$$B_m := \{(x,u) \in X \times U : f(x,u) > m\}$$

are weakly open for all real m, and $(S\bar{u},\bar{u}) \in B_m$. Therefore, if j is sufficiently large, $(x_{n_j},u_{n_j}) \in B_m$. Hence, $f(x_{n_j},u_{n_j}) \to \infty$, and from (4.7) it foolows that $f_o^* = -\infty$. Therefore, $\min (S) = - \inf (D) = \infty$. But this is not possible, since $I(u) < \infty$ for some $u \in U$.

b$_2$) Hence, $f(S\bar{u},\bar{u})$ must be finite. Using the same argument as in b$_1$), where

$$B_\varepsilon := \{(x,u) \in X \times U : f(x,u) > f(S\bar{u},\bar{u}) - \varepsilon\} \quad (\varepsilon > 0)$$

takes the place of B_m, we get $f(S\bar{u},\bar{u}) - \varepsilon < f(x_{n_j},u_{n_j})$, if j is sufficiently large. Therefore, from (4.7) it follows that

$$f(S\bar{u},\bar{u}) - \varepsilon < \gamma_{n_j} \|y_{n_j}\| + \varepsilon_{n_j} - f^*(y_{n_j},-S^*y_{n_j}) ,$$

and when j tends to infinity :

$$f(S\bar{u},\bar{u}) - \varepsilon \leq - f_o^* \quad \text{for all positive reals } \varepsilon .$$

Hence, $f(S\bar{u},\bar{u}) \leq - f_o^* \leq - \inf (D) = \min (S)$. Therefore, $f_o^* = \inf (D)$, and (S) achieves its minimum in $\bar{u}$.

4.5 C o r o l l a r y : Let the assumptions of 4.3 be satisfied.

4.5.1 : Assume that there are real constants $c > 0$, d such that

$$J(y) \geq c\|y\| - d \quad (y \in Y) .$$

Then the sequence $\{y_n\}$ is bounded, and the assertion of 4.4 holds.

4.5.2 : Assume that there is a real constant d such that the origin of X is an interior point of $A_d := \{x - Su : (x,u) \in X \times U, f(x,u) \leq d\}$ with respect to the norm topology of X (see 2.1). Then the assumption of 4.5.1 is satisfied, and the assertion of 4.4 holds.

Proof :

From the assumption of 4.5.1 and (4.7) it follows that

$$c\|y_n\| - d \leq J(y_n) \leq \gamma_n \|y_n\| + \varepsilon_n - m ,$$

where m is a lower bound for f (such a bound exists, since dom f is bounded). Therefore,

$$(c - \gamma_n)\|y_n\| \leq d - m + \varepsilon_n .$$

If n is sufficiently large, we get $(c/2)\|y_n\| \leq d - m + 1$, and the assertion of 4.5.1 follows.

From the assumption of 4.5.2 it follows that there is a constant $c > 0$ such that

$z \in A_d$ whenever $\|z\| \leqslant c$. Then, by definition of f^* ,

$$J(y) = f^*(y,-S^*y) = \sup_{(x,u)\in X\times U} [\langle x - Su,y\rangle - f(x,u)] \geqslant \sup_{\substack{(x,u)\in X\times U\\ f(x,u)\leqslant d}} [\langle x - Su,y\rangle - d]$$

$$\geqslant \sup_{\|z\|\leqslant c} \langle z,y\rangle - d = c\|y\| - d \qquad (y\in Y) .$$

5. Ordinary Control Problems

We consider an ordinary convex control problem :

$$(S_1)\begin{cases} \text{Minimize} \quad h(x(0),x(T)) + \int_0^T g(t,x(t),u(t))\,dt \quad \text{subject to} \\ \dot{x} = A(t)x + B(t)u \quad \text{in } [0,T] , \\ x : [0,T] \longrightarrow R^n \quad \text{absolutely continuous,} \\ u : [0,T] \longrightarrow R^m \quad \text{measurable, essentially bounded} \end{cases}$$

where $A(t)$ and $B(t)$ are (n,n)- and (n,m)-matrices respectively, the coefficients of which are integrable real functions with domain $[0,T]$ ($T > 0$ fixed),
$h : R^{2n} \longrightarrow R \cup \{\infty\}$ is a lower semicontinuous proper convex functional, and
$g : [0,T]\times R^{n+m} \longrightarrow R \cup \{\infty\}$ is a normal convex integrand in the sense of ROCKAFELLAR (68). h and g may attain the value ∞ . Therefore, as in problem (S), constraints for x and u are allowed.
Let $U := R^n\times L^\infty_m[0,T]$, $X := Q_n[0,T]\times R^n$ where $Q_n[0,T]$ denotes any linear function space of measurable functions containing all absolutely continuous functions $x : [0,T] \longrightarrow R^n$. Let

$$S(\xi_0,u) := (x,x(T)) \qquad ((\xi_0,u)\in U)$$

where x denotes the unique solution of

$$\dot{x} = A(t)x + B(t)u , \quad x(0) = \xi_0 \quad .$$

Let

$$f(x, \xi_1, \xi_0,u) := h(\xi_0, \xi_1) + \int_0^T g(t,x(t),u(t))\,dt \quad ((\xi_0,u)\in U,(x, \xi_1)\in X) \tag{5.1}$$

Then (S_1) has the form (S), if we define suitable topologies on U and X. - If $Q_n[0,T]$ is the space of all continuous n-vector valued functions, i.e. $X = C_n[0,T]\times R^n$, and if X and U are supplied with the norm topologies of $\| \ \|_\infty$, then, using some generalizations of results of ROCKAFELLAR (71), from 2.1 and 2.2 we get duality theorems for (S_1) as in ROCKAFELLAR (72) or, a little more general, in ORTLIEB (75) . We will not follow this way but guide ourselves along 2.3 . Therefore, let $X := L^\infty_n[0,T]\times R^n$, $Y := L^1_n[0,T]\times R^n$, $V := R^n\times L^1_m[0,T]$. $-S^* : Y \longrightarrow V$ is defined by :

$-S^*(y, \eta_1) = (\eta_0,v)$ if and only if $v(t) = B^T(t)p(t)$ a. e. , $\eta_0 = p(0)$,

where p denotes the unique solution of $\dot{p} = -A^T(t)p + y$, $p(T) = -\eta_1$.
If there are functions $(\bar{x},\bar{u})\in L_{n+m}[0,T]$ and $(\bar{y},\bar{v})\in L^1_{n+m}[0,T]$ such that $g(t,\bar{x}(t),\bar{u}(t))$ and $g^*(t,\bar{y}(t),\bar{v}(t))$ are integrable, then

$$f^*(y,\eta_1,\eta_0,v) = h^*(\eta_0,\eta_1) + \int_0^T g^*(t,y(t),v(t))\,dt \quad ((\eta_0,v)\in V,(y,\eta_1)\in Y) \tag{5.2}$$

(see ROCKAFELLAR (68), p.532). Therefore, we get the following dual problem of (S_1) :

$$(D_1)\begin{cases} \text{Minimize} \quad h^*(p(0),-p(T)) + \int_0^T g^*(t,y(t),B^T(t)p(t))\,dt \quad \text{subject to} \\ \qquad \dot{p} = -A^T(t) + y \quad \text{in } [0,T] \;, \\ p : [0,T] \longrightarrow R^n \quad \text{absolutely continuous,} \\ y : [0,T] \longrightarrow R^n \quad \text{integrable} \end{cases}$$

(see ROCKAFELLAR (70)) .

It is straightforward to get sufficient conditions such that (S_1), (D_1) satisfy the assumptions of 2.3.3 , for example. More general duality theorems concerning ordinary control problems of that kind are to be found in ROCKAFELLAR (70).
Applying the method given in 4. to (S_1), (D_1) it is of great importance to know the subgradients of the cost criterion of (D_1) :

5.1 **Lemma** : Let (S_1), (D_1) satisfy the assumptions of 2.3.3 . If $(y,\eta_1)\in Y$, the cost criterion of (D_1) is

$$J(y,\eta_1) = h^*(p(0),-p(T)) + \int_0^T g^*(t,y(t),B^T(t)p(t))\,dt$$

where $\dot{p} = -A^T(t)p(t) + y$, $p(T) = -\eta_1$. We get all subgradients (z,ζ) of J at (y,η_1) by finding $\bar{\xi}_0, \bar{\xi}_1 \in R^n$, $\bar{x}\in L_n[0,T]$, $\bar{u}\in L_m[0,T]$ such that $\bar{\xi}_0$, $\bar{\xi}_1$ minimize $h(\xi_0,\xi_1) - p^T(0)\xi_0 + p^T(T)\xi_1$, and

$$g(t,\bar{x}(t),\bar{u}(t)) - y^T(t)\bar{x}(t) - p^T(t)B(t)\bar{u}(t) = \min_{x,u}\,[\,g(t,x,u) - y^T(t)x + - p^T(t)B(t)u\,] \quad \text{a. e.} \;,$$

by determining the unique solution q of

$$\dot{q} = A(t)q + B(t)u \;, \quad q(0) = \bar{\xi}_0 \;,$$

and setting $z = \bar{x} - q$, $\zeta = \bar{\xi}_1 - q(T)$.

Hence, whether it is possible or not to find a subgradient of the dual cost criterion (see remark following 4.3) depends on the possibility of evaluating a kind of minimum principle and generalized transversality condition for (S).

6. Control Problems Governed by Partial Differential Equations

Optimal control problems governed by linear partial differential equations with convex cost criterion can be written in the form of problem (S), in general. Applying the method of 4. to such problems the main difficult is to get a representation of the control operator S (and its adjoint S^*) , which is easy to evaluate. For several control problems such a representation is known (see e.g. YEGOROV (69), BUTKOVSKIY (69),

GLASHOFF (75), GASHOFF/GUSTAFSSON (75), KRABS (75), SACHS (75) and some articles in this volume). Hence, the application of dual methods is possible for these problems.

7. Numerical Considerations

We were programming a dual method on a digital computer for several ordinary control problems and control problems of heat diffusion (GLASHOFF/GUSTAFSSON (75)) using Wolfe's method as algorithm 4.1 . All problems treated have nondifferentiable convex cost functionals. The speed of convergence observed thereby was different for different examples but very high for none of them. These experience do not come as a surprise, because, generally speaking, a dual method cannot be better than the minimization method used in it for solving auxiliary problems, and all known minimization methods for nondifferentiable convex cost functionals converge slowly indeed, particularly if the minimization problem has high dimension. Therefore, the success of dual methods depends strongly on a good choice of the subspaces M_n (see 4.3).

References

D.P. BERTSEKAS / S.K. MITTER (73) A Descent Numerical Method for Optimization Problems with Nondifferentiable Cost Functionals, SIAM J. Contr. 11, 637 - 653

A. BRØNDSTED (64) Conjugate Convex Functions in Topological Vector Spaces, Mat.-Fys. Medd. Danske Vid. Selsk. 34, 1 - 27

A.G. BUTKOVSKIY (69) Distributed Control Systems, Elsevier, New York - London - Amsterdam

U. DIETER (66) Optimierungsaufgaben in topologischen Vektorräumen I. Dualitätstheorie, Zeitschr. Wahrscheinl. verw. Gebiete 5, 89 - 117

N. DUNFORD / J.T. SCHWARTZ (57) Linear Operators. Part I. General Theory, Interscience, New York

K. GLASHOFF (75) Optimal Control of One Dimensional Linear Parabolic Differential Equation, to appear in BULIRSCH/OETTLI/STOER (eds.) Optimierungstheorie und optimale Steuerungen, Springer Lecture Notes in Mathematics 477, Berlin - Heidelberg - New York

K. GLASHOFF / S.-A. GUSTAFSSON (75) Numerical Treatment of a Parabolic Boundary-Value Control Problem, to appear in J. Opt. Th. Appl.

G. KÖTHE (66) Topologische lineare Räume, 2. Aufl., Springer, Berlin - Heidelberg - New York

W. KRABS (75) Ein Kontroll-Approximationsproblem für die schwingende Saite, Techn. Hochschule Darmstadt, Preprint 213

C. LEMARECHAL (74 a) An Extension of Davidon Methods to Nondifferentiable Functions, submitted to Math. Prog.

C. LEMARECHAL (74 b) Note on an Extension of Davidon Methods to Nondifferentiable Functions, Math. Prog. 7, 384 - 387

F. LEMPIO (71) Lineare Optimierung in unendlichdimensionalen Vektorräumen, Computing 8, 284 - 290

E.S. LEVITIN (69) A General Minimization Method for Unsmooth Extremal Problems, USSR Comp. Math. Math. Phys. 9, No 4, 63 - 93

D.G. LUENBERGER (69) Optimization by Vector Space Methods, Wiley, New York - London

D.G. LUENBERGER (73) Introduction to Linear and Nonlinear Programming, Addison-Wesley, Reading, Massachusetts

C.P. ORTLIEB (75) Konvexe Steuerungsprobleme mit Zustandsrestriktionen, submitted to Zeitschr. Ang. Math. Mech.

B.N. PSHENICHNIY (65 a) A Duality Principle for Convex Programming Problems, USSR Comp. Math. Math. Phys. 5, No 1, 131 - 143

B.N. PSHENICHNIY (65 b) Dual Methods in Extremal Problems I + II, Cybernetics 1, No 3, 91 - 99, No 4, 72 - 79

R.T. ROCKAFELLAR (66) Level Sets and Continuity of Conjugate Convex Functions, Trans. Amer. Math. Soc. 123, 46 - 63

R.T. ROCKAFELLAR (67) Duality and Stability in Extremum Problems.Involving Convex Functions, Pac. J. Math. 21, 167 - 187

R.T. ROCKAFELLAR (68) Integrals Which are Convex Functionals, Pac. J. Math. 24, 525 - 539

R.T. ROCKAFELLAR (69 a) Convex Analysis, Princeton University Press

R.T. ROCKAFELLAR (69 b) Convex Functions and Duality in Optimization Problems and Dynamics, in KUHN/SZEGÖ (eds.) Mathematical System Theory and Economics I, 117 - 141, Springer Lecture Notes in Operations Research and Mathematical Economics 11, Berlin - Heidelberg - New York

R.T. ROCKAFELLAR (70) Conjugate Convex Functions in Optimal Control and the Calculus of Variations, J. Math. Anal. Appl. 32, 174 - 222

R.T. ROCKAFELLAR (71) Integrals Which are Convex Functionals II, Pac. J. Math. 39, 439 - 469

R.T. ROCKAFELLAR (72) State Constraints in Convex Control Problems of Bolza, SIAM J. Contr. 10, 691 - 715

E. SACHS (75) Computation of Bang-Bang-Controls and Lower Bounds for a Parabolic Boundary-Value Control Problem, Techn. Hochsch. Darmstadt, Preprint 212

P. WOLFE (74 a) A Method of Conjugate Subgradients for Minimizing Nondifferentiable Functions, IBM Res. Rep. No RC 4845, Yorktown Heights, New York

P. WOLFE (74 b) Note on a Methods of Conjugate Subgradients for Minimizing Nondifferentiable Functions, Math. Prog. 7, 380 - 383

Y.V. YEGOROV (63) Some Problems in the Theory of Optimal Control, USSR Comp. Math. Math. Phys. 3, No 5, 1209 - 1232

A SUBGRADIENT ALGORITHM FOR SOLVING K-CONVEX INEQUALITIES

Stephen M. Robinson
Mathematics Research Center
University of Wisconsin-Madison
610 Walnut Street
Madison, WI 53706/USA

1. Introduction

In this paper we discuss an algorithm for numerical solution of a system of K-convex inequalities

$$f(x) \leqq_K 0 \qquad (1)$$
$$x \in C ,$$

where $f: D \subset \mathbb{R}^n \to \mathbb{R}^m$, C is a nonempty closed convex set in $\mathbb{R}^n$, K is a nonempty closed convex cone in $\mathbb{R}^m$, and where we write $y_1 \leqq_K y_2$ if $y_2 - y_1 \in K$. Under assumptions of convexity and regularity of f, we show that the algorithm converges, from an arbitrary starting point in C, to a solution of (1) at a rate which is at least linear. The algorithm requires the computation of a subgradient of (1) and the solution of a projection problem (a convex quadratic minimization problem) at each step. With additional differentiability assumptions on f, much faster convergence (e.g., quadratic) can be expected. The algorithm is an extension of a method proposed by Oettli [3], which in turn is related to earlier works such as those of Polyak [4] and Eremin [1], among others.

2. Assumptions; statement of the algorithm

The following assumptions will be made throughout the paper:

I) The set C is contained in the interior of the domain of definition D of the function f.

II) f is a non-vacuous closed K-convex function; that is, the set

$$\{(x, y) \mid x \in D, \ f(x) \leqq_K y\}$$

is nonempty, closed and convex in $\mathbb{R}^{n+m}$.

Sponsored by the United States Army under Contract No. DAAG29-75-C-0024 and by the National Science Foundation under Grant No. DCR74-20584.

III) The origin in $\mathbb{R}^m$ is a regular value of the multivalued function Φ from $\mathbb{R}^n$ to $\mathbb{R}^m$ given by

$$\Phi(x) := \begin{cases} f(x) + K, & x \in C \\ \emptyset, & x \notin C \end{cases} ;$$

that is, the projection of the graph of Φ into $\mathbb{R}^m$ covers some neighborhood of 0. (See [6] for more information about regular values of multivalued functions and for other results which we shall use in what follows.)

IV) The function f is subdifferentiable at every point $x \in C$: that is, for every such x there exists an $m \times n$ matrix $F'(x)$ (not necessarily unique) such that for each $y \in D$, $f(x) + F'(x)(y-x) \leqq_K f(y)$. Additional information about subdifferentials of K-convex functions can be found in [2].

With these assumptions in mind, we can state the algorithm as follows:

Step 0: Select any $x_0 \in C$ as a starting point; set $k := 0$.

Step 1: Having k and x_k, compute a subgradient $F'(x_k)$ of f at x_k; choose x_{k+1} to solve the convex quadratic minimization problem $\min\{\|x-x_k\| \mid f(x_k) + F'(x_k)(x-x_k) \leqq_K 0,\ x \in C\}$, where $\|\cdot\|$ is the Euclidean norm.

Step 2: If x_{k+1} belongs to the set

$$Z := \{x \in C \mid f(x) \leqq_K 0\},$$

or if x_{k+1} is otherwise satisfactory, stop; if not, set $k := k+1$ and go to Step 1.

It will be noted that this is a kind of "sublinearization" algorithm; if $C = D = \mathbb{R}^n$ and if f is differentiable then it reduces to the generalized Newton method for inequalities discussed by the author in [5]; local quadratic convergence was proved there under certain regularity assumptions. Since we do not assume differentiability here, we cannot expect to maintain the quadratic rate of convergence; however, we shall prove global linear convergence, and of course the quadratic rate can be expected to hold when the stronger hypotheses of [5] are met.

3. Convergence and rate of convergence

The convergence properties of the sequence $\{x_k\}$ produced by the algorithm are described in the following theorem.

THEOREM: Suppose Assumptions I-IV are satisfied. Then for any $x_0 \in C$ the algorithm given here, unless terminated satisfactorily in Step 2, yields an infinite sequence $\{x_k\}$ which converges to a point $x_\infty \in Z$. The rate of convergence is at least linear: that is, there exists some $\mu \in [0,1)$ such that for each k, $\|x_\infty - x_{k+1}\| \leqq \mu \|x_\infty - x_k\|$.

PROOF: We first prove that for a given $x_k \in C$, there is a point $x_{k+1} \in C$ satisfying the prescription given in Step 1 of the algorithm. To begin with, Assumption III implies that Z is nonempty; since $x \in Z$ if and only if $0 \in \Phi(x)$, we see that Z is closed and convex. For the given x_k, Assumption IV implies that $F'(x_k)$ exists, and the set $L_k := \{x \in C \mid f(x_k) + F'(x_k)(x-x_k) \leqq_K 0\}$ is easily shown to be nonempty (it contains Z) as well as closed and convex. It follows from elementary considerations that x_{k+1} exists as claimed. This shows that the algorithm is well defined and that, unless terminated satisfactorily in Step 2, it will produce an infinite sequence $\{x_k\}$. We assume now that such a sequence is generated, and proceed to investigate its properties.

As x_{k+1} is the closest point to x_k in L_k, we have by a well-known property of convex sets that for each point of L_k and, in particular, for each point $z \in Z$,

$$\langle z - x_{k+1}, x_k - x_{k+1} \rangle \leqq 0 ,$$

where $\langle \cdot , \cdot \rangle$ denotes the inner product. As it is an identity that

$$\|x_{k+1} - z\|^2 - \|x_k - z\|^2 + \|x_{k+1} - x_k\|^2 = 2\langle z - x_{k+1}, x_k - x_{k+1} \rangle ,$$

we see that for each $z \in Z$ and each k,

$$\|x_{k+1} - z\|^2 \leqq \|x_k - z\|^2 - \|x_{k+1} - x_k\|^2 . \tag{2}$$

It follows that $\{x_k\}$ is a Fejér sequence [1] with respect to Z (since by the assumption that $\{x_k\}$ was infinite we see that $x_k \notin Z$ and thus $x_{k+1} \neq x_k$). Evidently $\{x_k\}$

is bounded, and thus has a point of accumulation x_∞; we shall prove that $x_\infty \in Z$, which together with (2) will suffice to prove that the entire sequence $\{x_k\}$ converges to x_∞.

At this point we introduce the dual cone of K, written

$$K^* := \{y \in \mathbb{R}^m \mid \langle y, k\rangle \geqq 0 \text{ for each } k \in K\} ;$$

this is a closed convex cone in $\mathbb{R}^m$, and it is well known that any vector $w \in \mathbb{R}^m$ can be written uniquely as $w = w_+ + w_-$, with $w_+ \in K^*$, $w_- \in -K$, and $\langle w_+, w_-\rangle = 0$. We can now establish a lemma which will be useful for proving that $x_\infty \in Z$.

LEMMA: Suppose that f is K-Lipschitzian with constant L on some set $H \subset D$, in the sense of [6]: that is, for each $x, y \in H$ one has

$$f(x) - f(y) \in L\|x-y\| I_K ,$$

where

$$I_K := (B+K) \cap (B - K)$$

and B is the unit ball in $\mathbb{R}^m$. Let $x \in \text{int } H$; then for each subgradient $F'(x)$ and each $y \in K^*$,

$$\|y F'(x)\| \leqq L\|y\| .$$

PROOF of Lemma: Let v be any vector of unit length in $\mathbb{R}^n$, and let $\varepsilon > 0$ be small enough so that $x + \varepsilon v \in H$. We have

$$f(x+\varepsilon v) - f(x) \in L \varepsilon I_K , \tag{3}$$

but also

$$f(x) + F'(x)(\varepsilon v) \leqq_K f(x+\varepsilon v) ,$$

so that for some $k_0 \in K$,

$$f(x+\varepsilon v) - f(x) = F'(x)(\varepsilon v) + k_0 .$$

Now by (3) we have

$$f(x + \varepsilon v) - f(x) = L\varepsilon(b_1 - k_1)$$

where $b_1 \in B$ and $k_1 \in K$. As $y \in K^*$, we find that

$$\langle y, b_1 - k_1\rangle = \langle y, b_1\rangle - \langle y, k_1\rangle \leqq \langle y, b_1\rangle \leqq \|y\|$$

so that

$$\langle y, f(x+\varepsilon v) - f(x)\rangle \leqq L\varepsilon \|y\| \quad .$$

But

$$\begin{aligned} \langle y F'(x), v\rangle &= \varepsilon^{-1}\langle y, F'(x)(\varepsilon v)\rangle \\ &= \varepsilon^{-1}[\langle y, f(x+\varepsilon v) - f(x)\rangle - \langle y, k_0\rangle] \\ &\leqq \varepsilon^{-1}(L\varepsilon \|y\|) = L\|y\| \quad , \end{aligned}$$

and since v was arbitrary in the unit ball of $\mathbb{R}^n$ we have $\|y F'(x)\| \leqq L\|y\|$, which completes the proof of the lemma.

Returning to the proof of the theorem, we select a subsequence $\{x_{k_i}\}$ converging to x_∞; as C is closed we certainly have $x_\infty \in C$. But then by Assumption I, $x_\infty \in$ int D, so by Assumption II and by Theorem 5 of [6], we have that f is K-Lipschitzian with Lipschitz constant L on some neighborhood U about x_∞. With no loss of generality we can suppose that $\{x_{k_i}\} \subset U$. Now for any k, we have

$$f(x_k) + F'(x_k)(x_{k+1} - x_k) \leqq_K 0 \quad ,$$

so

$$\begin{aligned} \|f(x_k)_+\|^2 &= \langle f(x_k)_+, \ f(x_k)_+ + f(x_k)_-\rangle \\ &= \langle f(x_k)_+, f(x_k) + F'(x_k)(x_{k+1} - x_k)\rangle \\ &\quad - \langle f(x_k)_+, F'(x_k)(x_{k+1} - x_k)\rangle \end{aligned}$$

$$\leq - \langle f(x_k)_+, F'(x_k)(x_{k+1} - x_k) \rangle$$

$$\leq \| f(x_k)_+ F'(x_k) \| \, \| x_{k+1} - x_k \| .$$

If $x_k \in U$, then by the lemma we find that

$$\| f(x_k)_+ F'(x_k) \| \leq L \| f(x_k)_+ \| ,$$

and since $f(x_k)_+ \neq 0$ (since $x_k \notin Z$), we obtain

$$\| f(x_k)_+ \| \leq L \| x_{k+1} - x_k \| . \tag{4}$$

One has easily from (2) that $\| x_{k+1} - x_k \| \xrightarrow[k \to \infty]{} 0$, so for the subsequence $\{x_{k_i}\}$ it follows that

$$\lim_{i \to \infty} \| f(x_{k_i})_+ \| = 0 .$$

However, for each i we have $f(x_{k_i})_+ = f(x_{k_i}) - f(x_{k_i})_-$, and since $f(x_{k_i})_- \in -K$ the pairs $(x_{k_i}, f(x_{k_i})_+)$ belong to the graph of Φ. Assumption II, taken together with the fact that C is closed, implies that Φ is closed, so we must have $0 \in \Phi(x_\infty)$: that is $x_\infty \in Z$. It now follows from our earlier remarks that the entire sequence $\{x_k\}$ converges to x_∞. It remains to establish its rate of convergence.

We now recall that by Assumption III the origin in $\mathbb{R}^m$ is a regular value of Φ; this fact, together with the fact that $\{x_k\}$ is a bounded sequence, enables us to apply Theorems 1 and 2 of [6] to establish the existence of a constant M such that for each k,

$$d[x_k, Z] \leq M \, d[0, \Phi(x_k)] ,$$

where for a point p and a set A,

$$d[p, A] := \inf\{ \| p - a \| \mid a \in A \} .$$

However, we have seen that $f(x_k)_+ \in \Phi(x_k)$, so actually for $x_k \in U$,

$$d[x_k, Z] \leq M \| f(x_k)_+ \| \leq LM \| x_{k+1} - x_k \| ,$$

where we have used (4). We can also see from (2) that for any $z \in Z$ the sequence $\{\|x_k - z\|\}$ is monotonically decreasing; taking the limit we see that for any k, $\|x_\infty - z\| \leqq \|x_k - z\|$. Hence

$$\|x_k - x_\infty\| \leqq \|x_k - z\| + \|z - x_\infty\| \leqq 2\|x_k - z\| ,$$

and taking the infimum over all $z \in Z$ we have for $x_k \in U$,

$$\|x_k - x_\infty\| \leqq 2d[x_k, Z] \leqq 2LM\|x_{k+1} - x_k\| .$$

Using (2) again, we have for such x_k ,

$$\|x_{k+1} - x_\infty\|^2 \leqq \|x_k - x_\infty\|^2 - \|x_{k+1} - x_k\|^2$$

$$\leqq \|x_k - x_\infty\|^2 - (2LM)^{-2}\|x_k - x_\infty\|^2 ,$$

so for all sufficiently large k ,

$$\|x_{k+1} - x_\infty\| \leqq \nu\|x_k - x_\infty\| ,$$

with $\nu := [1 - (2LM)^{-2}]^{\frac{1}{2}} \in [0,1)$, which completes the proof of the theorem.

4. Computational example; remarks

Some computational results for a similar method (for differentiable functions) were given in [5]. Here we present only a simple example chosen to illustrate the convergence of the method from even very bad starting points. The example has $C = D = \mathbb{R}^5$, with $K = \mathbb{R}^4_+$ and $f\colon \mathbb{R}^5 \to \mathbb{R}^4$ given by

$$f_i(x) := \sum_{j=1}^{5} \alpha_{ij} x_j^2 - \beta_i, \quad 1 \leqq i \leqq 4 ;$$

the data are given in Table 1.

The algorithm was started at three "bad" initial points x_0; Table 2 gives these, as well as the number, k, of iterations required to reach a point x_k for which $\|x_{k+1} - x_k\|_\infty < 10^{-6}$.

i	β_i	α_{i1}	α_{i2}	α_{i3}	α_{i4}	α_{i5}
1	1	5	1	1	2	1
2	5	1	6	2	1	1
3	7	1	2	7	1	1
4	2	3	1	2	8	8

Table 1: Coefficients for f .

Starting point no.	$(x_0)_1$	$(x_0)_2$	$(x_0)_3$	$(x_0)_4$	$(x_0)_5$	No. of iterations, k
1	1	1	1	1	1	5
2	1	10	10^2	10^3	10^4	18
3	10^2	10^4	10^5	$-3\cdot 10^5$	10^6	24

Table 2: Starting points and numbers of iterations.

For starting points 2 and 3, the algorithm behaved initially in a characteristically linea fashion; as the iterates converged to the limit the quadratic rate became apparent, as would be expected since the function f is smooth.

In connection with the rate of convergence, it may be of interest to examine some what more closely the difference between the method outlined here and those discussed in [1] and in [4]. Those methods were primarily oriented toward the use of a single functional as an indicator of the degree to which a point x fails to solve the system (1) (or some specialization of it). In particular, in [1] the use of $\theta(x) := \max_i f_i(x)$ is proposed for the case in which $K = \mathbb{R}^m_+$; a similar device is used in [4]. As the computation of the next point is then carried out using information about θ and its derivative (or subgradient), all information about other components, say f_j, of f for which $0 < f_j(x) < \theta(x)$ is disregarded. By sublinearizing the entire system, the method proposed here takes account of such components as well as of those for which the maximum is attained. This additional information is, of course, obtained at the cost of additional computation in that an auxiliary minimization problem must be solved at each stage of the computation. However, the speed of convergence obtained in this way probably justifies the additional work involved (just as in the case of nonlinear equations, for which Newton's method, when usable, is generally preferable to a gradient method such as steepest descent).

In closing, we may remark on some possible extensions of the results presented here. First, it will be noted that except for showing that $Z \neq \emptyset$ we made no use of Assumption III in the proof of convergence of the algorithm; it was employed only in establishing the linear rate of convergence. Accordingly, we could have shown that the algorithm was implementable and convergent under Assumptions I, II and IV alone, with the additional hypothesis that $Z \neq \emptyset$. We have not bothered to do this primarily because of a hesitancy to propose, for any practical use, an algorithm whose rate of convergence is not even linear.

A much more useful extension would be to replace $\mathbb{R}^n$ and $\mathbb{R}^m$ by Hilbert spaces. All of the analysis in the proof of the theorem could be carried out in the Hilbert-space setting <u>except</u> for the crucial use of compactness to show the existence of x_∞. The need for compactness could have been avoided if we had assumed that Z had a nonempty interior; however, this assumption is so strong that we have not thought it of interest to state the result in detail. It would be very desirable to have a way of establishing the existence of x_∞ without requiring such an assumption.

REFERENCES

1. I. I. Eremin, "The relaxation method of solving systems of inequalities with convex functions on the left sides," Soviet Math. Doklady 6(1965), pp. 219-222.

2. V. L. Levin, "Subdifferentials of convex mappings and of compositions of functions," Siberian Math. J. 13(1972), pp. 903-909 (1973).

3. W. Oettli, "An iterative method, having linear rate of convergence, for solving a pair of dual linear programs," Math. Programming 3(1972), pp. 302-311.

4. B. T. Polyak, "Gradient methods for solving equations and inequalities," U.S.S.R. Computational Math. and Math. Phys. 4, #6(1964), pp. 17-32.

5. S. M. Robinson, "Extension of Newton's method to nonlinear functions with values in a cone," Numer. Math. 19(1972), pp. 341-347.

6. ________________, "Regularity and stability for convex multivalued functions," Technical Summary Report No. 1553, Mathematics Research Center, University of Wisconsin-Madison, 1975.

COMPUTATION OF BANG-BANG-CONTROLS AND LOWER BOUNDS FOR A PARABOLIC BOUNDARY-VALUE CONTROL PROBLEM

Ekkehard Sachs
Technische Hochschule Darmstadt
Fb Mathematik, AG 10
Schloßgartenstr. 7
D-6100 Darmstadt
West Germany

0. ABSTRACT

We consider the problem to minimize the deviation of the temperature distribution in a thin rod from a desired distribution. The system is governed by a diffusion equation and controlled by fuel flow which influences the temperature at the ends of the rod. We present a series of approximating problems and prove the convergence of their optimal values to the optimal value of the original problem. A method is given to solve the approximating problems by bang-bang-controls and we obtain upper and lower bounds of the optimal value. Inclusions of the optimal value of the original problem are computed.

1. INTRODUCTION

In this paper we consider the following one-dimensional heat-conducting system:
We study the temperature in a thin rod, which can be heated symmetrically at both end-points. The problem consists of minimizing the deviation of the temperature distribution in the rod from the desired distribution at a given time T.
With some simplifications we describe this problem mathematically in the following manner:
Let $z(x,t)$ be the *temperature distribution* in the rod depending on the space coordinate $x \in [0,1]$ and the time $t \in [0,T]$.

The process of heating is described by the diffusion equation

$$z_t(x,t) = z_{xx}(x,t) \tag{1}$$

for $x \in (0,1)$ and $t \in (0,T]$. The symmetry of the heating process, i.e.

$$z(-x,t) = z(x,t)$$

for $x \in [0,1]$, allows us to consider only the right part $[0,1]$ of the interval, but we have to add the following equation:

$$z_x(o,t) = 0 \tag{2}$$

for $t \in [0,T]$. The heat transfer from the surrounding medium to the rod is given by

$$z(1,t) + \frac{1}{b} z_x(1,t) = u(t) \tag{3}$$

where b is a constant heat transfer coefficient and u(t) is the *temperature of the medium.*
We take as an initial temperature distribution in the rod

$$z(x,0) = 0 \tag{4}$$

for $x \in [0,1]$ and the temperature distribution to be approximated is assumed as

$$z_T = \text{constant} .$$

In addition we assume that the temperature u(t) of the medium is controlled by *fuel flow* v(t) and that we can describe this by

$$u'(t) = a\{v(t) - u(t)\} \tag{5}$$

$$u(0) = 0 \tag{6}$$

for $t \in [0,T]$. As fuel flow cannot be unbounded, we consider controlsv(t) which define the set V:

$$V := \{ v \in L_\infty[0,T] \mid 0 \leqq v(t) \leqq 1 \}$$

A much more extensive discussion of the physical background and the corresponding mathematical equations can be found in the book of *BUTKOVSKIY* [*69*] or partially in the paper of *SAKAWA* [*64*] . Using techniques outlined in *KRABS/WECK* [*74*] we obtain by Fourier-expansion an explicit solution formula of (1) - (6). The solution z(v,x,t) which is uniquely determined (see e.g. *FRIEDMAN* [*64*]) is represented in the follwong lemmma:

LEMMA 1 : $z(v,\cdot,T)$ *is continuous on the interval* $[-1,1]$ *for each* $v \in L_\infty[0,T]$ *and can be represented as* $z(v,x,T) =$

$$\sum_{k=0}^{\infty} B_k \cos\mu_k x \int_0^T v(s) \exp\{-\mu_k^2(T-s)\}\, ds,$$

$$B_k := A_k \cdot a\mu_k^2 \cdot (\mu_k^2 - a)^{-1}, \quad k \in \mathbb{N},$$

$$B_o := a\left(\cos\sqrt{a} - \frac{\sqrt{a}}{b}\sin\sqrt{a}\right)^{-1}, \; \mu_o := a.$$

μ_k, $k\in\mathbb{N}$, are the positive solutions of $\mu \tan \mu = b$ and

$$A_k := \frac{2\sin\mu_k}{\mu_k + \sin\mu_k \cos\mu_k} \qquad k\in\mathbb{N}.$$

$$a \neq \mu_k^2 \qquad k \in \mathbb{N}$$

The continuity of $z(v,\cdot,T)$ can be proved by estimating the infinite series under consideration of

$$|B_k| \leqq \frac{2ab}{q}\mu_k^{-2}, \quad q := \min_{k\in\mathbb{N}} |1 - a\mu_k^{-2}| \quad \text{and}$$

$$(k-1)\pi \leqq \mu_k \leqq k\pi, \quad k \in \mathbb{N}.$$

We consider the following problem (*SAKAWA* [*64*]):

$$\int_0^1 |z(v,x,T) - z_T|\, dx \stackrel{!}{=} \min \qquad v \in V \qquad (P)$$

THEOREM 2 : *(GLASHOFF* [*75*]*) Problem (P) is solvable and the solution is uniquely determined as a bang-bang-control, i.e.* $v(t)\in\{0,1\}$ *a.e. on* $[0,T]$ *, with a finite number of jumps on every subinterval* $[0,T-\delta]$, $\delta>0$.

2. APPROXIMATING PROBLEMS

We approximate the infinite series by the first N terms, $N \in \mathbb{N}$, and consider the new problem (P_N) :

$$z_N(v,x,T) := \sum_{k=0}^{N} B_k \cos\mu_k x \int_0^T v(s)\exp\{-\mu_k^2(T-s)\}\, ds .$$

$$\int_0^1 |z_N(v,x,T) - z_T|\, dx \stackrel{!}{=} \min , \; v \in V \qquad (P_N)$$

The following theorem treats the *existence of solutions* :

<u>*THEOREM 3 :*</u> *For (P_N) there exists a solution.*

The proof is the same as for the existence of solutions of problem (P). This proof was given by *GLASHOFF* [*75*] as mentioned above.

Concerning the *estimation of the optimal values* of (P) and (P_N) we have the following theorem:

<u>*THEOREM 4 :*</u> *Let ρ and ρ_N resp. be the optimal values of (P) and (P_N) resp. Then for $N \geq 1$*

$$| \rho_{N+1} - \rho | \leq d \cdot N^{-\frac{7}{2}} \quad , \quad d = \frac{ab}{\sqrt{7q}} \pi^{-4} (1+b^{-1})$$

<u>Proof:</u> Using Hölder's inequality we obtain

$$|\rho_{N+1} - \rho| \leq \sup_{v \in V} \| \sum_{k=N+2}^{\infty} B_k \cos\mu_k \cdot \int_0^T v(s) \exp\{-\mu_k^2 (T-s)\} \, ds \|_2$$

Since the functions

$$\alpha_k \cdot \cos\mu_k x \quad , \quad \alpha_k := \frac{1}{2} (1 + b^{-1} \sin^2 \mu_k) \ , \quad k \in \mathbb{N}$$

define an orthonormal system in $L_2[0,1]$, we apply Parseval's inequality:

$$|\rho_{N+1} - \rho| \leq \sup_{v \in V} \left(\sum_{k=N+2}^{\infty} \alpha_k^2 B_k^2 \{ \int_0^T v(s) \exp\{-\mu_k^2 (T-s)\} ds \}^2 \right)^{\frac{1}{2}}$$

$$\leq \frac{1}{2} (1 + b^{-1}) \frac{2ab}{q} \left(\sum_{k=N+2}^{\infty} \mu_k^{-8} \right)^{\frac{1}{2}} \leq d \cdot N^{-\frac{7}{2}} \qquad \text{q.e.d.}$$

In the next theorem we *characterize the solution* under certain conditions. We define $Dv(x) := z_N(x,v,T)$ and

$$g_v(s) := \sum_{k=0}^{N} B_k \int_0^1 \operatorname{sgn}(Dv(x) - z_T) \cos\mu_k x \, dx \exp\{-\mu_k^2 (T-s)\} .$$

<u>*THEOREM 5 :*</u> *The solution $\hat{v} \in L_\infty[0,T]$ of (P_N) is characterized by*

$$\hat{v}(s) = (-\operatorname{sgn} g_{\hat{v}}(s))^+$$

for $s \in [0,T]$. $\hat{v}(s)$ has at most N jumps or $g_{\hat{v}}(s)$ vanishes on the whole interval. In the latter case the equation above gives no information on $\hat{v}$. But then $\hat{v}$ is also solution of the problem without any restrictions for $\hat{v}$ corresponding to (P_N).

In theorem 5 we set for $\sigma \in \mathbb{R}$ $\quad \sigma^+ := \max(0,\sigma)$ and

$$\operatorname{sgn}\sigma := \begin{cases} 1 & \text{if } \sigma>0 \\ -1 & \text{if } \sigma<0 \\ \text{not def.} & \sigma=0 \end{cases}$$

<u>Proof:</u> It is a well-known theorem *(DEMYANOV/RUBINOV [70])* for a Gâteaux-differentiable convex functional f defined on a convex set V of a linear space E that

$$f(\hat{v}) \leqq f(v) \qquad \text{for } v \in V$$

is equivalent to

$$f'_{\hat{v}}(\hat{v}) \leqq f'_{\hat{v}}(v) \qquad \text{for } v \in V \tag{7}$$

where $f'_{\hat{v}}(v)$ denotes the Gâteaux-derivative of f at $\hat{v}$ in the direction v.
In order to compute the Gâteaux-derivative of $\|Dv-z_T\|_1$ we use a theorem on the differentiability of L_1-norms *(YAMAMURO [74])*.
The Gâteaux-derivative of $\|Dv-z_T\|_1$ at $\hat{v}$ in the direction of v can be written as

$$\int_0^1 \operatorname{sgn}\,(D\hat{v}(x) - z_T)\; Dv(x)\, dx = \int_0^T g_{\hat{v}}(s)\; v(s)\, ds \; .$$

We define the problem

$$\int_0^T g_{\hat{v}}(s)\; w(s)\, ds \overset{!}{=} \min_{w \in V} \tag{$\widetilde{P}_N$}$$

which corresponds to (7). We call the solution $\hat{w}$ of $(\widetilde{P}_N)$ the *gradient-control* of $\hat{v}$. It is evident that $\hat{w}$ is characterized by

$$\hat{w}(s) = (\, -\operatorname{sgn} g_{\hat{v}}(s)\,)^+ \; .$$

Since $\hat{v}$ is optimal if and only if $\hat{v}$ is a solution of $(\widetilde{P}_N)$ the first part of the theorem is proved. Now we have to ask whether $g_{\hat{v}}(s)$ is defined.
As $\{\exp(-\mu_k^2(T-\cdot))\}_{k=0,..,N}$ forms a Chebyshev-system *(COLLATZ/KRABS [73])* and $g_{\hat{v}}(s)$ is a linear combination of those functions, $g_{\hat{v}}(s)$ either has at most N zeroes or is identically 0. In the latter case we have

$$\int_0^T g_{\hat{v}}(s)(v(s)-\hat{v}(s))\, ds = 0$$

for all $v \in L_\infty[0,T]$, i.e. $\hat{v}$ is a solution of the unrestricted problem.

q.e.d.

Let $\hat{v}$ be the optimal control of (P_N) and $\hat{w}$ the solution of $(\widetilde{P}_N)$, called the corresponding gradient-control. We assume that $\hat{v}$ and $\hat{w}$ are bang-bang-controls and $g_{\hat{v}}(s)$ does not vanish identically on $[0,T]$. Theorem 5 shows us that $\hat{v}$ is optimal if and only if the jumps of $\hat{v}$ and $\hat{w}$ are the same and $\hat{v}$ and $\hat{w}$ are both 1 or 0 on the first interval.

Therefore we have to compute such controls v whose corresponding gradient-controls w do not differ very much from v. For this reason we develop the following method:
For a given approximate control v and its gradient-control w (both are assumed to be bang-bang-controls) we compute the 'best' control v_α whose jumps are convex combinations of the jumps of v and w.

control v

gradient-control w

control v_α , $\alpha = \frac{1}{2}$

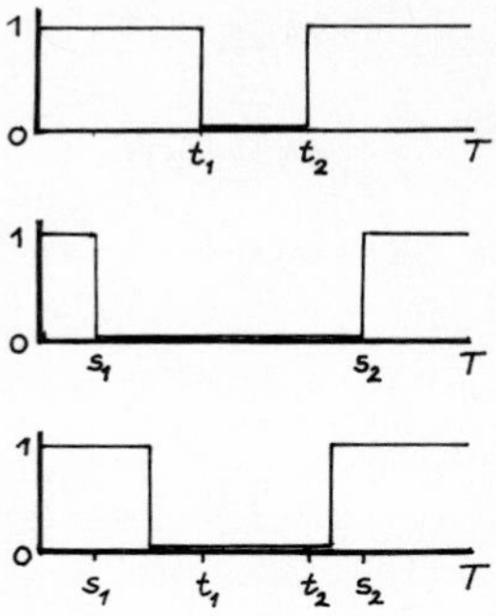

3. A METHOD FOR COMPUTING BANG-BANG-CONTROLS

At first we present the method in a rather abstract formulation but later we illustrate it by numerical examples:

STEP 0 : Start with $v_o = 1$.

Let v_i be a bang-bang-control at iteration i with jumps at

$$0 = t_o^i \leqq t_1^i \leqq \dots \leqq t_l^i \leqq t_{l+1}^i = T \ , \ l \leqq N \ .$$

STEP 1 : Compute the gradient- control $w_i = (-\operatorname{sgn} g_{v_i}(s))^+$.

If $g_{v_i}(s) = 0$, then v_i is optimal and stop the iteration.

Denote the jumps of w_i by

$$0 = s_o^i \leqq s_1^i \leqq \dots \leqq s_k^i \leqq s_{k+1}^i = T \ .$$

Theorem 5 implies $k \leqq N$.

STEP 2 : Define $t_j^i(\alpha) = \alpha s_j^i + (1-\alpha) t_j^i$, $j=1,\dots,\max(l,k)$ and

$$v_i^\alpha(t) = \begin{Bmatrix} 0.5(1-(-1)^j) \\ 0.5(1-(-1)^{j-1}) \end{Bmatrix} \quad \text{if} \quad v_i(0.5t_1^i) = \begin{Bmatrix} 1 \\ 0 \end{Bmatrix}$$

$$\text{and} \quad t \in \left[t_{j-1}^i(\alpha), t_j^i(\alpha) \right]$$

and compute $\hat{\alpha} \in [0,1]$ with $\|Dv_i^{\hat{\alpha}} - z_T\|_1 \leq \|Dv_i^{\alpha} - z_T\|_1 \quad \alpha \in [0,1]$.

STEP 3 : Define $v_{i+1} := v_i^{\hat{\alpha}}$ and go to step 1 .

REMARK 1 : *Sometimes it may occur that* $w_i(0.5\, s_1^i) \neq v_i(0.5\, t_1^i)$,

for example

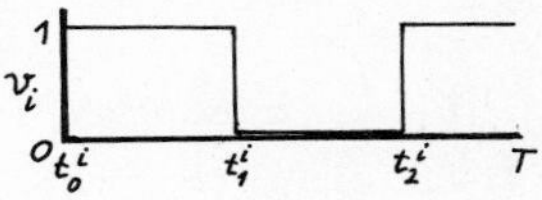

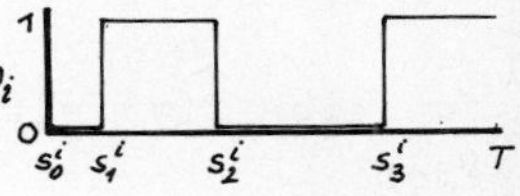

Then we interpret v_i *as a limit of the following controls*

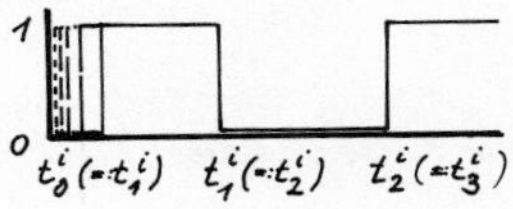

and we redefine $t_{j+1}^i := t_j^i$, $j=1,\dots,l$, $l:=l+1$, $t_1^i:=0$.

If after step 2 we have k=N+2 we better start again with $v_o=0$ hoping to avoid k=N+2.

REMARK 2 : *If in step 3* $k > l$, *we set* $t_j^i = T$, $l \leq j \leq k$.
Analogouly we define $s_j^i = T$, $l \geq j \geq k$, *if* $k < l$.

We give a condition under which the value $\|Dv - z_T\|_1$ decreases at every step. Let v_i and w_i be bang-bang-controls. We define

$t_j^i = T$, $N+1 \geq j \geq l+1$ and $s_j^i = T$, $N+1 \geq j \geq k+1$ and

$$\nu_j = v_i(\,0.5(t_j^i - t_{j-1}^i)),\ j=1,\dots,N+1.$$

THEOREM 6 : *If we have at the i-th iteration*

$$\sum_{k=0}^{N} B_k\, d_k \left(\sum_{j=1}^{N} \exp(-\mu_k^2 (T - t_j^i))(\nu_{j-1} - \nu_j)(t_j^i - s_j^i) \right) < 0 \qquad (8)$$

with $\qquad d_k := \int_0^1 \operatorname{sgn}\,(Dv_{i-1}(x) - z_T)\,\cos \mu_k x\, dx$

then there exists an $\varepsilon > 0$ *with*

$$\|Dv_i^{\alpha} - z_T\|_1 < \|Dv_{i-1} - z_T\|_1 \qquad \alpha \in (0,\varepsilon)$$

In order to prove the theorem we show by differential calculus that the Gâteaux-derivative $f_o'(1)$ of $f(\alpha) = \|Dv_i^{\alpha} - z_T\|_1$ at 0 in the direction of 1 is negative, which is equivalent to (8).

Before we begin to discuss the numerical results we look at the possibility to compute *lower bounds of the optimal value of* (P_N):

THEOREM 7 : *Let* $\bar{v} \in V$ *and* $g_{\bar{v}}(s) \neq 0$. *Then*

$$- z_T \int_0^1 \operatorname{sgn}(D\bar{v}(x)-z_T)\, dx + \int_0^T (-\operatorname{sgn}(g_{\bar{v}}(s)))^+ ds \leqq \rho_N \quad (9)$$

The function $f(v) := \| Dv - z_T \|_1$ is a convex function on a convex set V. Therefore we can use the following formula for lower bounds of the optimal value

$$f(\bar{v}) + \inf_{v \in V} f'_v(v-\bar{v}) \leqq \inf_{v \in V} f(v)$$

and obtain (9).

In connection with theorem 4 we have

THEOREM 8 : *Let* $\underline{\beta}_N$ *and* $\overline{\beta}_N$ *resp. be lower and upper bounds resp. of the optimal value of* (P_N). *Then we have the following inclusion of the optimal value of (P):*

$$\underline{\beta}_{N+1} - d \cdot N^{-\frac{7}{2}} \leqq \rho \leqq \overline{\beta}_{N+1} + d \cdot N^{-\frac{7}{2}}$$

4. NUMERICAL RESULTS

For our numerical results we define

$$a = 25, \quad b = 10, \quad T = 0.2, \quad z_T = 0.2, \quad N = 10$$

As indicated in step 0 of the algorithm we start with $v_0 = 1$. the corresponding gradient-control is $w_0 = 0$ and we apply remark 3:

Iteration 0

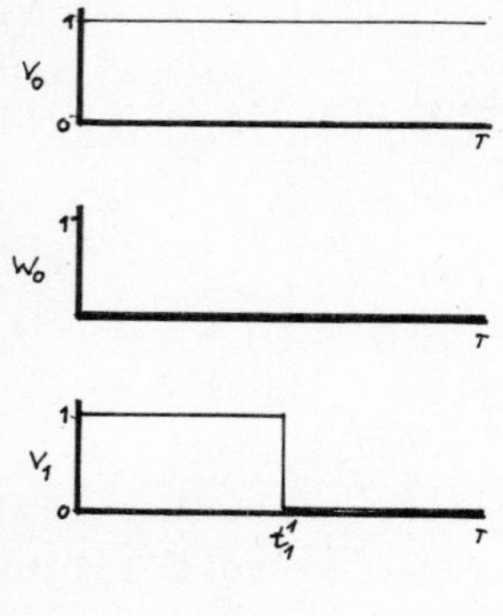

Iteration 1

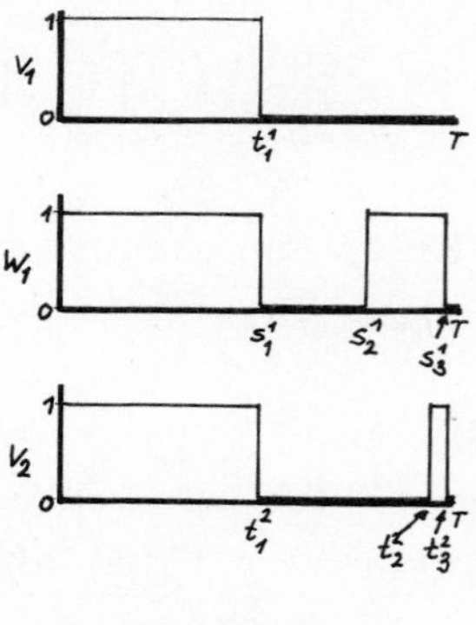

In order to compute v_2 we used the technique described in remark 4.

One of the main advantages of our method is the fact that the number of jumps need not to be fixed before starting the algorithm. The number is increased by the gradient controls as shown above. Table 2 shows that if v_i is in the neighbourhood of the optimal solution the strucure of v_i and w_i are the same and we need not pay attention to remark 4.

Table 1 : UPPER AND LOWER BOUNDS ON ρ

iteration	lower bound	upper bound
0	-6.17354'-2	20.01376'-2
1	2.54850'-2	4.01988'-2
2	3.04170'-2	3.73365'-2
3	3.27227'-2	3.71648'-2
7	3.63745'-2	3.67968'-2
11	3.63692'-2	3.66691'-2
15	3.56556'-2	3.66264'-2
19	3.56438'-2	3.65753'-2
22	3.65125'-2	3.65510'-2
23	3.59681'-2	3.65447'-2
24	3.65185'-2	3.65406'-2
25	3.61354'-2	3.65363'-2

Table 2: JUMPS OF THE APPROXIMATE CONTROLS (AND OF THE CORRESPONDING GRADIENT-CONTROLS)

iter.	t_1 (s_1)	t_2 (s_2)	t_3 (s_3)
0	0.2000 (0.0000)	0.2000 (0.2000)	0.2000 (0.2000)
1	0.1013 (0.1006)	0.2000 (0.1618)	0.2000 (0.1999)
2	0.1012 (0.0642)	0.1929 (0.2000)	0.2000 (0.2000)
3	0.1003 (0.0893)	0.1931 (0.1770)	0.2000 (0.1996)
7	0.0988 (0.0934)	0.1920 (0.1891)	0.1999 (0.1972)
11	0.0980 (0.1020)	0.1916 (0.1900)	0.1992 (0.1957)
15	0.0981 (0.1050)	0.1916 (0.1833)	0.1988 (0.1979)
19	0.0980 (0.1047)	0.1914 (0.1834)	0.1983 (0.1981)
22	0.0980 (0.0971)	0.1912 (0.1907)	0.1980 (0.1967)
23	0.0978 (0.1031)	0.1911 (0.1849)	0.1979 (0.1981)
24	0.0979 (0.0972)	0.1911 (0.1906)	0.1979 (0.1969)
25	0.0977 (0.1021)	0.1910 (0.1859)	0.1977 (0.1981)

The estimation of the optimal values which we proved in theorem 5 yields

$$|\rho_{11} - \rho| \leqq 2.994 \cdot 10^{-4}. \tag{10}$$

(10) implies

$$3.621 \cdot 10^{-2} \leq \rho \leq 3.684 \cdot 10^{-2} .$$

In order to get better inclusions of the optimal value of (P) we increase N and compute upper and lower bounds of (P_N) using $\hat{u}$ which has been the best approximating control:

$$3.65175 \cdot 10^{-2} \leqq \rho_{28} \leqq 3.65357 \cdot 10^{-2} .$$

With

$$|\rho_{28} - \rho| \leqq 9.6 \cdot 10^{-6}$$

we have the following result

$$3.650 \cdot 10^{-2} \leq \rho \leq 3.655 \cdot 10^{-2} .$$

The following results have been obtained in a similar way:

Table 3: UPPER AND LOWER BOUNDS ON ρ

Time T	lower bound	upper bound
0.15	6.630'-2	6.642'-2
0.20	3.650'-2	3.655'-2
0.25	1.426'-2	1.462'-2

The 'optimal' controls have the following structure :

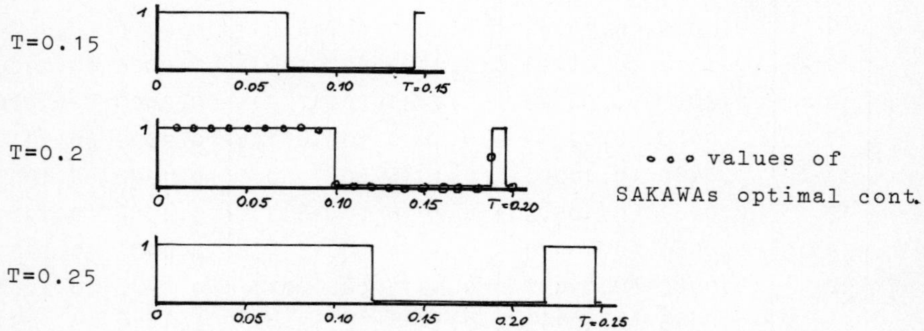

In the next figure we consider the error-curve $D\hat{v}(x) - z_T$.

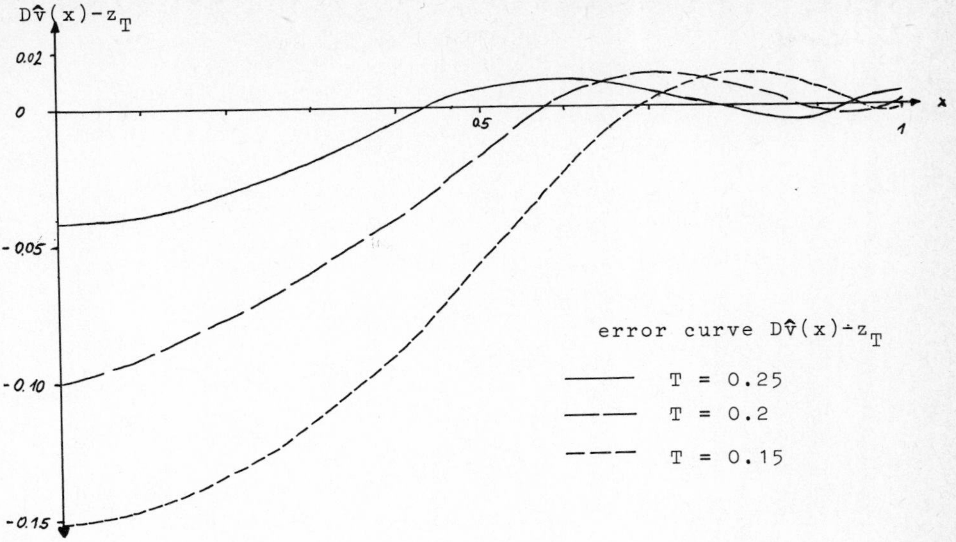

REFERENCES

BUTKOVSKIY, A.G.: Distributed control system. Elsevier: New York-London-Amsterdam, 1969.

COLLATZ, L. / KRABS, W.: Approximationstheorie, Teubner:Stuttgart, 1973.

DEMYANOV, V.F. / RUBINOV, A.M.: Approximate methods in optimization problems. Elsevier: New York-London-Amsterdam, 1970.

FRIEDMAN, A.: Partial differential equations of parabolic type, Prentice-Hall: Englewood Cliffs, 1964.

GLASHOFF, K.: Über Kontrollprobleme bei parabolischen Anfangs-Rand=wertaufgaben, Habilitationsschrift TH Darmstadt, 1975.

KRABS, W. / WECK, N.: Über ein Kontrollproblem in der Wärmeleitung, in Collatz/Wetterling: Numerische Methoden bei Optimierungs=aufgaben, S. 85 - 100, Birkhäuser: Basel, 1974.

SAKAWA, Y.: Solution of an optimal control problem in a distributed-parameter system, IEEE Transactions on automatic control, 420 - 426, 1964.

YAMAMURO, S.: Differential calculus in topological linear spaces, Springer: Berlin-Heidelberg-New York, 1974.

NUMERICAL SOLUTION OF SYSTEMS OF NONLINEAR INEQUALITIES

By Klaus Schittkowski

1. Introduction

In many cases we are concerned with the problem of solving a system of nonlinear inequalities, for example, if we have to construct a feasible initial point for an algorithm solving

$$\min \varphi(x)$$
$$x: f_i(x) \leq 0, \ i=1,\dots,m \ ,$$

that is a point $x_0 \in \mathbb{R}^n$ with $f_i(x_0) \leq 0$, $i=1,\dots,m$. Therefore, we will study now the following problem:

(1.1) Problem: Let $f:\mathbb{R}^n \to \mathbb{R}^m$ be continuously (Fréchet) differentiable. Find $x \in \mathbb{R}^n$ such that

$$f(x) \leq 0$$

Some iterative techniques have been developed till now, for example by Psenicnyi [11], Bulavskii [3], Daniel [4], Germanov and Spiridonov [7], Eremin [6], Mazurov [8], and Oettli [9]. Though in most practical cases we are interested in the solution of systems of nonlinear inequalities, we can extend this problem:

(1.2) Problem: Let $f:\mathbb{R}^n \to \mathbb{R}^m$ be continuously (Fréchet) differentiable and $K \subset \mathbb{R}^m$ be a closed, convex set. Find $x \in \mathbb{R}^n$ such that

$$f(x) \in K$$

Algorithms for this type were studied by Ben-Israel [1] and Robinson [12], but they can be realized numerically only if K is a simple set, for example a polyhedron. Proceeding from minimizing the distance function $d(f(x),K)$, the author [15] considered some possibilities to solve (1.2) even for such sets K, for which the projections can be calculated only approximatively.

Although the subsequent algorithms are programmable in general only for simple problems like (1.1), we will examine now the extension (1.2). It will be our special aim to develop modified Newton and quasi-Newton methods, that are such methods, which permit approximations of the Jacobian matrix.

In the following section we shall construct such algorithms, in the third some preliminary remarks and theorems will be mentioned, in order to obtain a loval convergence theorem of Kantorovich-type in the fourth section. Finally, we will consider some numerical results.

It should be noticed that we use in $\mathbb{R}^n$ and $\mathbb{R}^m$, respectively, always the Euclidean norm.

2. Construction of algorithms

For the first algorithm, we assume $m = n$, i.e. $f:\mathbb{R}^n \to \mathbb{R}^n$. We construct two sequences $\{x_k\}$ and $\{y_k\}$ such that $y_k \in K$ for all k.

<u>(2.1)Algorithm</u>:

0) Start: $x_o \in \mathbb{R}^n$, $y_o \in K$

For k=0,1,2,... compute x_{k+1}, y_{k+1} from x_k, y_k by

1) $x_{k+1} := x_k - C_k^{-1}(f(x_k) - y_k)$

2) Choose $y_{k+1} \in K$ with $\|f(x_{k+1}) - y_{k+1}\| \leq \|f(x_{k+1}) - y_k\|$

Here C_k is an approximation of the Jacobian matrix, what will be studied later in detail. This algorithm could be characterized in the following way: In each step we consider the system of nonlinear equations $f(x) - y_k = 0$ and perform one iteration step to solve this problem (with start x_k). The best y_k-selection is the projection of $f(x_k)$ on K, i.e.

$$\|f(x_k) - y_k\| = \min_{z \in K} \|f(x_k) - z\|$$

For the second algorithm, we assume that K is a closed, convex cone, and we examine in the domain of f the distance function

$$d(x,f^{-1}(K)) = \min \{\|x - z\|: f(z) \in K\}$$

This function is not known for any x, since then we would have a solution of (1.2). Therefore, we try to approximate $f^{-1}(K)$ linearly by

$$\min \{\|x - z\|: z: f(x) + Df(x)(z - x) \in K\}$$

So we get the following procedure:

(2.2) Algorithm:

Start: $x_0 \in \mathbb{R}^n$

For k=0,1,2,... compute x_{k+1} from x_k by

$$\|x_k - x_{k+1}\| = \min \{\|x_k - z\|: f(x_k) + C_k(z - x_k) \in K\}$$

If K is a polyhedron, we must solve in each step a linear least squares problem with linear constraints. For Newton's method, i.e. $C_k = Df(x_k)$, confer Robinson [12].

For both algorithms, it is necessary to generate matrices C_k, which should be, as mentioned above, approximations of the Jacobian matrices $Df(x_k)$. Now let us consider two different cases:

a) Modified Newton methods

Let $C_k := C(x_k)$, where C(x) is a matrix-valued mapping and an approximation of Df(x) in the following way:

(2.3) Assumption: There are constants δ, $\delta^* \in \mathbb{R}^+$ such that

$$\|Df(x) - C(x)\| \leq \delta + \delta^*\|x - x_0\| \quad \text{for all } x \in \mathbb{R}^n$$

x_0 is some initial point for algorithm (2.1) or (2.2). If C(x) = Df(x), we have Newton's method. In the other case, we can imagine for example that C(x) is determined by a secant method or a SOR-process, see Ortega, Rheinboldt [10].

b) Quasi-Newton methods

Quasi-Newton methods, well known from minimizing algorithms, are characterized by the following condition: Let $\{x_k\}$ be any sequence generated by (2.1) or (2.2), and C_0 an arbitrary initial matrix. Then we have for all k

$$C_{k+1}(x_{k+1} - x_k) = f(x_{k+1}) - f(x_k)$$

If we restrict ourself to rank-1 methods, we get easily the formula

$$C_{k+1} = C_k + \frac{(q_k - C_k p_k) d_k^T C_k}{p_k^T C_k^T d_k}$$

with $q_k := f(x_{k+1}) - f(x_k)$, $p_k := x_{k+1} - x_k$, and $d_k \in \mathbb{R}^m$ can be chosen arbitrarily up to the property that $d_k^T C_k p_k \neq 0$. For such updating procedures, Dennis [5] has shown the following useful result:

<u>(2.4) Lemma</u>: Suppose $Df(x)$ is Lipschitz-continuous, i.e. there is a $\gamma > 0$ with $\|Df(x) - Df(y)\| \leq \gamma \|x - y\|$ for all $x,y \in \mathbb{R}^n$.
Then for every rank-1 quasi-Newton method, we get the estimate

$$\|C_{k+1} - Df(x_{k+1})\| \leq \alpha_k \|C_0 - Df(x_0)\| + \gamma \sum_{j=0}^{k} \beta_k^j \|x_{j+1} - x_j\|$$

with $\alpha_k := \prod_{i=0}^{k} r_i$, $\beta_k^j := \prod_{i=-1}^{k-j-1} r_i \left(1 + \frac{1}{2} r_j\right)$,

$$r_i := \frac{\|p_i\| \|C_i^T d_i\|}{|d_i^T C_i p_i|} \quad , \quad r_{-1} := 1$$

Therefore, the following assumptions are convenient:

<u>(2.5) Assumption</u>: There are nonnegative constants δ and δ^* with

i) $\|C_0 - Df(x_0)\| \leq \delta$

ii) $\|C_k - Df(x_k)\| \leq \delta + \delta^* \sum_{j=0}^{k-1} \|x_{j+1} - x_j\|$ for all $k \in \mathbb{N}$

These estimates are valid for Broyden's rank-1 method [2]:

$$C_{k+1} = C_k + \frac{(q_k - C_k p_k) p_k^T}{p_k^T p_k}$$

For this correction, we have

$$\|C_k - Df(x_k)\| \leq \|C_0 - Df(x_0)\| + \frac{3}{2} \gamma \sum_{j=0}^{k-1} \|x_{j+1} - x_j\|$$

If $m = n$ and C_k is nonsingular, we can use the Sherman-Morrison-Woodbury formula to get

$$C_{k+1}^{-1} = C_k^{-1} - \frac{(C_k^{-1}q_k - p_k)p_k^T C_k^{-1}}{p_k^T C_k^{-1} q_k}$$

Therefore, it is not necessary to invert the matrices C_k of algorithm (2.1), since we are able to iterate the inverted matrices.

3. Some preliminaries

For the proof of the subsequent convergence theorem, we will use a special technique: We transfer the convergence studies from $\mathbb{R}^n$ into $\mathbb{R}$, cf. Ortega, Rheinboldt [10]. For this, we construct beside a given sequence $\{x_k\}$ a real, monotone increasing sequence $\{t_k\}$ such that $t_o = 0$, $\|x_1 - x_o\| \leq t_1$ and $\|x_{k+1} - x_k\| \leq t_{k+1} - t_k$ for all k. It is easy to see that the convergence of $\{t_k\}$ implies the convergence of $\{x_k\}$.

More specifically, we try to obtain a function $\nu:\mathbb{R}^3 \to \mathbb{R}$ so that

$$\|x_{k+1} - x_k\| \leq \nu(t_k - t_{k-1}, t_k, t_{k-1}) \quad \text{for all } k.$$

In this case, we define $t_{k+1} := t_k + \nu(t_k - t_{k-1}, t_k, t_{k-1})$. The following result will be very helpful, see Ortega, Rheinboldt [10], theorem (12.6.3):

<u>(3.1)Lemma</u>: Let $\nu(u,v,w) := \frac{1}{1 - p_4 v}[p_1 u^2 + (p_2 + p_3 w)u]$, $u,v,w \in \mathbb{R}$,

with $p_i \geq 0$, $i=1,\ldots,4$, $p_1 > 0$, $p_2 < 1$, $p_3 + p_4 = 2p_1$, and

$0 < \|x_1 - x_o\| \leq \frac{1}{4p_1}(1 - p_2)^2$.

Then $\{t_k\}$ is (strong) monotone increasing and

$$\lim_{k\to\infty} t_k = \frac{1}{2p_1}[(1 - p_2) - ((1 - p_2)^2 - 4\|x_1 - x_o\|p_1)^{1/2}]$$

For the convergence analysis of algorithm (2.2), we need some informations about convex processes, which were introduced by Rockafellar [14].

<u>(3.2)Definition:</u> A convex process T is a mapping of points in $\mathbb{R}^n$ into subsets of $\mathbb{R}^m$ with

a) $T(x + y) \supset T(x) + T(y)$ for all $x, y \in \mathbb{R}^n$

b) $T(\lambda x) = \lambda T(x)$ for all $\lambda > 0$ and $x \in \mathbb{R}^n$

c) $0 \in T(0)$

We say a convex process T is closed, if T(x) is closed for every $x \in \mathbb{R}^n$. For $P \subset \mathbb{R}^n$, we denote $T(P) := \cup \{T(x) : x \in P\}$. T is surjective, if $T(\mathbb{R}^n) = \mathbb{R}^m$. The mapping

$$T^{-1}(y) := \{x \in \mathbb{R}^n : y \in T(x)\}, \quad y \in \mathbb{R}^m,$$

is also a convex process and called the inverse process of T. Note that each convex process has an inverse.

Henceforth, suppose that $T(x) \neq \emptyset$ for all $x \in \mathbb{R}^n$. Then we define the norm of a convex process:

$$\|T\| := \sup_{\|x\| \leq 1} \inf_{y \in T(x)} \|y\|$$

We say a convex process T is normed, if $\|T\| < \infty$. Now, the following properties of the norm can easily be verified, see Robinson [13]:

<u>(3.3)Lemma</u>: Let λ be a real number and T, S be normed convex processes. Then $\|T\| \geq 0$, $\|\lambda T\| = |\lambda| \, \|T\|$, $\|S + T\| \leq \|S\| + \|T\|$

<u>(3.4)Lemma</u>: Let T be a closed convex process. Then for any $x_0 \in \mathbb{R}^n$ there is some $y_0 \in T(x_0)$ such that

$$\|y_0\| \leq \|T\| \, \|x_0\|$$

The following important perturbation theorem allows us to estimate the norm of an inverse convex process:

<u>(3.5)Theorem</u>: Suppose T and D are convex processes, T is surjective, T - D closed and $\|T^{-1}\| \, \|D\| < 1$. Then T - D is also surjective, $(T - D)^{-1}$ is normed, and we have

$$\|(T - D)^{-1}\| \leq \frac{\|T^{-1}\|}{1 - \|T^{-1}\| \|D\|}$$

For the proof, see Robinson [13], theorem 5. These statements are an extension of a well-known analogue theorem of the linear algebra:

(3.6) Theorem: Let T, D be (n,n)-matrices, T nonsingular and $\|T^{-1}\|\|D\| < 1$. Then $T - D$ is also nonsingular, and we get

$$\|(T - D)^{-1}\| \leq \frac{\|T^{-1}\|}{1 - \|T^{-1}\|\|D\|}$$

4. Convergence theorem

Now, let us analyse the convergence behavior of algorithm (2.1) and (2.2) with matrices C_k, which have been obtained either by a modified Newton or a quasi-Newton method. Then we establish the following local convergence statements of Kantorovich-type:

(4.1) Theorem: Suppose that the assumptions (2.3) and (2.5) are satisfied, respectively, and that the following conditions are valid:

i) $\|Df(x) - Df(y)\| \leq \gamma \|x - y\|$ for all $x, y \in \mathbb{R}^n$

ii) For algorithm (2.1), let C_o be nonsingular and $\beta := \|C_o^{-1}\|$, for (2.2), let the convex process $T_o(x) := C_o x - K$, $x \in \mathbb{R}^n$, be surjective and $\beta := \|T_o^{-1}\|$.

iii) The real numbers δ, δ^*, γ, β, and $\eta := \|x_1 - x_o\|$ may have the properties $\beta\delta < \frac{1}{3}$

$$0 < \eta \leq \frac{1 - 3\beta\delta}{2(1 - 2\beta\delta)}\,\alpha$$

with $\alpha := \dfrac{1 - 3\beta\delta}{\beta\gamma\sigma}$, $\sigma := 1 + 2\dfrac{\delta^*}{\gamma}$

Then the iterates x_k are well-defined, and we have

a) $x_k \in S(x_o, t^*) := \{x : \|x - x_o\| < t^*\}$, where t^* is denoted by

$$t^* := \alpha - \left(\alpha^2 - \frac{2\eta(1 - 2\beta\delta)}{\beta\gamma\sigma}\right)^{1/2}$$

b) There exists $\lim\limits_{k\to\infty} x_k =: x^*$

c) x^* solves (1.2), i.e. $f(x^*) \in K$

Proof: We define recurrently a real sequence $\{t_k\}$ by $t_o := 0$, $t_1 := \eta$, and $t_{k+1} := t_k + \nu(t_k - t_{k-1}, t_k, t_{k-1})$, where

$$\nu(u,v,w) := \frac{\beta}{1 - 2\beta\delta - \beta(\delta^* + \gamma)v}\left(\tfrac{1}{2}\gamma\sigma u + \delta + \delta^* w\right)u$$

This function satisfies the conditions of lemma (3.1), and we get

$$\lim_{k\to\infty} t_k = t^*$$

Now, we have to ensure that the sequence $\{x_k\}$ is well-defined and that $\|x_{k+1} - x_k\| \leq t_{k+1} - t_k$ for all k. This will be shown by induction.

Assumption ii) implies that the first step of procedure (2.1) or (2.2) is always performable, and, by definition, we have $\|x_1 - x_0\| = \eta = t_1 - t_0$.

Now, suppose that for some $k \in \mathbb{N}$, the iterates $x_1,\dots,x_k$ have been obtained from algorithm (2.1) or (2.2) such that $\|x_j - x_{j-1}\| \leq t_j - t_{j-1}$ for j=1,...,k. The induction step must be proved for each method separately.

Algorithm (2.1): Using the induction hypothesis and the definition of t^*, we get the estimates

$$\begin{aligned}\|C_k - C_0\| &\leq \|C_k - Df(x_k)\| + \|Df(x_k) - Df(x_0)\| + \|Df(x_0) - C_0\| \\ &\leq 2\delta + (\delta^* + \gamma)t_k \\ &< 3\delta + (\delta^* + \gamma)t^* \\ &\leq 3\delta + (\delta^* + \gamma)\alpha \\ &\leq 1/\beta\end{aligned}$$

Obviously, we have $\|C_k - C_0\|\|C_0^{-1}\| < 1$. The perturbation theorem (3.6) shows that C_k is nonsingular, i.e. the new iterate x_{k+1} is computable, and that

$$\|C_k^{-1}\| \leq \frac{\beta}{1 - 2\beta\delta - \beta(\delta^* + \gamma)t_k}$$

Moreover, using the induction hypothesis again, we obtain

$$\begin{aligned}\|x_{k+1} - x_k\| &= \|C_k^{-1}(f(x_k) - y_k)\| \\ &\leq \|C_k^{-1}\|\|f(x_k) - y_k\| \\ &\leq \|C_k^{-1}\|\|f(x_k) - y_{k-1}\| \\ &= \|C_k^{-1}\|\|f(x_k) - f(x_{k-1}) - C_{k-1}(x_k - x_{k-1})\|\end{aligned}$$

$$\leq \|C_k^{-1}\|(\|f(x_k) - f(x_{k-1}) - Df(x_{k-1})(x_k - x_{k-1})\| + \|Df(x_{k-1}) - C_{k-1}\|\|x_k - x_{k-1}\|)$$

$$\leq \|C_k^{-1}\|(\tfrac{1}{2}\gamma(t_k - t_{k-1})^2 + (\delta + \delta^* t_{k-1})(t_k - t_{k-1}))$$

$$\leq \frac{\beta}{1-2\beta\delta-\beta(\delta^*+\gamma)t_k}(\tfrac{1}{2}\gamma\sigma\ (t_k - t_{k-1}) + \delta + \delta^* t_{k-1})(t_k - t_{k-1})$$

$$= \nu(t_k - t_{k-1}, t_k, t_{k-1})$$

$$= t_{k+1} - t_k$$

Algorithm (2.2): In order to get analogous estimates for this method, we will use the idea of convex processes. Following Robinson [12], we define for each k convex processes by

$$T_k(x) := C_k x - K \qquad\qquad x \in \mathbb{R}^n$$
$$D_k(x) := (C_0 - C_k)x - \{0\}$$

T_0 and D_k are normed, T_0 is surjective, $T_0 - D_k = T_k$ closed, and we have, as in the previous part of this proof,

$$\|D_k\| = \|C_0 - C_k\| < 1/\beta$$

Clearly, $\|D_k\|\|T_0^{-1}\| < 1$, and all assumptions of the perturbation theorem (3.5) are satisfied. As (3.5) shows, the convex process T_k is surjective, i.e. the system $f(x_k) + C_k(z - x_k) \in K$ is solvable, and the new iterate x_{k+1} can be determined. Further, we obtain the relation

$$\|T_k^{-1}\| \leq \frac{\beta}{1 - 2\beta\delta - \beta(\delta^*+\gamma)t_k}$$

Now, define $u_k := f(x_{k-1}) - f(x_k) + C_{k-1}(x_k - x_{k-1})$. It is easy to see that the following two statements are equivalent:

1) $f(x_k) + C_k(z - x_k) \in f(x_{k-1}) + C_{k-1}(x_k - x_{k-1}) + K$

2) $z - x_k \in T_k^{-1}(u_k)$

The inverse process T_k^{-1} is closed, so we can apply lemma (3.4). There is a $z_k \in \mathbb{R}^n$, $z_k - x_k \in T_k^{-1}(u_k)$, such that $\|z_k - x_k\| \leq \|T_k^{-1}\|\|u_k\|$. Using these last two remarks and the induction hypothesis again, it follows that

$$\|x_{k+1} - x_k\| = \min\{\|x_k - z\|: f(x_k) + C_k(z - x_k) \in K\}$$

$$\leq \min\{\|x_k - z\|: f(x_k)+C_k(z-x_k) \in f(x_{k-1})+C_{k-1}(x_k-x_{k-1})+K\}$$

$$= \min\{\|x_k - z\|: z - x_k \in T_k^{-1}(u_k)\}$$

$$\leq \|x_k - z_k\|$$

$$\leq \|T_k^{-1}\|\|u_k\|$$

$$\leq \|T_k^{-1}\|(\|f(x_{k-1}) - f(x_k) + Df(x_{k-1})(x_k - x_{k-1})\| + \|C_{k-1} - Df(x_{k-1})\|\|x_k - x_{k-1}\|)$$

$$\leq \|T_k^{-1}\|(\tfrac{1}{2}\gamma(t_k - t_{k-1})^2 + (\delta + \delta^* t_{k-1})(t_k - t_{k-1}))$$

$$\leq \frac{\beta}{1-2\beta\delta-\beta(\delta^*+\gamma)t_k}(\tfrac{1}{2}\gamma\sigma\,(t_k - t_{k-1}) + \delta + \delta^* t_{k-1})(t_k - t_{k-1})$$

$$= \nu(t_k - t_{k-1}, t_k, t_{k-1})$$

$$= t_{k+1} - t_k$$

This completes the induction. As mentioned in the previous section, the sequence $\{x_k\}$ converges, i.e. there is a $x^* \in \mathbb{R}^n$ such that

$$\lim_{k\to\infty} x_k = x^*$$

Moreover, all x_k remain in $S(x_0, t^*)$, since $\|x_k - x_0\| \leq t_k < t^*$.

In order to prove that x^* solves (1.2), we consider again two cases:

Algorithm (2.1): From

$$\|f(x^*) - y_k\| \leq \|f(x^*) - f(x_k)\| + \|f(x_k) - y_k\|$$

$$\leq \|f(x^*) - f(x_k)\| + \|C_k\|\|x_{k+1} - x_k\|$$

$$\leq \|f(x^*) - f(x_k)\| + (\|C_k - C_0\| + \|C_0\|)\|x_{k+1} - x_k\|$$

$$\leq \|f(x^*) - f(x_k)\| + (1/\beta + \beta)\|x_{k+1} - x_k\|$$

we conclude $\lim_{k\to\infty} y_k = f(x^*)$ and $f(x^*) \in K$, since K is closed.

Algorithm (2.2): Since

$$f(x^*) - K = f(x^*) - f(x_{k+1}) + f(x_{k+1}) - K$$

$$\supset (f(x^*) - f(x_{k+1})) + (f(x_{k+1}) - f(x_k) - C_k(x_{k+1} - x_k)) - K$$

and, as above, $\{C_k\}$ is bounded, it follows that $0 \in f(x^*) - K$ or $f(x^*) \in K$.

5. Numerical example

In the previous sections we discussed various algorithms to solve (1.2). Now, we want to compare these methods numerically and illustrate their local convergence behavior by a simple, two dimensional problem. The computations have been performed on a TR 440 of the computing center of the Würzburg university.

Let $f(x) := \begin{pmatrix} 400(x_1^2 - x_2)x_1 + 2(x_1 - 1) \\ 200(x_2 - x_1^2) \end{pmatrix}$

and $K := \{(x_1, x_2) \in \mathbb{R}^2 : x_1 \leq 0,\ x_2 = 0\}$

Thus, the problem consists in solving a system of one nonlinear equality and one inequality:

$$400(x_1^2 - x_2)x_1 + 2(x_1 - 1) \leq 0$$

$$x_2 - x_1^2 = 0$$

The solution set is a part of a parabola:

$$f^{-1}(K) = \{(x_1, x_2) : x_1 \leq 1,\ x_2 = x_1^2\}$$

All methods were started at $x_0 := (-0.05, 0.25)$. x_0 doesn't satisfy the assumptions of theorem (4.1), but we realize in practice that the algorithms (2.1) and (2.2) are rather robust against deviations from these conditions.

In algorithm (2.1) we selected for all y_k the projection of $f(x_k)$ on K. For (2.2) it is necessary to solve in each step a linear least squares problem with linear constraints. For this, we used a numerical stable procedure of Stoer [16].

In the first column of the following tables, we have listed for each algorithm the distance of a point $f(x_k)$ from the set K, in the second the Euclidean norm of two successive iterates. In the last row, we find the calculating time in seconds.

Let us now consider the results of Newton's method, that is $C_k := Df(x_k)$. Table 1 shows the quadratic convergence of this method. Algorithm (2.2) needs less iteration steps than (2.1) to reach the stopping rule $\|x_{k+1} - x_k\| < 10^{-10}$, but requires more calculating time.

TABLE 1: Newton's method

K	(2.1) $d(f(x_k),K)$	(2.1) $\|x_{k+1}-x_k\|$	(2.2) $d(f(x_k),K)$	(2.2) $\|x_{k+1}-x_k\|$
0	$.50_{10}\ 2$	--	$.50_{10}\ 2$	
1	$.94_{10}-1$	.25	$.94_{10}-1$	.25
2	$.75_{10}-2$	$.63_{10}-2$	$.87_{10}-6$	$.46_{10}-3$
3	$.68_{10}-4$	$.60_{10}-3$	0	$.43_{10}-8$
4	$.56_{10}-5$	$.37_{10}-5$		$< {}_{10}-10$
5	$.91_{10}-10$	$.45_{10}-9$		
6		$< {}_{10}-10$		
ZEIT	0.0843		0.139	

Next, we consider the previously mentioned rank-1 method of Broyden. The initial matrix $C_0 := \begin{pmatrix} -100 & 0 \\ 0 & 200 \end{pmatrix}$ is a good approximation of the Jacobian $Df(x_0) = \begin{pmatrix} -95 & 20 \\ 20 & 200 \end{pmatrix}$. We get the following results:

TABLE 2: Broyden's rank-1 method

K	(2.1) $d(f(x_k),K)$	(2.1) $\|x_{k+1}-x_k\|$	(2.2) $d(f(x_k),K)$	(2.2) $\|x_{k+1}-x_k\|$
0	$.50_{10}\ 2$	--	$.50_{10}\ 2$	
1	$.15_{10}\ 3$	.99	$.15_{10}\ 3$	.99
2	.42	.75	.14	.74
3	$.45_{10}-2$	$.21_{10}-2$	$.48_{10}-4$	$.70_{10}-3$
4	$.27_{10}-4$	$.23_{10}-4$	$.14_{10}-10$	$.24_{10}-6$
5	$.40_{10}-6$	$.21_{10}-6$		$< {}_{10}-10$
6	$.39_{10}-8$	$.21_{10}-8$		
7		$< {}_{10}-10$		
ZEIT	0.124		0.206	

Specially for algorithm (2.1), we see the linear convergence of this quasi-Newton method. As above, (2.1) needs more iteration steps but much less calculating time than (2.2).

Analogue results have been obtained from all test functions studied by the author. Some more examples can be found in [15].

References

[1] A. Ben-Israel: On Newton's method in nonlinear programming. Proc. Princeton Symp. Math. Progr., H. Kuhn ed., Princeton 1967, 339-351

[2] C.G. Broyden: A class of methods for solving nonlinear simultaneous equations. Math. Comp. 19 (1965), 577-593

[3] V.A. Bulavskii: The solution of nonlinear inequalities and extremal problems by methods of increased accuracy. Optimizacija Vyp. 9, 26 (1973), 188-202

[4] J. Daniel: Newton's method for nonlinear inequalities. Num. Math. 21 (1973), 381-387

[5] J.E. Dennis: On the convergence of Broyden's method for nonlinear systems of equations. Math. Comp. 25 (1971), 559-567

[6] I. Eremin: Extension of the Motzkin-Agmon relaxation method. Usp. mat. Nauk, 20,2 (1965), 183-187

[7] M. Germanov, V. Spiridonov: A certain relaxation method for the solution of systems of nonlinear inequalities. B''lgar. Akad. Nauk. Otdel. Mat. Fiz. Nauk. Izv. Mat. Inst. 10 (1969), 113-128

[8] V. Mazurov: On the exponential method for solving a system of convex inequalities. Z. Vycisl. Mat. i Mat. Fiz. 6 (1966), 342-347

[9] W. Oettli: An iterative method, having linear rate of convergence, for solving a pair of dual linear programs. Math. Progr. 3 (1972), 302-311

[10] J.M. Ortega, W.C. Rheinboldt: Iterative solution of nonlinear equations in several variables. Academic Press, New York, London 1970

[11] B.N. Psenicnyi: A Newton method for the solution of systems of equalities and inequalities. Mat. Zametki, 8 (1970), 635-640

[12] S.M. Robinson: Extension of Newton's method for nonlinear functions with values in a cone. Num. Math. 19 (1972), 341-347

[13] S.M. Robinson: Normed convex processes. Trans. Amer. Math. Soc. 174 (1972), 127-140

[14] R.T. Rockafellar: Monotone processes of convex and concave type. Mem. Amer. Math. Soc., 77 (1967)

[15] K. Schittkowski: Algorithmen und Konvergenzsätze für Systeme nichtlinearer Funktionen mit Werten in einer konvexen Menge. Dissertation, Würzburg 1975

[16] J. Stoer: On the numerical solution of constrained least-squares problems. SIAM J. Numer. Anal. 8,2 (1971), 382-411

Institut für Angewandte Mathematik
Universität Würzburg
87 Würzburg
Am Hubland

LOWER BOUNDS AND INCLUSION BALLS FOR THE SOLUTION OF LOCALLY UNIFORMLY CONVEX OPTIMIZATION PROBLEMS

K. Schumacher
Universität Tübingen
7400 Tübingen, West Germany .

We study the minimization problem $f_0(x)=$ min! subject to constraints $f_i(x) \leq 0, i=1,2,\dots,k$, where each $f_i, i=0,1,\dots,k$, is a real convex and subdifferentiable functional on a closed convex set of a reflexive B-space. Supposing that at least one $f_i, i\in\{0,1,\dots,k\}$ is locally uniformly convex, lower bounds for the minimal value of f_0 are constructed which do not require the evaluation of the dual functional. It is also shown how in this case an inclusion for the unknown solutions can be obtained.

1. Preliminaries

Let $(B,\|\cdot\|)$ be a real Banach space with the dual space $(B^*, \|\cdot\|_*)$. Let $D \subseteq B$ be closed and convex. A real convex functional f on D is said to be subdifferentiable on D, if for all $x\in D$ there is a $g(x)\in B^*$ such that $f(y)-f(x)-g(x)(y-x) \geq 0$ for all $y\in D$. $\partial f(x)\subset B^*$, the so-called sub-differential of f at x, stands for the set of all subgradients g(x) of f at x. Let $I_0:=\{0,1,2,\dots,k\}\subset \mathbb{N}$, $I:I_0\setminus\{0\}$, and suppose $f_i:D\longmapsto \mathbb{R}$ are convex and subdifferentiable on D for all $i\in I_0$. We consider the optimization problem

(1) $$f_0(x)=\text{min!} \quad , \quad F(x) \leq 0 \quad , \quad x\in D,$$

with the abbreviations $F(x):=(f_i(x))_{i\in I}$ and "$\leq$" the componentwise partial ordering of $\mathbb{R}^k$. About (1) let us make the following hypotheses:

C1. $R:= \{x\in D \mid F(x) \leq 0\} \neq \emptyset$.

C2. There is at least one $j\in I_0$ and a function $h_j:D\times \mathbb{R}_+\longmapsto \mathbb{R}_+$ with the following properties:

i) $h_j(x,.)$ is continuous for all $x\in D$.

ii) $\lim_{t\to\infty} t^{-1} h_j(x,t) =\infty$ for all $x\in D$.

iii) For all $x,y\in D$ and for all $g_j(x)\in\partial f_j(x)$ holds

(2) $$f_j(y)-f_j(x)-g_j(x)(y-x) \geq h_j(x,\|y-x\|).$$

If in condition C2 the function h_j can be chosen such that $h_j(x,t)>0$ for all $x\in D$ and $t > 0$, then f_j is called locally uniformly convex on D. If in addition h_j is independent of $x\in D$, f_j is called uniformly convex on D. (see [2], sec. 1.6, [3], Ch.3.4).

Examples for C2:

1) Suppose D is bounded. Choose $d(x)\in[0,\infty)$ such that $\|y-x\|\le d(x)$ for all $y\in D$ and define with arbitrary $\beta>0$

$$h_j(x,t):=\left\{\begin{matrix} 0 & \text{for } 0\le t\le d(x) \\ \beta(t-d(x))^2 & \text{for } d(x)\le t \end{matrix}\right\} .$$

Then C2 is true for all f_j, $j\in I_0$.

2) Assume B is a Hilbert space and let $s_\mu : D\longmapsto \mathbb{R}$, $\mu=1,2,\ldots,r$, be twice Gâteaux-differentiable such that $m_\mu>0$ exists for all μ satisfying $s_\mu''(x)(y-x)(y-x)\ge m_\mu\|y-x\|^2$ for all $x,y\in D$. Then $s(x):=\max\{s_\mu(x)\,|\,\mu=1,\ldots,r\}$ is subdifferentiable and uniformly convex on D. $\partial s(x)$ is the convex hull of all gradients $s_\mu'(x)$ for which $s_\mu(x)=s(x)$ holds. (2) is fulfilled with $h(x,t):=\frac{1}{2}mt^2$, where $m:=\min\{m_\mu\,|\,\mu=1,\ldots,r\}$.

3) Let $(\Omega,\mathcal{A},\lambda)$ be a real measure space and let $B:=\mathcal{L}^p(\Omega)$, $p\ge 2$. Then $f(x):=\|x\|^p:=\int_\Omega |x(\omega)|^p d_\lambda\omega$ is uniformly convex on B. f has the gradient $g(x)\in B^*=\mathcal{L}^q(\Omega)$, $q:=p/(p-1)$, defined by $g(x)y:=p\int_\Omega x(\omega)|x(\omega)|^{p-2}y(\omega)d_\lambda\omega$, $y\in B$. (2) holds with $h(x,t):=p(p-1)\delta_p t^p$, $\delta_p:=\min_{\tau\in\mathbb{R}}\int_0^1(1-\vartheta)|\tau+\vartheta|^{p-2}d\vartheta>0$.

Let $J\subseteq I_0$ be the set of those numbers for which an h_j according to C2 exists. For $i\in I_0\setminus J$, we define $h_j:=0$. Let us introduce the abbreviations

$$\phi(x,u):=f_0(x)+u^TF(x),\quad u=(u_1,\ldots,u_k)\in\mathbb{R}^k,\quad \text{(Lagrangian)}$$

$$\partial F(x):=(\partial f_i(x))_{i\in I},$$

$$\partial\phi(x,u):=\partial f_0(x)+u^T\partial F(x),$$

$$h(x,u,t):=h_0(x,t)+\sum_{i=1}^k u_ih_i(x,t)\quad ,$$

$$P:=\begin{cases}\mathbb{R}^k_+ & \text{if } 0\in J\text{, otherwise:}\\ \{u\in\mathbb{R}^k_+\,|\,u_j>0 \text{ for at least one } j\in J\}\end{cases},$$

where $\mathbb{R}^k_+:=\{u\in\mathbb{R}^k\,|\,u\ge 0\}$.
Then we obtain from (2) for all $u\in\mathbb{R}^k_+$, for all $x,y\in D$, and every $g(x,u)\in\partial\phi(x,u)$ the inequality

(3) $\qquad \phi(y,u)-\phi(x,u)-g(x,u)(y-x)\ge h(x,u,\|y-x\|)$,

where h satisfies

(4) $\lim_{t\to\infty} t^{-1}h(x,u,t) = \infty$

for $u \in P$. This has the following consequence:

LEMMA.

If C1-2 hold, then problem (1) has a solution $\bar{x}$ and for all $u \in P$ the dual functional $S(u) := \min\{\phi(y,u) \mid y \in D\}$ exists.

Proof. The convexity and subdifferentiability of the f_i, $i \in I_0$, imply that $R_0 := \{y \in R \mid f_0(y) \leq f_0(x_0)\} \neq \emptyset$ is closed and convex for any $x_0 \in R$. We obtain from (3) for all $y \in R_0$, $g \in \partial\phi$, $u \in P$,

$$f_0(x_0) \geq \phi(y,u) \geq \phi(x_0,u) - \|g(x_0,u)\|_* \|y - x_0\| + h(x_0,u,\|y - x_0\|)$$

showing, by means of (4), the boundedness of R_0. Hence, (1) has a solution $\bar{x}$, since f_0 is weakly sequentially lower semicontinuous and R_0 is sequentially weakly compact by the well known theorem of EBERLEIN-SHMULYAN (see [2] ,sec. 1.4). Replacing R_0 by $\tilde{R}_u := \{w \in D \mid \phi(w,u) \leq \phi(w_0,u)\}$ for any $w_0 \in D$, we obtain in a similar manner the existence of $S(u)$ for $u \in P$. □

In order to get numerically effective lower bounds we suppose constraint qualifications which ensure the KUHN-TUCKER conditions.

C3. There is a solution $\bar{x}$ of (1) and a corresponding Lagrange-parameter $u \in \mathbb{R}^k_+$ such that

(5) $0 \in \partial\phi(\bar{x},\bar{u}) \quad , \quad \bar{u}^T F(\bar{x}) = 0$.

2. Inclusion theorems.

Inequality (3) leads us directly to lower bounds for the minimal value $f_0(\bar{x})$, which do not require the evaluation of $S(u)$. If some good estimates of the form (2) are known then $S(u)$ may be replaced by lower bounds, which can be more easily computed.

THEOREM 1.

Let C1-2 are fulfilled and let $\bar{x}$ be a solution of (1). Then

(6) $f_0(\bar{x}) \geq S(u) \geq \rho(x,u,g)$

holds for all $x \in D$, $u \in P$, and $g \in \partial\phi$, where

(7) $\rho(x,u,g) := \min_{t \geq 0} \{\Phi(x,u) - t\|g(x,u)\|_* + h(x,u,t)\}$.

<u>If in addition</u> C3 <u>holds, we obtain</u>

(8) $f_0(\bar{x}) = \sup \{ \rho(x,u,g) \mid x \in D,\ u \in P,\ g \in \partial\Phi \}$.

<u>Proof.</u> Because of $u^T F(\bar{x}) \leq 0$ for $u \geq 0$ we have $f_0(\bar{x}) \geq \Phi(\bar{x},u) \geq S(u)$. Given $x \in D$, $u \in P$, and $g \in \partial\Phi$, (3) yields $\Phi(y,u) \geq \rho(x,u,g)$ for all $y \in D$, showing (6). Now assume C3 and let $\bar{x}$ be a solution satisfying (5) with Lagrange-parameter $\bar{u} \geq 0$.

<u>Case i)</u> $\bar{u} \in P$. Choose $\bar{g} \in \partial\Phi$ with $\bar{g}(\bar{x},\bar{u})=0$. Since from (2) follows $h_j(x,0)=0$ for all $x \in D$, we have $\rho(\bar{x},\bar{u},\bar{g}) = f_0(\bar{x})$.

<u>Case ii)</u> $\bar{u} \notin P$. It follows $J \subseteq I$ and $\bar{u}_j=0$ for all $j \in J$. Let $j_0 \in J$ be fixed. Given $\varepsilon > 0$ define $u_\varepsilon \in P$ by $(u_\varepsilon)_i := \bar{u}_i$, $i \in I \setminus \{j_0\}$, and $(u_\varepsilon)_{j_0} := \varepsilon$. With $\bar{g}$ as in case i) we obtain $\bar{g}(\bar{x},u_\varepsilon) = \varepsilon\, \bar{g}_{j_0}(\bar{x})$ because of $\bar{u}_{j_0}=0$ and $\bar{g}(\bar{x},\bar{u})=0$. Therefore

$$\rho(\bar{x},u_\varepsilon,\bar{g}) = f_0(\bar{x}) + \varepsilon f_{j_0}(\bar{x}) + \varepsilon \min_{t \geq 0} \{-t\|\bar{g}_{j_0}(\bar{x})\|_* + h_{j_0}(\bar{x},t)\} ,$$

since $u_\varepsilon^T F(\bar{x}) = \varepsilon f_{j_0}(\bar{x})$. This proves (8). □

For locally uniformly convex functionals in a Hilbert space with h_j given by

(9) $h_j(x,t) := \frac{1}{2} m_j(x) t^2$, $m_j(x) > 0,\ x \in D,\ j \in J$,

we obtain instead of (7)

(10)
$$\rho(x,u,g) = \Phi(x,u) - \|g^T(x,u)\|^2/(2m(x,u)) ,$$
$$m(x,u) := m_0(x) + \sum_{i=1}^{k} u_i m_i(x), \quad m_i(x) := 0 \text{ for } i \in I \setminus J$$

(g^T stands for the RIESZ-representation of $g \in B^*$).

We turn now to the construction of inclusion balls for the unknown solutions $\bar{x}$ of (1). The following theorem can be applied without a priori information on the minimal value $f_0(\bar{x})$; however a good inclusion of $f_0(\bar{x})$ may be helpful for a good inclusion of $\bar{x}$.

<u>THEOREM 2.</u>

<u>Assume</u> C1-3 <u>holds and let</u> $\bar{x}$ <u>be a solution of</u> (1) <u>satisfying</u> (5) <u>with Lagrange-parameter</u> $\bar{u} \geq 0$. <u>For given</u> $x \in D$, $u \in P$, $g \in \partial\Phi$, and $\lambda \in \mathbb{R}_+$, <u>choose</u> $\delta : D \times P \times \mathbb{R}_+ \longrightarrow \mathbb{R}$ <u>and</u> $\alpha : \mathbb{R}_+ \xrightarrow{\sim} \mathbb{R}_+$ <u>such that</u>

(11) $\delta(x,u,\lambda) \geq (\lambda-1)(f_0(x)-f_0(\bar{x}))+ (\lambda\bar{u}-u)^T(F(x)-F(\bar{x}))$

and

(12) $h(\bar{x},\bar{u},t) \geq \alpha(t) \quad , \quad t\in\mathbb{R}^+.$

Let $t = r(x,u,\lambda,g)$ be the greatest real zero of the equation

(13) $\delta(x,u,\lambda)= -\|g(x,u)\|_* t+ \lambda\alpha(t)+h(x,u,t)$

(this zero always exists and is nonnegative). Then

(14) $\|x-\bar{x}\| \leq r(x,u,\lambda,g).$

Proof. Let $\bar{g}\in\partial\phi$ such that $\bar{g}(\bar{x},\bar{u})=0$ and $\bar{u}^T F(\bar{x})=0$. From (3) follows

(15) $\phi(x,\bar{u}) \geq \phi(\bar{x},\bar{u})+ \bar{g}(\bar{x},\bar{u})(x-\bar{x})+h(\bar{x},\bar{u},\|x-\bar{x}\|)=$

$= f_0(\bar{x}) +h(\bar{x},\bar{u},\|x-\bar{x}\|),$

(16) $\phi(\bar{x},u) \geq \phi(x,u)+g(x,u)(\bar{x}-x)+h(x,u,\|x-\bar{x}\|)$.

Multiplying (15) by $\lambda \geq 0$ and adding the obtained inequality to (16) yields

(17) $\delta(x,u,\lambda) \geq (\lambda-1)(f_0(x)-f_0(\bar{x}))+(\lambda\bar{u}-u)^T(F(x)-F(\bar{x})) \geq$

$\geq g(x,u)(\bar{x}-x)+\lambda h(\bar{x},\bar{u},\|x-\bar{x}\|)+h(x,u,\|x-\bar{x}\|) \geq$

$= -\|g(x,u)\|_* \tau + \lambda\alpha(\tau)+h(x,u,\tau)$

with $\tau:=\|x-\bar{x}\|$. Since $\lim_{t\to\infty} [-\|g(x,u)\|_* t+ \lambda\alpha(t)+h(x,u,t)] = \infty$,

there is a greatest zero $\tilde{t} \geq 0$ of (13). Hence, $\tau \leq \tilde{t}$.

□

Examples for the choice of λ, δ , and α.

1) Let $x\in R$ and real numbers s_0, s_1 be given with

(18) $s_1 \leq f_0(\bar{x}) \leq s_0 \leq f_0(x).$

Then

(19) $$\delta(x,u,\lambda):=\begin{cases}(\lambda-1)(f_0(x)-s_0)-u^TF(x) & \text{for } 0 \leq \lambda \leq 1\\ (\lambda-1)(f_0(x)-s_1)-u^TF(x) & \text{for } 1 \leq \lambda\end{cases}$$

fulfils (11). For instance, s_1 can be computed by use of theorem 1. $s_0\in[f_0(\bar{x}),f_0(x))$ can be easily computed if f_0 is twice Gâteaux-differentiable and an upper bound for the Hessian of f_0 on D is known (see [4], sec.3). If an inclusion (18) is not available, we may choose λ =1.

2) (12) holds trivially with $\alpha := 0$. In this case the best $\lambda \geq 0$ is that which minimizes $\delta(x,u,\lambda)$, e.g. $\lambda=0$ if δ is given by (19).

3) If f_0 is uniformly convex on D, we may choose $\alpha(t):=h_0(x,t)$, where h_0 is positive and independent of $x\in D$. Dividing equality (13) by $\lambda>0$ and passing to the limit $\lambda=\infty$ we obtain for $x\in R$ with the function δ from (19) the well known inclusion

(20) $$\alpha(\|x-\overline{x}\|) \leq f_0(x)-s_1$$

(e.g. see[2], Th. 1.6.3 and[5]).

Remark. It is easy to see that the inclusion (20) does not require C3. More general, if δ is defined by (19) and if $0 \leq \alpha(t) \leq h_0(x,t)$ holds, then inclusion (14) does not require C3. In this case the proof works with the optimality condition for $\overline{x}$, i.e. the existence of $\overline{g}_0(\overline{x})\in\partial f_0(\overline{x})$ such that $\overline{g}_0(\overline{x})(x-\overline{x}) \geq 0$ for all $x\in R$.

If B is a Hilbert space with inner product $\langle\cdot,\cdot\rangle$ and if the h_j, $j\in J$, have the form (9), (14) can be improved.

THEOREM 3.

Let B be a Hilbert-space. Suppose C1-3 hold where h_j, $j\in J$, have the special form (9). Let $\overline{x}$ be a solution of (1) with Lagrange-parameter $\overline{u}$ satisfying (5) and choose $\tilde{m} \geq 0$ such that $m(\overline{x},\overline{u}) \geq \tilde{m}$ (see (10)). Let $x\in D$, $u\in P$, $\lambda \geq 0$, and $g\in\partial\phi$, be given. Then, if $\delta(x,u,\lambda)$ fulfils (11) we have

(21) $$\|\overline{x}-a(x,u,\lambda,g)\| \leq r(x,u,\lambda,g) \quad ,$$

(22) $$a(x,u,\lambda,g):=x-\frac{g(x,u)^T}{\lambda\tilde{m}+m(x,u)}$$

(23) $$r(x,u,\lambda,g):=\left(\frac{\|g(x,u)^T\|^2}{(\lambda\tilde{m}+m(x,u))^2}+\frac{2\delta(x,u,\lambda)}{\lambda\tilde{m}+m(x,u)}\right)^{\frac{1}{2}} .$$

Proof. From (17) follows

$\delta(x,u,\lambda) \geq \langle g(x,u)^T, \overline{x}-x\rangle + \frac{1}{2}(\lambda\tilde{m}+m(x,u))\|\overline{x}-x\|^2$. An easy manipulation shows that this inequality is equivalent with $\|\overline{x}-a(x,u,\lambda,g)\|^2 \leq \tilde{r}(x,u,\lambda,g)^2$ □

If the solution $\overline{x}$ of problem (1) is unique then an inclusion for $\overline{x}$ does not only provide an error estimate for an approximation x but also may lead to a decision whether the boundary of D is active. Namely if the inclusion ball lies in the interior of D then the boundary can not be active.

3. The computation of appropriate parameters.

In this section we discuss a method for the choice of parameters $u \in P$ and subgradients $g \in \partial\phi$ in order to obtain good lower bounds and inclusion balls. Let us make the stronger hypothesis that f_0 or at least one essential restriction f_j is uniformly convex:

C4. Let one of the following two conditions be satisfied:

i) $h_0(x,t) \geq \alpha_0(t) \geq 0$ for all $x \in D$, where $\alpha_0(t) > 0$ for $t > 0$,
$\lim_{t \to \infty} t^{-1} \alpha_0(t) = \infty$.

ii) There is at least one $j \in I$ such that

(24) $-\infty \leq \inf\{f_0(x)/x \in D, f_i(x) \leq 0,\ i \in I \setminus \{j\}\} < \inf\{f_0(x)/x \in R\}$,

where the corresponding h_j satisfies $h_j(x,t) \geq \alpha_j(t) \geq 0$ for all $x \in D$ and α_j has the same properties as α_0.

Conditions C1-4 imply uniqueness of $\overline{x}$. Let $\tilde{I} \subseteq I$ be the subset of indices of active constraints f_j .

DEFINITION

Let $\overline{x}$ be a solution of (1) satisfying C3 and let $\{x_n\}_{n \in \mathbb{N}} \subset D$ with $\lim_{n \to \infty} x_n = \overline{x}$. The sequence $\{x_n\}$ is called an admissible approach of $\overline{x}$, if there are $g_i(\overline{x}) \in \partial f_i(\overline{x})$, $g_i(x_n) \in \partial f_i(x_n)$ for all $i \in I_0$ and if $\overline{u} \in \mathbb{R}^k_+$ such that $g(\overline{x},\overline{u}) = 0$, $\overline{u}^T F(\overline{x}) = 0$, and $\|g_i(x_n) - g_i(\overline{x})\|_* \xrightarrow[n \to \infty]{} 0$ for all $i \in \{0\} \cup \tilde{I}$.

Given a forcing function $p: \mathbb{R}^2_+ \overset{\backsim}{\longrightarrow} \mathbb{R}_+$ (i.e. $p(0,0)=0$ and for arbitrary $\gamma_m \geq 0$, $\eta_m \geq 0$, $m \in \mathbb{N}$, $p(\gamma_m,\eta_m) \longrightarrow 0$ implies $\gamma_m \longrightarrow 0$ and $\eta_m \longrightarrow 0$), we assume that we are able to execute the following program:

A. For arbitrary $x \in D$ and $\varepsilon > 0$ find $u \in P$ and $g \in \partial\phi$ such that

$$\gamma \leq p(\|g(x,u)\|_*, |u^T F(x)|) \leq \gamma + \varepsilon ,$$

(25)

$$\gamma := \inf\{p(\|g(x,u)\|_*, |u^T F(x)|) \mid u \in \mathbb{R}^k_+,\ g \in \partial\phi\}$$

We can prove:

THEOREM 4.

Suppose C1-4 and let $\{x_n\}_{n \in \mathbb{N}} \subset D$ be an admissible approach of the solution $\overline{x}$ of (1). Given a sequence $\{\varepsilon_n\}_{n \in \mathbb{N}} \subset R_{++}$, $\varepsilon_n \longrightarrow 0$, compute

$g^{(n)} \in \partial\phi$ <u>and</u> $u^{(n)} \in P$ <u>by</u> <u>A</u> <u>for</u> $\varepsilon := \varepsilon_n$ <u>and</u> $x := x_n$. <u>Then</u>

$$\lim_{n\to\infty} \rho(x_n, u^{(n)}, g^{(n)}) = f_0(\bar{x}) \quad . \tag{26}$$

<u>Moreover,</u> <u>if</u> <u>the</u> <u>function</u> δ <u>defined</u> <u>in</u> (11) <u>satisfies</u> $|\delta(x_n, u^{(n)}, 1)| \leq \leq \Psi(|u^{(n)T} F(x_n)|)$, <u>with</u> $\Psi: \mathbb{R}_+ \hookrightarrow \mathbb{R}_+$, $\Psi(0)=0$, <u>we</u> <u>obtain</u>

$$\lim_{n\to\infty} r(x_n, u^{(n)}, 1, g^{(n)}) = 0. \tag{27}$$

<u>Proof.</u> Let $g_i(\bar{x})$, $g_i(x_n)$, and $\bar{u} \geq 0$ be chosen as in the above definition. Since $\bar{u}_i = 0$ for $i \in I \setminus \tilde{I}$ we have

$$\| g(x_n, \bar{u}) \|_* \leq \| g_0(x_n) - g_0(\bar{x}) \|_* + \sum_{i\in\tilde{I}} \bar{u}_i \| g_i(x_n) - g_i(\bar{x}) \|_*$$

showing $g(x_n, \bar{u}) \longrightarrow 0$. Because of $f_i(\bar{x}) = 0$ for $i \in \tilde{I}$ we have

$$f_i(x_n) \geq g_i(\bar{x})(x_n - \bar{x}) \text{ and } -f_i(x_n) \geq g_i(x_n)(\bar{x} - x_n), \ i \in \tilde{I} .$$

Therefore $f_i(x_n) \rightarrow 0$, $i \in \tilde{I}$, which implies $\bar{u}^T F(x_n) \longrightarrow 0$. We conclude that the infima γ_n of (25), taken for $x_n \in D$, tend to zero for $n \rightarrow \infty$. Hence, $p(\| g^{(n)}(x_n, u^{(n)}) \|_*, |u^{(n)T} F(x_n)|) \longrightarrow 0$. Thus by the forcing function property

$$\lim_{n\to\infty} g^{(n)}(x_n, u^{(n)}) = 0 \quad , \qquad \lim_{n\to\infty} u^{(n)T} F(x_n) = 0. \tag{28}$$

Next we will show $\underline{\lim}_{n\to\infty}\, u^{(n)}_j > 0$ under the hypothesis that f_j, $j \in I$, is a uniformly convex essential restriction (see C4). For this purpose choose $y \in D$ such that $f_0(y) < f_0(\bar{x})$ and $f_i(y) \leq 0$ for all $i \in I \setminus \{j\}$. Employing $f_0(x_n) - f_0(y) \leq g_o^{(n)}(x_n)(x_n - y)$ and $f_i(x_n) \leq f_i(x_n) - f_i(y) \leq g_i^{(n)}(x_n - y)$, $i \in I \setminus \{j\}$, we get

$$\begin{aligned} f_0(x_n) - f_0(y) &\leq g^{(n)}(x_n, u^{(n)})(x_n - y) - u^{(n)T} F(x_n) + \\ &+ u_j^{(n)}(f_j(x_n) - g_j^{(n)}(x_n)(x_n - y)) \quad \leq \\ &\leq g^{(n)}(x_n, u^{(n)})(x_n - y) - u^{(n)T} F(x_n) + u_j^{(n)} f_j(y). \end{aligned} \tag{29}$$

From (28) and (29) we obtain $f_0(\bar{x}) - f_0(y) \leq f_j(y) \underline{\lim}_{n\to\infty} u_j^{(n)}$, which proves our statement since $f_j(y) > 0$ by (24) (Hint: $f_0(x_n) \rightarrow f_0(\bar{x})$ follows from $\{x_n\}_{n\in\mathbb{N}}$ admissible).

C4 and the above consideration guarantee the existence of $n_0 \in \mathbb{N}$ and of $\tilde{\alpha}: R_+ \xrightarrow{\nearrow} R_+$ with $\tilde{\alpha}(t) > 0$ for $t > 0$ and $\lim_{t\to\infty} t^{-1}\tilde{\alpha}(t) = \infty$ such that $h(x_n, u^{(n)}, t) \geq \tilde{\alpha}(t)$ for $n \geq n_0$. Since by a standard argument

$$0 \leq \max_{t\geq 0} \{t\|g^{(n)}(x_n,u^{(n)})\|_* - h(x_n,u^{(n)},t)\} \leq \max_{t\geq 0} \{t\|g^{(n)}(x_n,u^{(n)})\|_* - \tilde{\alpha}(t)\}$$

$\longrightarrow 0$ for $n \longrightarrow \infty$ and $\Phi(x_n, u^{(n)}) \longrightarrow f_0(\bar{x})$, we have proved (26).

Since $\delta(x_n,u^{(n)},1) + t_n \|g^{(n)}(x_n,u^{(n)})\|_* \geq \tilde{\alpha}(t_n)$ with $t_n := r(x_n,u^{(n)},1,g^{(n)})$ we obtain from (28) by another standard argument $t_n \longrightarrow 0$. □

Obviously $r(x_n,u^{(n)},\lambda_n,g^{(n)}) \longrightarrow 0$ also holds if λ_n is chosen to minimize $r(x_n,u^{(n)},\lambda,g^{(n)})$ over the range $\lambda \geq 0$. (27) remains true if r is replaced by $\tilde{r}$ from theorem 3, since $\tilde{r}$ is not greater then the corresponding r due to theorem 2. In this case a suitable choice for the forcing function p corresponding to (23) is

(30) $$p(\gamma,\eta) := \gamma^2 + c\eta , \quad c > 0 .$$

4. A numerical example.

Let $B = D = \mathbb{R}^6$, k=3, and let

$f_0(x) := a^T x$, $f_1(x) := b^T x + \beta$, $f_2(x) := x^T C x + \gamma$, $f_3(x) = \sum_{i=1}^{6} e^{x_i} + \delta$,

with

$a = (-\frac{41}{6}, 1, -\frac{2}{3}, -5, 4, \frac{2}{3})$, $b = (5,-8,-10,0,-18,-10)$, $\beta = 41$,

$c_{11} = c_{33} = c_{66} = \frac{5}{3}$, $c_{13} = c_{36} = \frac{1}{3}$, $c_{16} = -\frac{1}{3}$, $c_{22} = 1$, $c_{44} = c_{55} = 2$,

$c_{ij} = c_{ji}$, $c_{ij} = 0$ otherwise,

$\gamma = -\frac{32}{3}$, $\delta = -6e$, $e = 2.7182818...$.

Problem (1) has the solution $\bar{x} = (1,1,1,1,1,1,)$, $f_0(x) = -6.833333...$. The eigenvalues of C are bounded below by 1. Thus we can choose $h_i :\equiv 0$ for i=0,1,3 and $h_2(x,t) := 0.5\, t^2$. The parameter $u = (u_1,u_2,u_3)$ is determined from the minimization of the quadratic functional

$$\|f_0'(x) + \sum_{i=1}^{3} u_i f_i'(x)\|^2 + 4\left|\sum_{i=1}^{3} u_i f_i(x)\right|$$

over the range $u \geq 0$ for fixed $x \in D$ ($\|\cdot\|$ is the euclidian norm). The following table shows for various approximations $x \in R$ the lower bounds $\rho(x,u,g)$ and the radius $\tilde{r}(x,u,1,g)$ computed with $\delta(x,u,1) := -u^T F(x)$. $\tilde{r}$ is compared with the exact error $\delta := \|\bar{x} - a(x,u,1,g)\|$:

	I	II	III	IV
x	0.88	0.98	0.9625	0.9713
	1.05	1.02	1.0692	1.0356
	0.95	0.97	1.2005	1.0689
	0.79	0.91	0.8846	0.9892
	1.19	1.02	0.8543	0.9478
	1.02	1.01	0.9875	0.9821
$f_0(x)$	-4.11	-6.12	-6.6556	-6.8140
$f_1(x)$	-4.12	-0.42		
$f_2(x)$	-0.35	-0.39	$< 10^{-4}$	$< 10^{-4}$
$f_3(x)$	-7.05	-0.85		
ρ	-7.14	-6.87	-7.4605	-6.8884
$\tilde{r}$	2.03	1.12	1.2750	0.3768
δ	0.40	0.18	0.9977	0.2710

The discrepancy between $\tilde{r}$ and δ for the approximations I and II is caused by the somewhat arbitrary choice of δ .

5. References.

[1] J.Céa, Optimisation théorie et algorithmes, Dunod Paris 1971(chap.5).

[2] J.W.Daniel, The approximate minimization of functionals, Prentice-Hall Inc. 1971 (sec. 1.5-1.6).

[3] J.M.Ortega and W.C. Rheinboldt, Iterative solution of nonlinear equations in several variables, Academic Press New York-London 1970 (ch.3.4, 4.2, 4.3).

[4] P.Wolfe, Convergence theory in nonlinear programming, in J.Abadie, Integer and nonlinear programming, North-Holland Publishing Company Amsterdam-London 1970 (ch.1).

[5] B.Poljak, Existence theorems and convergence of minimizing sequences in extremum problems with restrictions, Dokl.Akad.Nauk SSSR 166, 287-290; Soviet Math.Dokl.7,72-75,1966.

[6] K.Schumacher, Untersuchungen zur Theorie der Gradientenverfahren, Thesis, Tübingen 1974.

K.Schumacher
Lehrstuhl f. Biomathematik
Universität Tübingen

7400 Tübingen/West Germany
Auf der Morgenstelle 28

ÜBER VEKTORMAXIMIERUNG UND ANALYSE DER GEWICHTUNG VON SUBZIELEN

Klaus Spremann

Institut für Wirtschaftstheorie und
Operations Research der Universität
D-75 Karlsruhe 1, Kaiserstraße 12

Bei nichtlinearen Vektormaximierungsproblemen in reellen Banachräumen soll für den schwachen Lösungsbegriff der lokalen-Slater-Kegel-Effizienz eine notwendige Bedingung hergeleitet werden. Verwendung findet dabei die Methode der lokalen Approximation von Majorantenmenge und Menge zulässiger Entscheidungen durch disjunkte Kegel, deren Konvexität eine Regularitätsbedingung garantiert, sowie deren Separation mit topologischen Trennungssätzen, wie sie auf A.Y.Dubovitskii und A.A.Miljutin zurückgeht. Schließlich werden die Anwendungsmöglichkeiten unserer Resultate bei der Analyse der Zielgewichtung bzw. der Explizierung der Präferenz eines rationalen Entscheidungsträgers aufgezeigt.

1. EINFÜHRUNG

Gegeben seien ein reeller Banachraum X, eine nichtleere Teilmenge Ω von X, ein reeller normierter Raum V, eine Abbildung $F: X \to V$ sowie ein Ordnungskegel P in V mit nichtleerem Inneren, also $P \subset V$, $\mathbb{R}_+ P \subset P$, $P+P \subset P$, $P \cap -P = \{0\}$ und $\text{int}(P) \neq \emptyset$. Zur Interpretation sehe man X als Raum der "Entscheidungsvariablen" an; Ω ist die zulässige Menge; jede Entscheidung $x \varepsilon X$ bewirke den "Output" $F(x) \varepsilon V$ und der Ordnungskegel P legt durch

$$v^1 \leqq v^2 \;:\Longleftrightarrow\; v^2 - v^1 \varepsilon P$$
$$v^1 \leq v^2 \;:\Longleftrightarrow\; v^2 - v^1 \varepsilon P \setminus \{0\}$$
$$v^1 < v^2 \;:\Longleftrightarrow\; v^2 - v^1 \varepsilon \text{int}(P)$$

transitive, antisymmetrische, kompatible Relationen auf V fest, die dazu dienen, verschiedene Outputs zu "vergleichen". Das Quintupel

$$(X, \Omega, V, F, P)$$

wird als "Problem" der Vektormaximierung, kurz VMP, angesehen, indem man Teilmengen von Ω als (vollständige) "Lösung von VMP auszeichnet :

Definition 1 : Für ein VMP (X,Ω,V,F,P) heißt

$$E := \{\tilde{x}\in\Omega \mid \forall\, x\in\Omega : F(x) - F(\tilde{x}) \notin P\setminus\{0\}\}$$

Menge der kegeleffizienten Entscheidungen,

$$S := \{\tilde{x}\in\Omega \mid \forall\, x\in\Omega \quad F(x) - F(\tilde{x}) \notin \mathrm{int}(P)\}$$

Menge der Slater-kegeleffizienten Entscheidungen und

$$LS := \{\tilde{x}\in\Omega \mid \text{es existiert eine Umgebung } \tilde{U} \text{ von } \tilde{x} \text{ in } X \text{ so, daß für alle } x\in\Omega\cap\tilde{U} \quad F(x)-F(\tilde{x}) \notin \mathrm{int}(P) \text{ gilt}\}$$

Menge der lokal-Slater-kegeleffizienten Entscheidungen.

Bemerkung 1 : Die Inklusion $LS \supset S \supset E$ ist offensichtlich. Die Verwendung von allgemeinen Ordnungskegeln, die vom "positiven Orthanten" in V verschieden sein können, geht in der Theorie der Vektormaximierung wohl auf P.L.Yu zurück. Wie bei deren Behandlung als nichtlineares Programmierungsproblem erfordern Aufgabenstellungen der Kontrolltheorie die Untersuchung von $X \neq \mathbb{R}^n$. L.Hurwicz behandelte zuerst den Fall $V \neq \mathbb{R}^\ell$. Die Bedingung $\mathrm{int}(P) \neq \emptyset$ schließt die Verwendung der kanonischen Positivitätskegel in den ℓ_p- und L_p-Räumen aus. Als Kegel in einem reellen Vektorraum wird jede nichtleere Teilmenge K bezeichnet, für die

$$\mathbb{R}_{++}K \subset K \quad , \quad \mathbb{R}_{++} := \{r\in\mathbb{R} \mid r>0\}$$

gilt. Ein Kegel ist konvex genau dann, wenn $K+K \subset K$ gilt. In topologischen linearen Räumen ist mit K auch $\mathrm{int}(K)$ ein Kegel, sofern $\mathrm{int}(K)=\emptyset$.

Für Vektormaximierungsprobleme der Art $(\mathbb{R}^n, \Omega, \mathbb{R}^\ell, F, \mathbb{R}^\ell_+)$ wird E als Lösungsmenge angesehen, denn genau die Elemente von E werden durch keine zulässige Entscheidung in den "Subzielen" $F(.)_i : X \to \mathbb{R}$, $i=1,\dots,\ell$, dominiert, d.h.

$$\tilde{x}\in E \iff \begin{array}{l} \tilde{x}\in\Omega \text{ und es gibt kein } x\in\Omega \text{ mit :} \\ F(x)_i \geqq F(\tilde{x})_i \quad \text{für alle } i=1,\dots,\ell \text{ und} \\ F(x)_j > F(\tilde{x})_j \quad \text{für ein } j\in\{1,\dots,\ell\} \ . \end{array}$$

"Although there is no universally accepted solution concept for decision problems with multiple noncommensurable objectives, one would agree that a 'good' solution must not be dominated by other feasible alternatives" bemerkt P.L.Yu. Zur "Berechnung" von E hat man zunächst "algorithmisch auswertbare" Bedingungen aufgestellt, die notwendig oder sogar hinreichen für das Enthaltensein in E sind. Als Leitgedanke dabei dienen Sattelpunktseigenschaften oder Stationarität einer Lagrangefunktion. Aus den Arbeiten von H.W.Kuhn, A.W.Tucker, L.Hurwicz und A.M.Geoffrion etwa sei der Fall angeführt, bei dem Ω Lösungsmenge eines Systems von $m \in \mathbb{N}$ Ungleichungen $G(x)_\kappa \geqq 0$, $\kappa=1,\ldots,m$ ist:

Satz 1 : Für ein VMP der Art $(\mathbb{R}^n, \{x\in\mathbb{R}^n \mid G(x) \in \mathbb{R}^m_+\}, \mathbb{R}^\ell, F, \mathbb{R}^\ell_+)$ wobei $G: \mathbb{R}^n \to \mathbb{R}^m$; bei dem die Abbildungen F und G stetig differenzierbar sind, betrachte man zu einem $\tilde{x} \in X$ die Menge

$$M(\tilde{x}) := \{\mu \in \mathbb{R}^\ell_+ \setminus \{0\} \mid \text{es gibt ein } \lambda \in \mathbb{R}^m \text{ so, daß: } F'(x)\mu + G'(x)\lambda = 0 \text{ und } G(x)\lambda = 0\}.$$

Dann gilt:

a) Sind G und F konkav, $\tilde{x} \in E$ und erfüllt G an der Stelle $\tilde{x}$ die Slaterbedingung $G(X) \cap \mathbb{R}_{++} \neq \emptyset$, dann ist $M(\tilde{x})$ nicht leer;

b) Sind G und F konkav und $M(\tilde{x}) \cap \mathbb{R}_{++} \neq \emptyset$, dann folgt $\tilde{x} \in E$;

c) Erfüllt G an der Stelle x entweder die "constraint qualification" von Kuhn-Tucker oder die von Arrow-Hurwicz-Uzawa und gilt $\tilde{x} \in E$, dann folgt $M(\tilde{x}) \neq \emptyset$.

Beweis: Mit dem "Trick" von L.Hurwicz kann man das Vektormaximumproblem als Programmierungsproblem deuten. Für die genaue Zusammenstellung der Beweisführung vgl. A.Takayama.

Bemerkung 2 : So wie die Sensitivitätsanalyse bei nichtlinearen Programmierungsproblemen zeigt, daß sich die Lagrangemultiplikatoren λ_i, $i=1,\ldots,m$ als "Kosten" für (aktive) Restriktionen interpretieren lassen (siehe etwa D.G.Luenberger), erfahren die μ_i, $i=1,\ldots,\ell$ ihre Deutung als Outputpreise bzw. als Bewertung der "Subziele" $F(.)_i$, $i=1,\ldots,\ell$.

Im Hinblick auf den Einsatz von VMP als Planungsinstrument und zur Hilfe bei der Entscheidungsfindung, hat man versucht, geeignete echte Teilmengen von E als Lösung von VMP anzusehen. Mit den Begriffen der "proper efficiency" und der "properly efficient solution" zogen H.W.Kuhn und A.W.Tucker bzw. A.M.Geoffrion solche Entscheidungen von E außer Betracht, bei deren "Variation" in zulässigen Richtungen einem Zuwachs der

Werte eines Subzieles $F(.)_j$ nur ein "Verlust von höherer Ordnung" in den Werten der anderen Subziele $F(.)_i$, $i \in \{1,\ldots,\ell\} \setminus \{j\}$ gegenübersteht. Hierzu gehören auch spezielle Bedingungen an F und G, die sichern, daß E einelementig ist (vgl. P.Bod) sowie der Ansatz des "goal programming", bei dem man sich zu einem VMP (X,Ω,V,F,P) eine Abstandsfunktion $d(.,.) : V \times V \to \mathbb{R}$ und ein $\bar{v} \in V$ vorgibt und die Lösungsmenge der Approximationsaufgabe "minimiere $\{d(F(x),\bar{v}) \mid x \in \Omega\}$" als Lösungsmenge von VMP wählt, sofern sie in E enthalten ist.

Bei der in dieser Arbeit gestellten Aufgabe geht es nun nicht um eine möglichst enge Fassung des Lösungsbegriffs für ein VMP und die Berechnung der Lösungsmenge (die dann eine Teilmenge von E wäre), sondern es wird eine zulässige Entscheidung gegeben sein, die sich nur unter einer Verallgemeinerung des Lösungsbegriffs als Lösung des VMP auffassen läßt, und die es zu "analysieren" gilt. Die Beispiele in §3 begründen, weshalb diese zulässige Entscheidung Element von LS aus Def.1 ist. Nun kann Satz 1 so verallgemeinert werden, daß auch für Elemente $\tilde{x} \in LS$ Existenz und Stationarität einer Lagrangefunktion bewiesen werden kann. In dieser Lagrangefunktion sind die in $F'(\tilde{x}) : X \to V$ zusammengefassten linearisierten Änderungen der Werte der Subziele durch ein positives $\mu \in V^*$ aggregiert, das sich als "Preissystem" oder Bewertung interpretieren läßt und so Aufschluß auf die nicht offenliegende Präferenz des Entscheidungsträgers (der $\tilde{x}$ aus Ω gewählt hat) liefert: Mit der Wahl von $\tilde{x}$ aus Ω handelt der Entscheidungsträger so, als ob es gelte, das reelle Funktional $x \mapsto \mu F(x)$ auf Ω zu maximieren oder wenigstens einen stationären Punkt zu finden.

2. LOKALE-SLATER-KEGEL-EFFIZIENZ

Zur Herleitung der notwendigen Bedingung für lokal-Slater-kegeleffiziente Entscheidungen werden Begriffe benötigt, wie sie von P.P.Varaiya verwendet werden und die auf die Separationsmethode von A.Y.Dubovitskii und A.A.Miljutin zurückgehen, bei der zulässige Menge und Majorantenmenge lokal durch disjunkte Kegel approximiert werden, die unter Regularität konvex und dann trennbar sind.

Definition 2 : Sei X ein topologischer linearer Raum, Ω eine nichtleere Teilmenge von X und $\tilde{x} \in \Omega$. Der Durchschnitt aller abgeschlossenen Kegel in X die $\Omega - \tilde{x}$ enthalten, wird "closed cone of Ω at $\tilde{x}$" genannt und mit $C(\Omega,\tilde{x})$ bezeichnet. Unter dem "local closed cone of Ω

at $\tilde{x}$" versteht man die Menge

$$LC(\Omega,\tilde{x}) := \bigcap_{U \in \mathfrak{U}} C(\Omega \cap U, \tilde{x}) ,$$

wobei $\mathfrak{U}$ das System der Umgebungen von $\tilde{x}$ in X ist.

Bemerkung 3 : Die Inklusionen $C(\Omega,\tilde{x}) \supset LC(\Omega,\tilde{x}) \supset \{0\}$ sind klar. Zur Definition des Tangentialkegels von M.R.Hestenes : es gilt

$$LC(\Omega,\tilde{x}) \subset \{k \in X \mid \text{es gibt eine Folge } \{x_\nu\}_{\nu\in\mathbb{N}} \text{ in } \Omega \text{ sowie eine Folge } \{r_\nu\}_{\nu\in\mathbb{N}} \text{ nichtnegativer reeller Zahlen mit } \{x_\nu\} \to \tilde{x} , \{r_\nu(x_\nu-\tilde{x})\} \to k \}$$

immer und Mengengleichheit in topologischen linearen Räumen mit abzählbarer Nullumgebungsbasis: Insbesondere in normierten Räumen stimmt der "local closed cone" mit dem Tangentialkegel von Hestenes überein.

Satz 2 : Für ein VMP (X,Ω,V,F,P) sei $\tilde{x} \in LS$ und F an der Stelle $\tilde{x}$ differenzierbar und K eine nichtleere konvexe Teilmenge von $LC(\Omega,\tilde{x})$. Dann existiert ein $\mu \in V^*$ mit der Positivitätsbedingung $\mu(\text{int}(P)) > 0$ so, daß für alle $k \in K$: $\mu(F'(\tilde{x})k) \leqq 0$ gilt.

Beweis : Zunächst wird $F'(\tilde{x})(LC(\Omega,\tilde{x})) \cap \text{int}(P) = \emptyset$ gezeigt, dann der Trennungssatz von M.Eidelheit (vgl.G.Köthe) angewendet. Also angenommen, es gebe ein $\bar{k} \in LC(\Omega,\tilde{x})$ mit $F'(\tilde{x})\bar{k} \in \text{int}(P)$. Zu $\bar{k}$ gibt es nach Bem.3 und der Def.2 eine Folge $\{x_\kappa\}_{\kappa\in\mathbb{N}} \to \tilde{x}$ mit $x_\kappa \in \Omega$ sowie eine Folge $\{r_\kappa\}_{\kappa\in\mathbb{N}}$ nichtnegativer reeller Zahlen mit der Konvergenz $\{r_\kappa(x_\kappa-\tilde{x})\} \to \bar{k}$. Aus $\bar{v} := F'(\tilde{x})\bar{k} \in \text{int}(P)$ folgt $\bar{v} \neq 0$ und daraus $\bar{k} \neq 0$ und weiter $x_i \neq \tilde{x}$, $r_i \neq 0$ für alle (genügend große) $i \in \mathbb{N}$. Die Abschätzung : Für alle $\gamma > 0$ gibt es ein $n(\gamma) \in \mathbb{N}$ so, daß

$$\| r_\kappa(F(x_\kappa)-F(\tilde{x})) - \bar{v} \| \leqq \text{(Dreiecksungleichung)} \leqq$$

$$\leqq \| r_\kappa(F(x_\kappa) - F(\tilde{x}) - F'(\tilde{x})(x_\kappa-\tilde{x}))\| + \| r_\kappa F'(\tilde{x})(x_\kappa-\tilde{x}) - \bar{v} \| \leqq$$

$$\leqq r_\kappa \cdot \| x_\kappa \ \tilde{x} \| \cdot \gamma + \| F'(\tilde{x}) \| \cdot \| r_\kappa(x_\kappa-\tilde{x}) - \bar{k} \| \leqq$$

(da $\{r_\nu \| x_\nu-\tilde{x} \|\} \to \|\bar{k}\|$ und folglich diese Folge beschränkt ist)

$$\leqq \left(\sup\{r_\nu \cdot \| x_\nu-\tilde{x} \| \mid \nu \in \mathbb{N}\} + \sup\{ \| r_\nu(x_\nu-\tilde{x}) - \bar{k} \| \mid \nu \in \mathbb{N}\}\right) \cdot \gamma$$

für alle $\kappa \geqq n(\gamma)$ gilt, zeigt, daß in jeder γ - Umgebung von $\bar{v} \in \text{int}(P)$ Punkte $r_\kappa(F(x_\kappa) - F(\tilde{x}))$, $\kappa \geqq n(\gamma)$, liegen; folglich auch in einer Umgebung, die ganz in $\text{int}(P)$ liegt. Aus $r_i(F(x_i) - F(\tilde{x})) \in \text{int}(P)$ für ein

$i \in \mathbb{N}$ folgt mit der Kegeleigenschaft von $int(P)$ (vgl.Bem.1) aber $F(x_i) - F(\tilde{x}) \in int(P)$, ein Widerspruch zur lokalen-Slater-Kegeleffizienz von $\tilde{x}$, denn man kann wegen $\{x_\nu\} \to \tilde{x}$ ein x_i aus der mit "lokal" angesprochenen Umgebung von $\tilde{x}$ wählen. Man hat also zwei konvexe Mengen P und K mit der Eigenschaft $K \cap int(P) = \emptyset$ und $int(P) = \emptyset$. Der topologische Trennungssatz besagt die Existenz einer P und K trennenden abgeschlossenen Hyperebene, die keinen inneren Punkt von P enthält. Dazu äquivalent ist die Existenz eines stetigen linearen Funktionals mit den in der Aussage des Satzes genannten Eigenschaften. Q.E.D.

Zur Anwendung von Satz 2 wäre es gut, wenn der Kegel $LC(\Omega,\tilde{x})$ selbst bereits konvex wäre, wenn er etwa eine Hyperebene oder ein von einer Hyperebenen begrenzter Halbraum wäre und wenn man ihn "geeignet charakterisieren" könnte. Aus der Theorie der nichtlinearen Programmierung in Banachräumen sind derartige Bedingungen bekannt. In sie gehen Regularitätsvoraussetzungen ein. Im Hinblick auf Kontrollprobleme soll der Fall

(†) $\begin{cases} \Omega = \{x \in X \mid x \in A_X \text{ und } G(x) \in A_Y\} \\ \text{wobei } A_X \text{ eine konvexe Teilmenge von } X \text{ mit} \\ \text{nichtleerem Inneren ist; } G \text{ ist eine Abbildung} \\ \text{von } X \text{ in einen reellen Banachraum } Y \text{ und} \\ A_Y \text{ ist eine konvex Teilmenge von } Y . \end{cases}$

behandelt werden.

<u>Korollar zu Satz 2</u> : Für ein VMP (X,Ω,V,F,P) wo die zulässige Menge Ω durch (†) gegeben ist, sei $\tilde{x} \in LS$, F an der Stelle $\tilde{x}$ differenzierbar, G an der Stelle $\tilde{x}$ stetig differenzierbar sowie $G'(\tilde{x}) \in \mathcal{L}(X,Y)$ surjektiv. Dann existiert ein $\mu \in V^*$ mit der Positivitätsbedingung $\mu(int(P)) > 0$ sowie ein $\lambda \in Y^*$ mit:

für alle $y \in LC(A_Y, G(\tilde{x}))$ gilt: $\lambda(y) \geqq 0$ und
für alle $k \in \mathbb{R}_+(int(A_X) - \tilde{x})$ gilt:

$$\mu(F'(\tilde{x})k) + \lambda(G'(\tilde{x})k) \leqq 0.$$

Ist speziell $A_X = X$ und $A_Y = \{0\}$, hat man für alle $k \in X$:

$$\mu(F'(\tilde{x})k) + \lambda(G'(\tilde{x})k) = 0 ,$$

also $\mu(F'(\tilde{x})k) = 0$ für alle $k \in Kern(G'(\tilde{x}))$.

Zum Beweis des Korollars zu Satz 2 benötigen wir drei Hilfssätze:

Lemma 1 (von L.A.Ljusternik über Tangentialmannigfaltigkeiten) : Seien X und Y reelle Banachräume, $G : X \to Y$ eine Abbildung, die an einer Stelle $\tilde{x} \in \Omega := \{x \in X \mid G(x)=0\}$ stetig differenzierbar ist und ferner $G'(\tilde{x}) \in \mathcal{L}(X,Y)$ surjektiv; dann gilt $LC(\Omega,\tilde{x}) = \mathrm{Kern}(G'(\tilde{x}))$.

Beweis von Lemma 1 : Die Inklusion $LC(\Omega,\tilde{x}) \subset \mathrm{Kern}(G'(\tilde{x}))$ gilt auch ohne die Vollständigkeit von X und Y, ohne stetige Differenzierbarkeit und ohne die Regularitätsvoraussetzung der Surjektivität von $G'(\tilde{x})$, denn zu jedem $k \in LC(\Omega,\tilde{x})$ existieren Folgen $\{x_\kappa\} \to \tilde{x}$, mit $G(\tilde{x}_\kappa)=0$, und $\{r_\kappa\}_{\kappa\in\mathbb{N}}$, $r_\kappa \in \mathbb{R}_+$ mit $\{r_\kappa(x_\kappa-\tilde{x})\} \to k$, also :

$$\begin{aligned} \|G'(\tilde{x})k\| &= \|r_\kappa(G(x_\kappa)-G(\tilde{x}) - G'(\tilde{x})k\| \leqq \\ &\leqq \|r_\kappa G(x_\kappa) - r_\kappa G(\tilde{x}) - r_\kappa G'(\tilde{x})(x_\kappa-\tilde{x})\| + \|G'(\tilde{x})\|\cdot\|r_\kappa(x_\kappa-\tilde{x}) - k\| \leqq \\ &\leqq (\sup\{r_\nu\|x_\nu-\tilde{x}\| \mid \nu\in\mathbb{N}\})\cdot(1+\|G'(\tilde{x})\|)\cdot\gamma \quad , \text{ für } \kappa \geqq n(\gamma) \in \mathbb{N} . \end{aligned}$$

Für alle $\gamma > 0$ gilt also $\|G'(\tilde{x})k\| < \gamma$, was $k \in \mathrm{Kern}(G'(\tilde{x}))$ bedeutet und womit die erste Inklusion gezeigt ist.
Besonders wichtig (um die Orthogonalität von $\mu F'(\tilde{x})$ auf dem Tangentialkegel auf den Kern von $G'(\tilde{x})$ übertragen zu können) ist aber die andere Inklusion, $LC(\Omega,\tilde{x}) \supset \mathrm{Kern}(G'(\tilde{x}))$. Um sie zu zeigen, konstruieren wir zu einem beliebigen $k \in \mathrm{Kern}(G'(\tilde{x}))$ Folgen $\{x_\kappa\}$ in Ω und $\{r_\kappa\}$ in $\mathbb{R}_+$ und zeigen, daß sie die Eigenschaften $\{x_\kappa\} \to \tilde{x}$ und $\{r_\kappa(x_\kappa-\tilde{x})\} \to k$ haben, die $k \in LC(\Omega,\tilde{x})$ sichern: Als stetiger linearer Operator zerlegt $G'(\tilde{x}) : X \to Y$ den Banachraum X in die direkte Summe $X = H^1 \oplus H^2$ aus dem abgeschlossenen Unterraum $H^1 := \mathrm{Kern}(G'(\tilde{x}))$ und $H^2 :\cong X/H^1$, also ist jedes $x_\kappa \in \Omega \subset X$ eindeutig zerlegbar, $x_\kappa = \tilde{x} + h_\kappa^1 + h_\kappa^2$ mit $h_\kappa^1 \in H^1$ und $h_\kappa^2 \in H^2$. Dann bedeutet $x_\kappa \in \Omega$ gerade $G(x_\kappa) = G(\tilde{x}+h_\kappa^1+h_\kappa^2) = 0$ oder $\Phi(h_\kappa^1,h_\kappa^2) = 0$, wenn man $\Phi : H^1 \times H^2 \to Y$ durch $\Phi(h^1, h^2) := G(\tilde{x}+h^1+h^2)$ definiert. Die Abbildung Φ ist wegen der Voraussetzungen an G in einer offenen Teilmenge von $(0,0) \in H^1 \times H^2$ stetig differenzierbar, es gilt $\Phi(0,0) = 0$ und $\nabla_2\Phi(0,0)$ ist surjektiv. Der Satz über implizite Funktonen sichert die Existenz einer γ-Umgebung von $0 \in H^1 \subset X$ und einer dort stetig differenzierbaren Abbildung Ψ aus H^1 in H^2 mit der Eigenschaft : $\Phi(h^1, \Psi(h^1)) = 0$ für alle $h^1 \in H^1$ mit $\|h^1\| \leqq \gamma$.
Nun wählen wir zu $k \in \mathrm{Kern}(G'(\tilde{x}))$ für alle $\kappa \in \mathbb{N}$ $x_\kappa := \tilde{x} + \frac{\gamma}{\kappa}k + \Psi(\frac{\gamma}{\kappa}k)$ und $r_\kappa := \kappa/\gamma$; damit sind $x_\kappa \in \Omega$ und $\{x_\kappa\} \to \tilde{x}$ klar; weiter folgt $\{r_\kappa(x_\kappa-\tilde{x})\} \to k$ aus $\|r_\kappa(x_\kappa-\tilde{x}) - k\| = \|\frac{\kappa}{\gamma}\cdot\Psi(\frac{\gamma}{\kappa}k)\| \to 0$ für $\kappa \to \infty$ da $\Psi(0)=0$, $\Psi'(0) = -\nabla_2\Phi(0,0)^{-1} \circ \nabla_1\Phi(0,0) = 0$, wie sich sofort aus

$\nabla_1\Phi(O,0) = G'(\tilde{x})|_{H^1} = 0 \in \mathcal{L}(H^1,Y)$ ergibt. Q.E.D.

Lemma 2 : Seien X und Y reelle Banachräume, $G : X \to Y$ eine Abbildung die an einer Stelle $\tilde{x} \in \Omega := \{x \in X \mid G(x) \in A_Y\}$ stetig differnzierbar ist wobei A_Y eine Teilmenge von Y ist. Dann gilt

$$LC(\Omega,\tilde{x}) = \{k \in X \mid G'(\tilde{x})k \in LC(A_Y,G(\tilde{x})\} .$$

Ist darüberhinaus A_Y konvex, dann ist auch $LC(\Omega,\tilde{x})$ konvex.

Beweis von Lemma 2 : Zunächst die Mengeninklusion "$\subset$" : zu $k \in LC(\Omega,\tilde{x})$ existieren Folgen $\{x_\kappa\}_{\kappa \in \mathbb{N}}$ und $\{r_\kappa\}_{\kappa \in \mathbb{N}}$ mit $x_\kappa \in \Omega$, $r_\kappa \geq 0$ sowie den Konvergenzeigenschaften $\{x_\kappa\} \to \tilde{x}$ und $\{r_\kappa(x_\kappa - \tilde{x})\} \to k$. Daraus folgt für die durch $y_\kappa := G(x_\kappa)$ definierte Folge $\{y_\kappa\}$ in A_Y : $\{y_\kappa\} \to G(\tilde{x})$, $\{r_\kappa(y_\kappa - G(\tilde{x}))\} = \{r_\kappa(G(x_\kappa) - G(\tilde{x}))\} \to \lim \{r_\kappa G'(\tilde{x})(x_\kappa - \tilde{x})\} = G'(\tilde{x})k$,was $G'(\tilde{x})k \in LC(A_Y,G(\tilde{x}))$ bedeutet und die erste Inklusion des Lemmas zeigt. Für die zweite Inklusion erinnern wir an den Beweis von Lemma 1; an die Zerlegung $X = H^1 \oplus H^2$ mit $H^1 := \mathrm{Kern}(G'(\tilde{x}))$ und $H^2 \cong X/H^1$. Der auf dem Produktraum von $H^1 \times Y$ und H^2 durch

$$\Phi((h^1,\Delta y),h^2) := G(\tilde{x} + h^1 + h^2) - G(\tilde{x}) - \Delta y$$

definierte Operator $\Phi : (H^1 \times Y) \times H^2 \to Y$ erfüllt an der Stelle ((O,O),O) die Voraussetzungen des Satzes über implizite Funktionen, denn es gilt

$\Phi((0,0),0) = 0$,

Φ ist an dieser Stelle stetig differenzierbar,

$\nabla_1\Phi((0,0),0) = (G'(\tilde{x})|_{H^1}, -Id_Y) \in \mathcal{L}(H^1 \times Y,Y)$,

$\nabla_2\Phi((0,0),0) = G'(\tilde{x})|_{H^2} \in \mathcal{L}(H^2,Y)$ ist surjektiv.

Also gibt es eine Abbildung $\psi : (H^1 \times Y) \cap U_\gamma \to Y$ wobei U_γ eine γ-Umgebung von (O,O) ist, mit den angeführten Eigenschaften.
Sei nun $k \in \{k \in X \mid G'(\tilde{x})k \in LC(A_Y,G(x))\}$, $k = k^1 + k^2$ mit $k^i \in H^i$, i=1,2 Ist $k^2 = O$, also $k \in \mathrm{Kern}(G'(\tilde{x}))$, haben wir mit Lemma 1 sofort

$$k \in LC(\{x \in X \mid G(x) = G(\tilde{x})\}, \tilde{x}) \subset LC(\Omega,\tilde{x})$$

und wir wären fertig; deshalb können wir im folgenden $k^2 \neq 0$ annehmen. Zu k existieren Folgen $\{y_\kappa\} \to G(\tilde{x})$, $y_\kappa \in A_Y$ und $\{r_\kappa\}$ mit $r_\kappa \geq 0$ und der Konvergenz von $\{r_\kappa(y_\kappa - G(\tilde{x})\}$ gegen $G'(\tilde{x})k = G'(\tilde{x})k^2$. Die Zahlenfolge $\{r_\kappa\}$ divergiert, andernfalls würde wegen $\{y_\kappa\} \to G(\tilde{x})$ sofort $G'(\tilde{x})k = 0$ folgen im Widerspruch zu $k^2 \neq 0$ und der Injektivität von $G'(\tilde{x})|_{H^2}$. Ohne die Konvergenz von $\{r_\kappa(y_\kappa - G(\tilde{x}))\}$ gegen $G'(\tilde{x})k$ zu ändern, ersetzen wir diejenigen r_κ , $\kappa \in \mathbb{N}$ mit $r_\kappa < 1$ durch 1; ebenso dürfen wir $\sup(\{\| y_\nu - G(\tilde{x})\| \mid \nu \in \mathbb{N}) < 1$ annehmen. Für alle $\kappa \in \mathbb{N}$ definieren wir

$$\begin{aligned} k_\kappa^1 &:= \frac{\gamma}{r_\kappa} k^1 , \\ \Delta y_\kappa &:= \gamma \cdot (y_\kappa - G(\tilde{x})) , \\ x_\kappa &:= \tilde{x} + k_\kappa^1 + \Psi(k_\kappa^1, \Delta y_\kappa) . \end{aligned}$$

Dann gilt zunächst $\Phi((k_\kappa^1, \Delta y_\kappa), \Psi(k_\kappa^1, \Delta y_\kappa)) = 0$, also $G(x_\kappa) = y_\kappa$, d.h. $x_\kappa \in \Omega$, $\kappa \in \mathbb{N}$, ferner $\{x_\kappa\} \to \tilde{x}$ denn die Folge $\{\Delta y_\kappa\}$ konvergiert gegen 0 und auch $\{k_\kappa^1\}$ konvergiert gegen 0 , da die Folge der Zahlen r_κ divergiert. Schließlich konvergiert

$$\{(r_\kappa/\gamma)\cdot(x_\kappa - \tilde{x})\}_{\kappa\in\mathbb{N}} = \{k_\kappa^1 + \frac{r_\kappa}{\gamma}\cdot\Psi(k_\kappa^1, \Delta y_\kappa)\}_{\kappa\in\mathbb{N}}$$

wegen der stetigen Differenzierbarkeit von Ψ gegen

$$\lim_{\kappa\to\infty} (k_\kappa^1 + \frac{r_\kappa}{\gamma}\cdot(\Psi(0,0) + \Psi'(0,0)(k_\kappa^1, \Delta y_\kappa))) = k^1 + k^2 = k ,$$

da $\Psi(0,0)=0$ und $\Psi'(0,0) = -(G'(\tilde{x})|_{H^2})^{-1} \circ (G'(\tilde{x})|_{H^1}, -Id_Y)$ gilt. Also ist

$$\begin{aligned} \lim \frac{r_\kappa}{\gamma}\cdot\Psi'(0,0)(k_\kappa^1, \Delta y_\kappa) &= \lim 0 + \frac{\gamma}{r_\kappa}(G'(\tilde{x})|_{H^2})^{-1}(\Delta y_\kappa) = \\ = (G'(\tilde{x})|_{H^2})^{-1}(G'(\tilde{x})k^2) &= k^2 \end{aligned}$$

womit $k \in LC(\Omega,\tilde{x})$ gezeigt ist.

Um den Beweis von Lemma 2 zu vollenden, folgern wir noch aus der Konvexität von A_Y die von $LC(\Omega,\tilde{x})$. Ist nun C eine konvexe Teilmenge eines normierten Raumes und c Element der abgeschlossenen Hülle cl(C) von C , dann ist $LC(C,c) = cl(\mathbb{R}_+ (cl(C)-c))$; vgl. P.J.Laurent. Klarerweise ist $\mathbb{R}_+(cl(C)-c)$ konvex und dann ist es auch der Abschluß dieser Menge. Somit folgt aus der Konvexität von A_Y die Konvexität des Kegels $LC(A_Y, G(\tilde{x}))$ und mit der Linearität folgt die Konvexität seines Urbildes $LC(\Omega,\tilde{x})$. Q.E.D.

Während Lemma 1 den "local closed cone" der zulässigen Menge Ω im Punkt $\tilde{x}$ für den Fall angibt, bei dem Ω Lösungsmenge einer Gleichung $G(x) = 0$ ist, behandelt Lemma 2 den Fall $\Omega = \{x \in X \mid G(x) \in A_Y\}$. Für (†) , also $\Omega = \{x \in X \mid x \in A_X \text{ und } G(x) \in A_Y\}$, können wir nicht $LC(\Omega,\tilde{x})$ genau charakterisieren, wohl aber eine konvexe Teilmenge Q dieses Kegels berechnen, die im Hinblick auf die beabsichtigte Herleitung des Korollars aus Satz 2 ausreichend "groß" ist :

Lemma 3 : Seien A und A_X Teilmengen eines reellen normierten Raumes X und $\tilde{x} \in X$. Ferner sei A_X konvex und habe nichtleeres Inneres. Der Kegel $LC(A,\tilde{x})$ sei konvex. Dann gilt

$$LC(A \cap A_X, \tilde{x}) \supset Q := LC(A,\tilde{x}) \cap \mathbb{R}_+(\mathrm{int}(A_X) - \tilde{x})$$

und Q ist ein konvexer Kegel.

Beweis von Lemma 3 : Sei $q \in Q$. Aus $q \in \mathbb{R}_+(\mathrm{int}(A_X) - \tilde{x})$ folgt die Existenz eines $\alpha > 0$ mit $\bar{q} := \alpha q \in \mathrm{int}(A_X)$ und $\bar{q} \in LC(A,\tilde{x})$. Für die wegen $\bar{q} \in LC(A,\tilde{x})$ existierenden Folgen $\{\bar{x}_\kappa\}$ und $\{\bar{r}_\kappa\}$ von Elementen von A bzw. nichtnegativer reeller Zahlen mit den beiden Konvergenzeigenschaften $\{\bar{x}_\kappa\} \to \tilde{x}$ und $\{\bar{r}_\kappa(\bar{x}_\kappa - \tilde{x})\} \to \bar{q}$ gilt wegen $\bar{q} + \tilde{x} \in \mathrm{int}(A_X)$ ab einem gewissen $n \in \mathbb{N}$ für $\kappa \geqq n$ auch

$$\tilde{x} + \bar{r}_\kappa(\bar{x}_\kappa - \tilde{x}) \in \mathrm{int}(A_X)$$

und wegen der Konvexität von A_X auch $\bar{x}_\kappa \in A_X$, also $\bar{x}_\kappa \in \Omega \cap A_X$. Die Existenz dieser Folge $\{\bar{x}_\kappa\}_{\kappa \geq n}$ zeigt $\bar{q} \in LC(A \quad A_X,\tilde{x})$ und wegen der Kegeleigenschaft auch $q \in LC(A \cap A_X,\tilde{x})$. Die Konvexität und Kegeleigenschaft von Q ist offensichtlich. Q.E.D.

Beweis des Korollars zu Satz 2 : Der Satz 2 besagt $\mu(F'(\tilde{x})(k)) \leqq 0$ für alle $k \in Q$, wobei Q sich wie in Lemma 3 berechnet Für die in diesem Lemma mit A bezeichnete Menge wähle man $\{x \in X \mid G(x) \in A_Y\}$ und wende Lemma 2 an, woraus sich die Konvexität von $LC(A,\tilde{x})$ ergibt. Nun bedeutet $k \in Q$ gerade $T(k) \in (LC(A_Y, G(\tilde{x})), \mathbb{R}_+(\mathrm{int}(A_X) - \tilde{x}))$ wenn man die Abbildung $T : X \to Y \times X$ durch $T(k) := (G'(\tilde{x})k, k)$ definiert. Man hat also

$$\mu(F'(\tilde{x}))(k) \leqq 0 \,, \quad \text{für alle } k \in X \text{ mit}$$
$$T(k) \in (LC(A_Y, G(\tilde{x})), \mathbb{R}_+(\mathrm{int}(A_X) - \tilde{x})) \quad \text{und}$$
$$\mu(F'(x))(k) = 0 \,, \quad \text{für alle } k \in \mathrm{Kern}(T) \,.$$

Nun ist zwar T nicht surjektiv, aber $\mathrm{Bild}(T)$ ist eine abgeschlossene Teilmenge von $Y \times X$, denn allein schon aus der Stetigkeit von $G'(\tilde{x})$ und der Vollständigkeit von X und Y folgt dies mit dem Satz über den abgeschlossenen Graphen, vgl. J.Dieudonné. Nun ist der Orthogonalraum von $\mathrm{Kern}(T)$, in dem $\mu(F'(\tilde{x})) \in X^*$ liegt, gleich dem Bild des zu adjungierten Operators T^* , wegen der Abgeschlossenheit von $\mathrm{Bild}(T)$

und der Vollständigkeit von X sowie $X \times Y$. Also existiert ein Paar $(\lambda,\zeta) \varepsilon Y^* \times X^*$ mit $\mu(F'(\tilde{x})) + \lambda(G'(x)) + \zeta = 0$ und

$$(\lambda,\zeta)(T(k)) \geqq 0 \,, \quad \text{für alle } k \varepsilon X \text{ mit}$$
$$T(k) \ \varepsilon \ (LC(A_Y, G(\tilde{x})), \mathbb{R}_+(\mathrm{int}(A_X)-\tilde{x}))$$

woraus sich sofort die Hauptaussage des Korollars zu Satz 2 ergibt. Zur Stationarität der Lagrangefunktion für den Spezialfall $A_X=X$, $A_Y=\{0\}$ hat man lediglich Lemma 1 anzuwenden. Q.E.D.

3. ANWENDUNGEN

Zunächst ein einfaches Beispiel, das auf ein VMP $(X,\Omega,\mathbb{R},G,\mathbb{R}^\ell_+)$ führt. In einer Ungewißheitssituation betrage die Auszahlung, auf die eine zulässige Entscheidung $x \varepsilon \Omega$ führt, gerade $G(x)_i$, wobei ungewiß ist, welcher "Umweltzustand" $i \varepsilon \{1,\dots,\ell\}$ eintreten wird. Eine Entscheidung $\tilde{x}$ ist mit Sicherheit echt besser als ein $x \varepsilon X$, wenn

$$\Delta v \ := \ G(\tilde{x}) - G(x) \ \varepsilon \ \mathrm{int}(\mathbb{R}^\ell_+) \ = \ \mathbb{R}^\ell_{++}$$

gilt, und nicht schon dann, wenn $\Delta v \ \varepsilon \ \mathbb{R}^\ell_+ \setminus \{0\}$ der Fall ist : als Menge "rationaler" Entscheidungen kann also nicht E sondern muß S angesehen werden, zumal ein Entscheidungsträger das Eintreten gewisser Umweltzustände aufgrund eines subjektiven Wahrscheinlichkeitsbegriffs für unmöglich halten könnte.

Angenommen, er habe sich in einer derartigen Ungewißheitssituation auf eine Entscheidung $\tilde{x} \varepsilon S$ festgelegt. Satz 2 besagt dann, daß der Entscheidungsträger sich so entschieden hat, wie wenn ein Risikofall vorläge, er subjektive Wahrscheinlichkeiten p_j für das Eintreten von j, $j=1,\dots,\ell$ habe und es gelte, die Erwartungswertfunktion

$$x \mapsto \sum_{i=1}^{\ell} p_i G(x)_i$$

auf Ω zu maximieren bzw. einer ihrer stationären Punkte zu finden. Dabei läßt sich für diese subjektiven Wahrscheinlichkeiten p_j die Abschätzung

$$\inf\{ \frac{\mu_j}{\Sigma \mu_i} \mid \mu \varepsilon M(\tilde{x}) \} \ \leqq \ p_j \ \leqq \ \sup\{ \frac{\mu_j}{\Sigma \mu_i} \mid \mu \varepsilon M(\tilde{x}) \}$$

angeben, wobei ähnlich wie in Satz 1 die Menge $M(\tilde{x})$ durch

$$M(\tilde{x}) := \{ \mu \varepsilon \mathbb{R}^\ell \mid \text{es gibt eine nichtleere konvexe Teilmenge } K \text{ von } LC(\Omega,\tilde{x}) \text{ so, daß } \mu^\dagger F'(\tilde{x}) K \leqq 0 \text{ gilt } \}$$

definiert ist.

Ist $F'(\tilde{x})(LC(\Omega,\tilde{x}))$ eine Hyperebene in $V = \mathbb{R}^{\ell}$, ist diese Analyse der Entscheidung $\tilde{x}$ scharf, da dann die p_j eindeutig bestimmt sind.

Bemerkung 4 : Besonders bei Problemstellungen aus dem ökonomischen Bereich ist es sinnvoll anzunehmen, der Entscheidungsträger kenne Ω und F nur lokal um eine (vielleicht in einer Vorperiode realisierten) Entscheidung. In diesem Fall würde er gegebenfalls "lokale Verbesserungen" vornehmen und sich auf eine Entscheidung $\tilde{x}$ festlegen, für die nicht mehr $\tilde{x} \in S$, aber noch $\tilde{x} \in LS$ angenommen werden kann.

Bemerkung 5 : Es genügt nicht, die Aktionen durch Vektoren des $\mathbb{R}^{\ell}$ zu bewerten. Ein im Zeitintervall $[a,b]$ ablaufender Prozeß könnte zu einem ungewissen Zeitpunkt inspiziert werden. Hier wäre etwa $V = \mathcal{C}_o[a,b]$ denkbar. Ein weiteres Beispiel dazu ist die Analyse der Bewertung zukünftiger Ereignisse : Sei $X = (L_2[a,b])^{n+m}$ der Raum der Zustands- und Steuervariablen eines Kontrollproblems; Ω beschreibe den Prozeß und Beschränkungen; das zu maximierende reellwertige Zielfunktional sei

$$x \mapsto \int_a^b \beta(t) g(x(t),t)\, dt$$

wobei $\beta(.)$ eine Diskontierfunktion ist, $\beta \in B := (\mathcal{C}_o[a,b])_+ \setminus \{0\}$. Dieses Kontrollproblem habe die Lösungsmenge L_β. Folgendes Problem der Analyse ist nun gestellt : gegeben sei ein

$$\tilde{x} \in \underline{L} \qquad \text{mit} \qquad \underline{L} := \bigcup_{\beta \in B} L_\beta \quad ,$$

gesucht ist ein $\tilde{\beta} \in B$ mit $\tilde{x} \in L_{\tilde{\beta}}$. Nun gilt $\underline{L} \subset LS$, wobei LS die Menge aller lokal-Slater-kegeleffizienten $x \in X$ des VMP

$$(X , \Omega , \mathcal{C}_o[a,b] , F , (\mathcal{C}_o[a,b])_+)$$

mit $F(x)(t) := g(x(t),t)$ für $t \in [a,b]$, ist. Also ist ein $\tilde{x} \in LS$ gegeben. Das Funktional μ aus Satz 2 gibt Information über die $\tilde{x}$ entsprechende Diskontierfunktion $\tilde{\beta}(.)$.

Bemerkung 6 : Die in Satz 2 und dem Korollar zu Satz 2 angegebenen notwendigen Bedingungen für die lokale-Slater-Kegeleffizienz sind auch hinreichend, d.h. charakterisierend, bei Vektormaximierungsproblemen VMP (X,Ω,V,F,P) mit $(\dagger)$, wenn die folgende Konvexitätsbedingung erfüllt ist :

$$(\dagger\dagger)\quad \begin{cases} F : X \to V & \text{ist konkav,} \\ \{x \in X \mid G(x) \in A_Y\} & \text{ist konvex und} \\ \underline{\Omega} := \{x \in X \mid G(x) \in A_Y\} \cap \operatorname{int}(A_X) \neq \emptyset . \end{cases}$$

Der Beweis verwendet an wesentlicher Stelle Proposition (1.3.2) von P.J.Laurent, mit dem es möglich ist, $\Omega \subset cl(\underline{\Omega})$ zu folgern und ist hier nicht weiter ausgeführt.

4. LITERATUR

P.BOD : Über "indifferente" Optimierungsaufgaben. Methods of Operations Research XVI(1973)4o-5o.

J.DIEUDONNE : Foundations of Modern Analysis. Pure and Applied Mathematics 10 , Academic Press (1960)p.88.

A.Y.DUBOVITSKII and A.A.MILJUTIN : Extremum Problems in the Presence of Restrictions. USSR Computational Mathematics and Mathematical Physics 5(1965)3,1-8o.

A.M.GEOFFRION : Proper Efficiency and the Theory of Vector Maximization. Journal of Mathematical Analysis and Applications 22(1968)618-630.

L.HURWICZ : Programming in Linear Spaces; in K.J.Arrow, L.Hurwicz and H.Uzawa : Studies in linear and non-linear Programming, Stanford University Press (1958)38-102.

G.KÖTHE : Topologische Lineare Räume I . Die Grundlehren der Mathematischen Wissenschaften in Einzeldarstellungen 107 , Springer-Verlag (1966)S.191.

H.W.KUHN and A.W.TUCKER : Nonlinear Programming. Proc.2nd Berkely Symposium on Mathematical Statistics and Probability, University of California Press (1951)481-492.

P.J.LAURENT : Approximation et optimisation. Collection Enseignement des sciences 13, Hermann, Paris (1972)p.8-10.

L.A.LJUSTERNIK : Sur les extrémes relatifs des fonctionelles. Recueil Mathématique de la Société Mathématique de Moscou 41 (1934)390-401.

D.G.LUENBERGER : Introduction to Linear and Nonlinear Programming. Addison-Wesley (1973)230-231.

Y.NAGAHISA and Y.SAKAWA : Nonlinear Programming in Banach Spaces. Journal of Optimization Theory and Applications 4(1969)3,182-190.

M.Z.NASHED : Differentiability and Related Properties of Nonlinear Operators - Some Aspects of the Role of Differentials in Nonlinear Functional Analysis; in L.B.Rall (ed.): Nonlinear Functional Analysis and Applications, Academic Press (1971) 103-309.

A.TAKAYAMA : Mathematical Economics. The Dryden Press (1974) 112-116.

P.P.VARAIYA : Nonlinear Programming in Banach Space. SIAM J.Appl. Math. 15(1967)2, 284-293.

P.L.YU : Cone Convexity, Cone Extreme Points, and Nondominated Solutions in Decision Problems with Multiobjectives. Journal of Optimization Theory and Applications 14(1974)3,319-377.

Preference Optimality

(An optimality concept in multicriteria problems)

W. Stadler

Summary

Any concept of optimality on a set is based on an 'ordering' of the set. In its most general form such an 'ordering' will be termed a 'preference' on a set and it will simply be a binary relation whose purpose it is to introduce a hierarchy among the elements of the set. In finite-dimensional multicriteria decision problems one basically deals with a mapping $g(.):\mathcal{D} \to \mathcal{A}$, where $\mathcal{D}$ is some decision set and $\mathcal{A} = \{a \in R^N: a=g(d), d \in \mathcal{D}\}$ is the set of corresponding criterion values, e.g. $\mathcal{D}$ may be a functional constraint set in programming or a set of admissible controls in control theory. The optimality concept introduced here consists of a preference $\precsim$ on $\mathcal{A}$ and the definition of preference optimal decisions $d^* \in \mathcal{D}$ as those for which $g(d^*)=a^*$ is a least or a minimal element of $\mathcal{A}$ with respect to $\precsim$.

1. Introduction

Any definition of an optimal element of a set rests on some notion of 'order' among the elements of the set. In the usual programming and control problems one maximizes or minimizes a single criterion function. In similar problems with several criterion functions Pareto optimality has become an engrained concept. In the first case one uses the 'natural' order on the reals and defines optimality in terms of least elements with respect to that order, and in the second case one uses the 'natural' order on R^N and defines optimality in terms of minimal elements with respect to this partial order. The word 'natural' here is based on the fact that the real number system has served as the prototype and everything that refers to it is 'natural'.

Whereas 'order' has become the dominant word in this context in mathematics, the word 'preference' is more generally used in economics. Unfortunately, each word has acquired an almost fixed meaning among mathematicians and economists. Thus, 'order' to the mathematician usually implies at least a 'partial order' and 'preference' is often taken to be at least a 'partial preorder' by economists due largely to the treatment of the subject by Debreu in his classic work "Theory of Value" [2]. But even with respect to these latter terms there seems to be no general consensus; for example, Kelley [4] calls any transitive relation a partial order or a quasi-ordering, Royden [5] defines a partial order as a transitive and antisymmetric relation and Fishburn [3] takes a quasi-order to be a reflexive and transitive relation. In the light of all these special names it is easy to lose sight of the fact that, above all, these relations serve to introduce a hierarchy among the elements of a set. Since the word 'preference' has, at least to some people, the more general connotation, it will be used here.

Definition 1.1. Strict preference and indifference.

(i) Let R_1 be a binary relation on a set M. R_1 is a strict preference on M iff R_1 serves to provide a hierarchy among elements of M. R_1 is then denoted by $\prec$.

(ii) Let R_2 be a binary relation on a set M. R_2 is an indifference on M iff R_2 serves to provide a sense of equality among elements of M. R_2 is then denoted by $\sim$.

Definition 1.2. Preference. Let R be a binary relation on a set M. R is a preference on M iff $R = R_1 \cup R_2$ is the disjoint union of a strict preference R_1 and an indifference R_2. R is then denoted by $\precsim$.

Once a preference has been established on a set then it also makes sense to talk about optimal elements of the set with respect to this preference.

Definition 1.3. Least and minimal elements. Let $\precsim$ be a preference on a set M. Let $x^0, x^1 \in M$. Then

(i) x^0 is a least element of M with respect to $\precsim$ on M iff $x^0 \precsim x \;\forall x \in M$.

(ii) x^1 is a minimal element of M with respect to $\precsim$ on M iff $x \precsim x^1 \Rightarrow x \sim x^1$ for every x^1-comparable $x \in M$. (In the following terms such as 'x^1-comparable' will be omitted. It will be understood that 'for every' in this context refers to only those elements which are comparable to the element under consideration.)

An analogous definition may be given for greatest and maximal elements. Some additional properties of preferences will be of use later on; also, some preferences with special properties have already been cited. Their names and some of these properties are collected in the next definitions.

Definition 1.4. Order relations. Consider the following properties for a binary relation R defined on a set M with $x,y,z \in M$:

(i) $xRx \ \forall x \in M$ (reflexivity);

(ii) 'xRy and yRz' $\Rightarrow$ 'xRz' (transitivity);

(iii) 'xRy and yRx' $\Rightarrow$ '$x=y$' (antisymmetry);

(iv) 'xRy' $\Rightarrow$ '$\neg yRx$' (asymmetry);

(v) 'xRy' $\Rightarrow$ 'yRx' (symmetry);

(vi) for every $x,y \in M$ either xRy or yRx or both (connexity).

Collectively one then defines:

(1) (i) and (ii) together as a partial preorder;

(2) (i), (ii) and (iii) together as a partial order;

(3) (i), (ii) and (v) together as an equivalence;

(4) the inclusion of (vi) in (1) as a complete preorder;

(5) the inclusion of (vi) in (2) as a linear order.

Definition 1.5. Properties of preferences. M is an arbitrary set unless mentioned otherwise. Let $\precsim$ be a preference on M.

(i) Continuity. $\precsim$ is continuous iff the sets $\{x \in M: x \precsim y\}$ and $\{x \in M: x \succsim y\}$ are closed in M for every $y \in M$.

(ii) Differentiability. $\precsim$ on a set $M \subseteq R^n$ is of class C^k on M iff $\{(x,y) \in M \times M: x \sim y\}$ is a C^k-hypersurface in R^n.

(iii) Monotonicity. Let $M \subseteq R^n$. The derived strict preference $\prec$ on M is

(a) weakly monotone iff for $x,y \in M$ one has $x \ll y \Rightarrow x \prec y$;

(b) monotone iff for $x,y \in M$ one has $x < y \Rightarrow x \prec y$.

(For $x,y \in R^N$: $x \leq y \Leftrightarrow x_i \leq y_i, \quad \forall i \in I = \{1,\dots,N\}$;
$x < y \Leftrightarrow x_i \leq y_i, \quad x \neq y, \ \forall i \in I$;
$x \ll y \Leftrightarrow x_i < y_i, \quad \forall i \in I$.)

In the theory of economic equilibria as in most other calculations involving preferences the existence of a utility function representing the preference plays a vital role.

Definition 1.6. Utility function. Let $\precsim$ be a preference on a set M. A real valued function $\phi(.): M \to R$ is a utility function for $\precsim$ on M iff for every $x,y \in M$:

(i) $x \precsim y \Leftrightarrow \phi(x) \leq \phi(y)$,

(ii) $x \prec y \Leftrightarrow \phi(x) < \phi(y)$,

(iii) $x \sim y \Leftrightarrow \phi(x) = \phi(y)$.

The above notion of preference is now used to introduce an optimality concept in multicriteria decision making, along with corresponding discussions of existence, necessary and sufficient conditions. The treatment here is necessarily somewhat brief. The subject was introduced by the author in [6] with a discussion of the fundamental aspects, existence and of necessary conditions; sufficient conditions were given in [7] and finally a connected treatment together with some applications of Pareto-Optimality in mechanics was given in [8]. As a matter of fact, by first reading [8] which was completed before the other two, one gets an impression of the evolution of the subject to its present form.

2. The Multicriteria Problem

The basic problem treated here consists of a decision set $\mathcal{D}$ and a mapping $g(.):\mathcal{D}\to\mathcal{A}\subseteq R^N$, where $\mathcal{A}=\{a\in R^N: a=g(d), d\in\mathcal{D}\}$ is the so-called attainable criteria set or the set of values of $g(.)$ when d is restricted to $\mathcal{D}$. No consideration is given to problems involving more general vector spaces such as arise with countably many criterion functions or a continuum of them.

For such a problem there are two basically different approaches to an optimality concept. One may introduce a preference either on the decision space $\mathcal{D}$, a usual approach among economists or one may introduce a preference on $\mathcal{A}$, the approach which is used in programming and control theory. The approach here will be to introduce a preference $\precsim$ on $\mathcal{A}$ and thereafter to define optimality in terms of either least or minimal elements of $\mathcal{A}$ with respect to $\precsim$. In all cases, of course, one wishes to define the optimal decision.

Definition 2.1. Preference optimal decisions. Let $\precsim$ be a preference on $\mathcal{A}$. A decision $d^*\in\mathcal{D}$ is preference optimal iff $g(d^*)$ is either a minimal or a least element of $\mathcal{A}$ with respect to $\precsim$.

In view of the often mentioned connection of the multicriteria problem in this form with cooperative game theory and the calculation of economic optima it may be helpful to state just how this type of problem enters into an essentially classic treatment of the latter two concepts. Possibly the simplest illustrations thereof are the concept of the core and a pure exchange economy, respectively.

With a set I of N consumers associate consumption sets $X_i\subseteq R^n$, preferences $\precsim_i$ on each X_i and an initial resource $\omega^i\in R^n$. Corresponding to a consumption $x^i\in X_i$ for every $i\in I$ one defines a set A of feasible allocations $x=(x^1,\dots,x^N)\in X=\prod_{i=1}^N X_i$ by

$$A=\{x\in X: \sum_{i=1}^N x^i = \sum_{i=1}^N \omega^i\},$$

and an optimal allocation $x^*\in A$ as one for which

$$x^*\precsim x \Rightarrow x^*\sim x \quad \forall x\in A,$$

where the preference $\precsim$ is defined by

$$x \precsim y :\Leftrightarrow x^i \precsim_i y^i \quad \forall i \in I,\ x,y \in A.$$

If one has utility functions $\phi_i(.):X_i \to R \forall i \in I$, one may introduce a mapping $\phi(.):X \to R^N$ by

$$\phi(x) = (\phi_1(x^1),\phi_2(x^2),\ldots,\phi_N(x^N))$$

and an optimal allocation $x^* \in A$ then is one for which

$$\phi(x^*) \leq \phi(x) \Rightarrow \phi(x^*) = \phi(x), \quad \forall x \in A.$$

This is an ordinal optimality concept, since the definition is independent of any monotone increasing transformations of the utility functions $\phi_i(.)$.

The fundamental aspect of cooperative game theory is the formation of coalitions S, non-empty subsets of I, and the resultant allocations when members of a coalition cooperate in reaching goals in competition with other possible coalitions. For each quantity let a subscript s denote its restriction to $S \subset I$. Then an S-feasible set is given by

$$A_s = \{x_s \in X_s : \sum_{i \in S} x^i = \sum_{i \in S} \omega^i\},$$

along with a mapping $\phi_s(.):X_s \to R^S$ given by

$$\phi_s(x_s) = (\phi_i(x^i))_{i \in S}.$$

Particular effort has been expended in determining those coalitions which "block" an allocation x, or

S blocks $x \in A$ iff there exists an $\bar{x}_s \in A_s$ such that

$$x_s \prec_s \bar{x}_s, \quad \forall x_s \in A_s,$$

or equivalently, such that

$$\phi_s(x_s) < \phi_s(\bar{x}_s), \quad \forall x_s \in A_s,$$

with $\prec_s$ defined by $x_s \prec_s \bar{x}_s$ iff $x^i \prec_i \bar{x}^i$, $i \in S$. Thus, an allocation $\tilde{x}$ is not blocked by a coalition S iff

$$\phi_s(\tilde{x}_s) \leq \phi_s(x_s) \Rightarrow \phi_s(\tilde{x}_s) = \phi_s(x_s), \quad \forall x_s \in A_s.$$

The set of all $\tilde{x} \in A$ for which there does not exist a blocking coalition is called the core.

Now, let $\mathcal{P}(I)$ be the power set of I and let $K = \cup\{A_s : S \in \mathcal{P}(I)\}$ with $A_\emptyset = \emptyset$ and define a correspondence $v(.):\mathcal{P}(I) \to \phi(K)$ by

$$v(S) = \{\phi^S \in R^S : \phi^S = \phi^S(x^S),\ x^S \in A_s\}.$$

The couple $(v(.),I)$ is called a cooperative game.

It is apparent that the only way in which the previously introduced multicriteria problem enters into this discussion is in the calculation of the so-called Pareto-optimal allocations - either as the optimal allocations of a pure exchange economy or as those allocations which are not blocked by a coalition S.

In conclusion, the usual optimality concept in economics consists of the definition of an optimal or blocking allocation in the decision spac which is then induced on the utility space. In this context it thus follows that the definition of an optimal allocation in terms of the utility functions must be independent of monotone increasing transformations. This is precisely the reason why the present approach would not generally appear in economics. All these difficulties are of course circumvented if one uses cardinal utilities (which requires quite some modification of the original utility concept as arising from a system of preferences) or if one simply does not associate with each consumer a utility function but instead a cardinal function which is fixed for once and for all. This will be done here.

3. Existence, Necessary and Sufficient Conditions

The discussion here consists of results. All the proofs and more detailed discussions along with some illustrative examples may be found in the previously indicated references.

Theorem 3.1. Let $\mathcal{A} \subset R^N$ be compact and assume that the preference $\precsim$ on $\mathcal{A}$ is continuous and a complete preorder on $\mathcal{A}$. Then there exists a preference optimal decision $d^* \in \mathcal{D}$.

The proof is simple and it is based on the existence of a continuous utility function over $\mathcal{A}$.

The next theorem deals with conditions on $\precsim$ subject to which the optimal decisions with respect to $\precsim$ on $\mathcal{A}$ are also Pareto-optimal.

Theorem 3.2. Let $\precsim$ be a preference on $\mathcal{A} \subset R^N$ and assume that the derived strict preference $\prec$ is monotone and asymmetric on $\mathcal{A}$ and that the derived indifference $\sim$ is symmetric on $\mathcal{A}$. Then a preference optimal decision $d^* \in \mathcal{D}$ is also a Pareto-optimal decision.

Consequently, any conditions which are necessary for Pareto-optimal decisions are also necessary with respect to a preference optimal decision of the indicated type. For example, one thus has necessary conditions for preference optimal decisions with respect to the lexicographic order $\precsim_L$ on $\mathcal{A}$. This latter fact is of interest, since it is a result which is independent of the existence of a utility function $\phi(.)$ over $\precsim$ on $\mathcal{A}$, whereas the remaining results do make use of at least the existence of such a utility function.

By far the easiest deduction of necessary conditions for Pareto-optimality is based on the following beautifully simple observation which is attributed to Leonid Hurwicz.

Lemma 3.3. If $d^* \in \mathcal{D}$ is a Pareto-optimal decision then d^* minimizes $g_j(d)$ subject to $g_i(d) \leq g_i(d^*)$, $i \in I$, $i \neq j$, $d \in \mathcal{D}$ where the choice of j is arbitrary.

Thus, one may derive necessary conditions from the usual necessity theorems in optimal control and programming by simply adjoining some additional inequality constraints. In essence the resulting necessary conditions differ from those with a single criterion function $G(.):\mathcal{D} \to R$ only in that $G(.)$ is now replaced by $G(.) = cg(.)$, where $c > 0$ when the conditions are based on the previous statements. Similar conditions are obtained when there exists a C^1-utility function for $\precsim$ on $\mathcal{A}$, although the statement concerning c may be altered, in particular $c=0$ becomes a possibility. The following theorems are a collectivization of the results of the previously cited references.

Theorem 3.4. Let $\precsim$ be a preference on $\mathcal{A}^o$(open)$\supset \mathcal{A}$ and assume that there exists a C^1-utility function $\phi(.)$ for $\precsim$ on $\mathcal{A}^o$. Then if $d^* \in \mathcal{D}$ is a preference optimal decision with respect to $\precsim$ on $\mathcal{A}$ and if $a^* = g(d^*)$ the vector c satisfies the compatibility condition

$$c = k\nabla\phi(a^*),\ k>0.$$

Naturally, this theorem may be modified to include hypotheses which guarantee the existence of a differentiable utility function. A theorem assuring the existence of a C^2-utility function was given by Debreu [1].

As a consequence of using necessary conditions which involve the parameter c the corresponding optimal decision candidates and criterion values become functions of $c \in C$, where the set C largely depends on the optimality concept which was used. For example, $C=\{c \in R^N : c \geq 0\}$ when the preference is the natural order on R^N($c \geq 0$ because the abnormal case where the criterion functions do not enter the Hamiltonian has already been incorporated). This dependence on c will be denoted by $d(c) \in \mathcal{D}$ and $g(d(c)) = a(c) \in \mathcal{A}$.

Theorem 3.5. Let $\precsim$ be a monotone preference on $\mathcal{A}$ and assume that there exists a utility function $\phi(.)$ for $\precsim$ on $\mathcal{A}$. Then, if $c^* \in C$ is such that $d^* = d(c^*) \in \mathcal{D}$ is a preference optimal decision and $a(c^*) = g(d(c^*))$ is the corresponding criterion value, one has

$$\phi(a(c^*)) \leq \phi(g(d(c)))\ \forall c \in C.$$

Obviously, Theorem 3.2 may be used to generate sufficient conditions for Pareto optimality. This gives rise to the following two corollaries.

Corollary 3.6. Let $\precsim$ be a monotone preference on $\mathcal{A}^o$(open) $\supset \mathcal{A}$. Assume that a utility function $\phi(.)$ is known for $\precsim$ on $\mathcal{A}^o$ and that $a^* = g(d^*)$, $d^* \in \mathcal{D}$, satisfies $\phi(a^*) \leq \phi(a)\ \forall a \in \mathcal{A}$. Then d^* is Pareto-optimal.

Corollary 3.7. Let $\mathcal{A}^0$(open)$\supset\mathcal{A}$ and let $\phi(.):\mathcal{A}^0\to R$ be a C^1-function which satisfies $\nabla\phi(a)>>0$ on $\mathcal{A}^0$. Then any decision $d^*\in\mathcal{D}$ with $a^*=g(d^*)$ which satisfies $\phi(a^*)\leq\phi(a)\forall a\in\mathcal{A}$ is a Pareto-optimal decision.

As a last theorem a meta-theorem for sufficiency in preference optimality is given. For the theorem, let Th denote any sufficiency theorem which assures that $G(.):\mathcal{D}\to R$ satisfies $G(d^*)\leq G(d)\forall d\in\mathcal{D}$.

Theorem 3.8. Let $\precsim$ be a preference on $\mathcal{A}^0$(open)$\supset\mathcal{A}$ and let $a^* = g(d^*)$ with $d^*\in\mathcal{D}$. Assume:

(i) It is known that there exists a differentiable utility function $\phi(.)$ for $\precsim$ on $\mathcal{A}^0$ such that $\phi(.)$ is pseudoconvex with respect to $\mathcal{A}$ at $a^*\in\mathcal{A}$ and such that $\nabla\phi(a^*) \neq 0$.

(ii) A normal vector $n(a^*)$ to the indifference set $I(a^*)=\{b\in\mathcal{A}^0:a^*\sim b\}$ at a^* is known and it has the same orientation as $\nabla\phi(a^*)$.

(iii) All the hypotheses of Th are satisfied with $G(.)= n(a^*)g(.)$.

Then d^* is a preference optimal decision.

4. Illustrative Example

The example is presented in what might be a useful agricultural context. The equations are academic, however, since they are intended to illustrate computational procedure rather than quantitatively meaningful results.

Assume that a farmer wishes to raise two different species of livestock, designated 1 and 2. Let x_1 represent the amount of time spent in groooming and in caring for the species and x_2 the amount of allocated fodder. Let $g(x_1)$ and $g(x_2)$ be the respective numbers of the species capable of thriving for given amounts x. Now, the farmer may control the amounts x and because of the possibly varying benefit of the species to him, he may consider some combinations of amounts of the species more desirable than others thus giving rise in a natural way to a preference over g_1g_2 - space.

Let the preference be described in the following manner with $P=\{y\in R^2:y>>0\}$, consider the family of hyperbolas $y_2=\frac{k}{y_1}$, $k>0$, $y=(y_1,y_2)\in P$. Let the indifference surfaces of the preference be the members of this family; that is, if $z=(z_1,z_2)$ is any point in P and k is such that z may be written as $z=(z_1,\frac{k}{z_1})$; i.e., $k=z_1z_2$ then

$$I_k(z)\triangleq\{x\in P:x_2=\frac{k}{x_1}\}$$

is the indifference class of z. With

$$I_k^+(z)\triangleq\{x\in P:x_2\geq\frac{k}{x_1}\}$$

the statement

$$\forall z\in P,\ x\succsim z \text{ iff } x\in I_k^+(z),$$

along with a similar one for $x \nprec z$ defines a monotone and continuous preference of class C^∞ on P which is a complete preorder on P. The statement of the problem then is taken to be: Obtain preference optimal decisions for

$$g_1(x)=(x_1x_2)^{1/2} \text{ and } g_2(x)=x_1-x_2$$

subject to the inequality constraints

$$f_1(x)=x_2-x_1<0, \; f_2(x)=x_1-1\leq 0, \; f_3(x)=-x_2<0.$$

Solution: From the necessary conditions implied by Theorem 3.2 one has

$$-(\tfrac{1}{2}c_1\sqrt{\frac{x_2}{x_1}} + c_2, \; \tfrac{1}{2}c_1\sqrt{\frac{x_1}{x_2}} - c_2) + (-\lambda_1 + \lambda_2, \lambda_1 - \lambda_3) = 0.$$

Since $f_1(.)$ and $f_3(.)$ are never active and since an interior point cannot be a solution, it follows that

$$-\tfrac{1}{2}\, c_1 \sqrt{x_2} - c_2 = -\lambda_2,$$

$$-\tfrac{1}{2}\, c_1 \frac{1}{\sqrt{x_2}} + c_2 = 0,$$

with the result $x_2 = \frac{1}{4}(\frac{c_1}{c_2})^2$ and with corresponding criterion values

$$y_1 = \frac{1}{2}\,\frac{c_1}{c_2} \text{ and } y_2 = 1 - (\frac{c_1}{2c_2})^2.$$

Now, $c=(c_1,c_2)$ still needs to satisfy the compatibility condition, Theorem 3.4. Any normal vector with the same orientation as $\nabla\phi(a^*)$ will do. Take

$$n(y) = (\frac{a}{y_1^2}, 1)$$

as the definition of a normal vector field on P. This results in

$$\tilde{n}(c) = (\frac{2c_2}{c_1}[1-(\frac{c_1}{2c_2})^2], 1)$$

and a compatibility condition

$$c_1 = \frac{2c_2}{c_1}[1-(\frac{c_1}{2c_2})^2] \text{ and } c_2 = 1.$$

Solving for c_1 and c_2 yields

$$x^* = (1,\tfrac{1}{3}) \text{ and } y^* = (\frac{1}{\sqrt{3}}, \frac{2}{3}),$$

and one may use sufficiency conditions to deduce that these are indeed preference optimal.

Thus, if x_1 is in terms of hours per day, x_2 in hundreds of pounds of food per day and y_1, y_2 are in hundreds of members of the species, one may conclude that one hour of grooming and care together with $33\frac{1}{3}$ lbs of food per day will result in approximately 58 members of the first species and 67 of the second.

5. Conclusions and Suggestions

As a consequence of the present state of the art of multicriteria theory one should look for ways of specifying c in the linear combination cg(.) of the criterion functions. When a utility function for $\precsim$ on $\mathcal{A}$ is known explicitly then one may use standard programming methods to do so; on the other hand, it would be highly desirable to devise numerical algorithms which deal with the preference $\precsim$ directly. To my knowledge, no such algorithms exist to date.

Acknowledgements

I wish to express my gratitude to the "Mathematisches Forschungsinstitut Oberwolfach" for the opportunity to present this material under the title "Ordnung Muss Sein."

References

[1] Debreu, G.: "Smooth Preferences", Econometrica, Vol. 40, No. 4, 1972.

[2] Debreu, G.: Theory of Value, Wiley and Sons, N. Y., 1959.

[3] Fishburn,P.C.: Utility Theory for Decision Making, John Wiley and Sons, New York, 1970.

[4] Kelley, J.L.: General Topology, Van Nostrand, Princeton, New Jersey, 1955.

[5] Royden, H.L.: Real Analysis, The Macmillan Company, London, 1968.

[6] Stadler, W.: Preference Optimality in Multicriteria Control and Programming Problems, SIAM Journal on Control

[7] Stadler, W.: Sufficient Conditions for Preference Optimality, JOTA, Vol. 18, No. 1, January 1976.

[8] Stadler, W.: Preference Optimality and Applications of Pareto Optimality, Multicriteria Decision Making, edited by Marzollo, A. and Leitmann, G., CISM Courses and Lectures, Springer Verlag, New York, 1975.

Department of Mechanical Engineering
College of Engineering
University of California
Berkeley, Calif. 94720

Decomposition Procedures for Convex Programs

J. Stahl, Budapest

Introduction

In [4] several procedures were presented for finding a saddle point of a continuous convex-concave function, and it was shown that some decomposition methods for convex programs can be considered as special cases of these procedures. This paper follows the same way of thinking.

Under certain conditions the programming problem

$$\begin{gathered} f_1(x_1,x_2) \le 0 \\ f_2(x_1,x_2) \le 0 \\ x_1 \in X_1 \\ \min F(x_1,x_2) \end{gathered}$$

is equivalent to finding the saddle point value of the function $\varphi(x_1,y_1)$, where $\varphi(x_1,y_1)$ is the optimal value of the programming problem

$$\begin{gathered} f_2(x_1,x_2) \le 0 \\ \min \left(F(x_1,x_2) + y_1\, f_1(x_1,x_2)\right) . \end{gathered}$$

$x_1 \in X_1$, and y_1 is nonnegative. In the first part of the paper we introduce two procedures for calculating saddle point values, and the second part contains applications of these procedures to obtain decomposition methods for some programming problems.

The presentation of the procedures for calculating saddle point values is optional, i.e., the decomposition procedures can be derived directly. But the saddle-point procedures are of independent interest and they give a certain basis. Such a basis may help to understand what our decomposition procedures actually do, and many of the existing decomposition methods can also be imbedded into this framework.

1. Procedures for finding saddle point value

Let the function $\varphi(x,y)$ be continuous on $X \times Y$ where X and Y are compact subsets of Euclidean spaces of appropriate dimensions, and let us assume that $m^* = \max_Y \min_X \varphi(x,y) = \min_X \max_Y \varphi(x,y) = M^*$.

If the elements $y^1, y^2, \ldots, y^n \in Y$ and $x^1, x^2, \ldots, x^n \in X$ are given, let us consider the following two programming problems:

(1)
$$M \geq \varphi(x, y^i) \qquad (i=1,2,\ldots,n)$$
$$x \in X$$
$$\min M$$

and

(2)
$$m \leq \varphi(x^i, y)$$
$$y \in Y$$
$$\max m \; .$$

φ being continuous, X and Y compact, both these problems have optimal solutions which will be denoted by (M^n, x^{n+1}) and (m^n, y^{n+1}) respectively.

Lemma 1. $M^n \leq M^{n+1} \leq M^* = m^* \leq m^{n+1} \leq m^n$

and $$\lim M^n = \lim m^n \; (= M^* = m^*) \; .$$

So we can easily have an iterative procedure for finding the saddle point value of φ.

If $\varphi(x,y)$ is convex-concave - in which case $m^* = M^*$ - then (1) and (2) are convex problems. Since for these problems the Slater condition is obviously fulfilled, sets of multipliers $\mu^{1n}, \mu^{2n}, \ldots, \mu^{nn}$ and $\lambda^{1n}, \lambda^{2n}, \ldots, \lambda^{nn}$ resp. satisfying the Kuhn-Tucker saddle point conditions exist.

Lemma 2. Any accumulation point of the sequence $\left\{\left(\sum_i \lambda^{in} x^i, \sum_i \mu^{in} y^i\right)\right\}$ is a saddle point of φ.

Let now the function φ be continuously differentiable. In this case one can replace (1) and (2) by the - "weakened" - problems

(3)
$$M \geq \varphi(x^i, y^i) + \nabla_x \varphi(x^i, y^i)(x - x^i) \qquad (i=1,2,\ldots,n)$$
$$x \in X$$
$$\min M$$

and

(4)
$$m \leq \varphi(x^i, y^i) + \nabla_y \varphi(x^i, y^i)\,(y - y^i) \qquad (i=1,2,\ldots,n)$$
$$y \in Y$$
$$\max m \; ,$$

where $x^1, x^2, \ldots, x^n \in X$ and $y^1, y^2, \ldots, y^n \in Y$ are given.

All the arguments following (1) and (2) can be repeated, i.e., denoting by $(\overline{M}^n, x^{n+1})$ and $(\overline{m}^n, y^{n+1})$ the optimal solutions of these problems we have

Lemma 3. $\overline{M}^n \le \overline{M}^{n+1}$ $(\le M^{n+1}) \le M^* = m^*(\le m^{n+1}) \le \overline{m}^{n+1} \le \overline{m}^n$

and $\lim \overline{M}^n = \lim \overline{m}^n$ $(= M^* = m^*)$.

Lemma 4. Any accumulation point of $\left\{\left(\sum_i \lambda^{in} x^i, \sum_i \mu^{in} y^i\right)\right\}$ is a saddle point of φ, where now the λ-s and μ-s are optimal multipliers of (3) and (4).

Obviously, to obtain the convergence of this iterative procedure it is enough to assume the existence of bounded subgradients. The importance of replacing (1) and (2) by (3) and (4) will be clear in the cases below, where having already known the x^i and y^i we have no explicit expressions for the terms $\varphi(x,y^i)$ and $\varphi(x^i,y)$, but we can calculate $\nabla_x\varphi(x^i,y^i)$ and $\nabla_y\varphi(x^i,y^i)$ (or subgradients).

Let us consider the programming problem

(5)
$$\begin{aligned} f_1(x_1,x_2) &\le 0 \\ f_2(x_1,x_2) &\le 0 \\ x_1 &\in X_1 \\ \min F(x_1,x_2) &, \end{aligned}$$

where f_1, f_2 and F are convex functions and the set X_1 is convex and compact.

Assuming the existence of an optimal solution to this problem and of such an $x_1^o \in X_1$ for which $f_1(x_1^o,x_2^o) < 0$ and $f_2(x_1^o,x_2^o) \le 0$ with some x_2^o, we have

$$\varphi^* = \max_{\overline{Y}_1} \min_{X} \overline{\varphi}(x_1,x_2,y_1)$$

where

$$\overline{Y}_1 = \{y_1 : y_1 \ge 0\} ,$$
$$X = \{(x_1,x_2) : x_1 \in X_1, f_2(x_1,x_2) \le 0\} ,$$
$$\overline{\varphi}(x_1,x_2,y_1) = F(x_1,x_2) + y_1 f_1(x_1,x_2) ,$$

and φ^* is the optimal value of (5).

Assuming for any $(x_1,y_1) \in X_1 \times Y_1$ the existence of an optimal solution to the (convex) problem

(6)
$$\begin{gathered} f_2(x_1,x_2) \le 0 \\ \min \{F(x_1,x_2) + y_1 f_1(x_1,x_2)\} \end{gathered}$$

it follows that

$$\varphi^* = \max_{\overline{Y}_1} \min_{X} \overline{\varphi}(x_1,x_2,y_1) = \max_{Y_1} \min_{X_1} \varphi(x_1,y_1) \quad ,$$

where Y_1 is an appropriately chosen compact subset of $\overline{Y}_1$ and $\varphi(x_1,y_1)$ is the optimal value of (6).

Lemma 5. $\varphi(x_1,y_1)$ is a convex-concave function on $X_1 \times Y_1$.

Lemma 6. Let the functions $F(x_1,x_2)$, $f_1(x_1,x_2)$ and $f_2(x_1,x_2)$ be continuous, and for any $x_1 \in X_1$ let furthermore be $X_2(x_1) = \{x_2 : f_2(x_1,x_2) \leq 0\} \subseteq X_2$, where X_2 is bounded. Then $\varphi(x_1,y_1)$ is continuous on $X_1 \times Y_1$.

Concluding this part we see that under the assumptions above the optimal value of (5) is equal to the saddle-point value of the continuous convex-concave function φ. If we want to apply now the first procedure for calculating saddle-point value we have just the problem already mentioned. Namely, we have no explicit formula for the terms $\varphi(x_1,y_1^i)$ and $\varphi(x_1^i,y_1)$. In the following we will discuss cases where the second procedure can be applied.

2. Decomposition procedures

The gradients of φ can be determined using the results of [2].

Lemma 7. Let us assume the following. The functions $F(x_1,x_2)$, $f_1(x_1,x_2)$ and $f_2(x_1,x_2)$ are twice continuously differentiable and they are strictly convex in x_2. For any $(x_1,y_1) \in X_1' \times Y_1'$, where $X_1 \times Y_1 \subseteq \text{int}\,(X_1' \times Y_1')$, there exist optimal solutions and multipliers to (6) satisfying the Kuhn-Tucker conditions and strict complementarity - i.e., if some component of f_2 is tight at the optimal solution, then this component of y_2 is positive - and the rows of the Jacobian $\frac{\partial f_2}{\partial x_2}$ corresponding to the binding constraints are linearly independent at the optimal point. Then for any $(x_1,y_1) \in X_1 \times Y_1$

$$\nabla_{x_1} \varphi(x_1,y_1) = \nabla_{x_1} F(x_1,x_2) + y_1 \frac{\partial f_1}{\partial x_1}(x_1,x_2) + y_2 \frac{\partial f_2}{\partial x_1}(x_1,x_2)$$

and

$$\nabla_{y_1} \varphi(x_1,y_1) = f_1(x_1,x_2) \ .$$

Now applying the second procedure of part 1 we obtain the following decomposition method for solving (5).

Assuming that $x_1^1, x_1^2, \ldots, x_1^n \in X_1$ and $y_1^1, y_1^2, \ldots, y_1^n \in Y_1$ are already given and $x_2^1, x_2^2, \ldots, x_2^n$ and $y_2^1, y_2^2, \ldots, y_2^n$ are the optimal solutions and multipliers satisfying the Kuhn-Tucker conditions of the corresponding problems (6). In the next iterative step one has to solve the two separate master problems below:

$$(7)\quad M \geq F(x_1^i, x_2^i) + y_1^i\, f_1(x_1^i, x_2^i) + \left(\nabla_{x_1} F(x_1^i, x_2^i) + y_1^i \frac{\partial f_1}{\partial x_1}(x_1^i, x_2^i) + y_2^i \frac{\partial f_2}{\partial x_1}(x_1^i, x_2^i)\right)(x_1 - x_1^i) \quad (i=1,2,\ldots,n)$$

$$x_1 \in X_1$$

$$\min M$$

and

$$(8)\quad m \leq F(x_1^i, x_2^i) + y_1\, f_1(x_1^i, x_2^i) \quad (i=1,2,\ldots,n)$$

$$y_1 \in Y_1$$

$$\max m\ .$$

Having obtained the optimal solution (M^n, x_1^{n+1}) and (m^n, y_1^{n+1}) to (7) and (8) resp. one solves the subproblem

$$(9)\quad f_2(x_1^{n+1}, x_2) \leq 0$$

$$\min \left(F(x_1^{n+1}, x_2) + y_1^{n+1}\, f_1(x_1^{n+1}, x_2)\right)\ .$$

Denoting the optimal solution and multipliers of this last problem by x_2^{n+1} and y_2^{n+1} resp. the iterative step is completed.

It is obvious, that introducing into (2) and (4) the constraint $m \leq \varphi(x^o, y)$ does not change the validity of the statements of the corresponding lemma 5, where $x^o \in X$ is arbitrary. Choosing now $x^o = x_1^o$ and adding to (8) the constraint $m \leq F(x_1^o, x_2^o) + y_1\, f_1(x_1^o, x_2^o)$ also does not change the validity of the procedure. Applying a theorem of Dantzig [1] now we can replace in (8) the constraint $y_1 \in Y_1$ by $y_1 \geq 0$, since the sequence $\{y^i\}$ obtained by solving the problems (8) will belong to a bounded set. Of course, this is only a formal thing. Now (8) is a linear programming problem and the other master will be too, if the set X_1 is given by linear constraints.

Furthermore, denoting by $\lambda^{on}, \lambda^{1n}, \ldots, \lambda^{nn}$ an optimal solution to the dual of the augmented (8), $\left(\sum_{i=0}^{n} \lambda^{in}\, x_1^i,\ \sum_{i=0}^{n} \lambda^{in}\, x_2^i\right)$ is obviously a feasible solution to problem (5), and $F\left(\sum_{i=0}^{n} \lambda^{in}\, x_1^i,\ \sum_{i=0}^{n} \lambda^{in}\, x_2^i\right) \leq m^n$.

Since from our earlier assumptions follows the continuity (even the differentiability) of all the terms in the expressions for $\nabla_{x_1}\varphi$ and $\nabla_{y_1}\varphi$, we have now the following

Theorem 1. Let us assume the following. Problem (5) has an optimal solution, and φ^* is the optimal value of this problem. There exists an $x_1^o \in X_1$ for which $f_1(x_1^o,x_2^o) < 0$ and $f_2(x_1^o,x_2^o) \le 0$ with some x_2^o. The assumptions of lemma 7 are satisfied. Then the iterative procedure, given by solving the problems (7),

$$(8') \qquad m \le F(x_1^i,x_2^i) + y_1\, f_1(x_1^i,x_2^i) \qquad (i=o,1,2,\dots,n)$$
$$y_1 \ge 0$$
$$\max m ,$$

and (9), solves (5); i.e., the sequence $\{M^i\}$ is monotonically increasing and converges to φ^*, $(\bar{x}_1^n,\bar{x}_2^n) = \left(\sum_{i=o}^{n} \lambda^{in} x_1^i,\ \sum_{i=o}^{n} \lambda^{in} x_2^i\right)$ is a feasible solution to (5), where the λ^{in} -s are the optimal multipliers of (8'), and $F(\bar{x}_1^n,\bar{x}_2^n) \le m^n$, where the sequence $\{m^i\}$ is monotonically decreasing and converges to φ^*.

The rather strong assumptions of lemma 7 can be somewhat weakened in the case, when

$$(10) \qquad \begin{aligned} F(x_1,x_2) &= F_1(x_1) + F_2(x_2) \\ f_1(x_1,x_2) &= f_{11}(x_1) + f_{12}(x_2) \\ f_2(x_1,x_2) &= f_{21}(x_1) + f_{22}(x_2) \end{aligned}$$

where all the functions on the right hand sides are continuous convex and the functions $F_1(x_1)$, $f_{11}(x_1)$ and $f_{21}(x_1)$ are differentiable.

Let us assume again the existence of an optimal solution to (5) and that for any $(\bar{x}_1,\bar{y}_1) \in \bar{X}_1 \times \bar{Y}_1$ the set of the feasible solutions of the problem

$$(11) \qquad f_{22}(x_2) \le -f_{21}(\bar{x}_1)$$
$$\min F_2(x_2) + \bar{y}_1\, f_{12}(x_2)$$

is a subset of a bounded set and that (11) has an optimal solution and multipliers satisfying the Kuhn-Tucker saddle point conditions.

Lemma 8. If x_1, $x_1^i \in X_1$ and y_1, $y_1^i \in Y_1$ then

$$\varphi(x_1^i,y_1) \le \varphi(x_1^i,y_1^i) + (y_1 - y_1^i)\, f_1(x_1^i,x_2^i)$$

and

$$\varphi(x_1,y_1^i) \geq \varphi(x_1^i,y_1^i) + \left(\nabla_x F_1(x_1^i) + y_1^i \frac{\partial f_{11}}{\partial x_1}(x_1^i) + y_2^i \frac{\partial f_{21}}{\partial x_1}(x_1^i)\right)(x_1-x_1^i),$$

where $\varphi(x_1^i,y_1^i)$ is the optimal value of (11) and y_2^i is the system of the optimal multipliers belonging to (11) in the case when $\bar{x}_1 = x_1^i$ and $\bar{y}_1 = y_1^i$. (Actually, the first inequality does not require the separability of the functions).

To be able to apply the second procedure described in part 1 we still have to assure the boundedness of $\nabla_x F_1(x_1^i) + y_1^i \frac{\partial f_{11}}{\partial x_1}(x_1^i) + y_2^i \frac{\partial f_{21}}{\partial x_1}(x_1^i)$. Again, recalling to the existence of (x_1^0,x_2^0) and incorporating into the appropriate problem the same constraint as earlier we have no problem with the y_1^i-s. Assuming that the functions $F_1(x_1)$, $f_{11}(x_1)$ and $f_{21}(x_1)$ are continuously differentiable the only terms we have to deal with are the y_2^i-s. A simple possibility is to assume that for every $x_1 \in X_1$ there exist such an $x_2^0(x_1)$ for which $f_{21}(x_1) + f_{22}(x_2^0(x_1)) < 0$. Then one has

$$\|y_2^i\| \leq c \cdot \frac{F_2(x_2^0(x_1^i)) + y_1^i f_{12}(x_2^0(x_1^i)) - F_2(x_2^i) - y_1^i f_{12}(x_2^i)}{\| f_{22}(x_2^0(x_1^i))\|}$$

where c is some constant, and from this inequality the boundedness of y_2^i follows.

Now gathering all these assumptions for case (10) a theorem similar to the earlier one can be stated.

For case (10) a decomposition procedure is provided in [3]. The only difference between this procedure and the one suggested here is in the problem of determining x_1^{n+1}. In [3] there are constraints of the form

$$M \geq F_2(x_2^i) + y_1^i f_{12}(x_2^i) + y_2^i f_{22}(x_2^i) + F_1(x_1) + y_1^i f_{11}(x_1) + y_2^i f_{21}(x_1) .$$

If we start out from the approximation of $\varphi(x_1,y_1^i)$ given by

$$\varphi(x_1,y_1^i) \geq \varphi(x_1^i,y_1^i) + F_1(x_1) - F_1(x_1^i) + y_1^i(f_{11}(x_1) - f_{11}(x_1^i)) + y_2^i(f_{21}(x_1) - f_{21}(x_1^i))$$

and which is better than the one stated in lemma 7, the decomposition procedure in [3] can also be set up into the frames given by the second procedure of part 1. Obviously, its validity can now be proven under the same assumptions as in our case, even the assumption of continuous differentiability of the functions $F(x_1)$, $f_{11}(x_1)$ and $f_{21}(x_1)$ is not necessary.

At last we consider the linear case, i.e. when all the functions in (5) are linear. In this case one does not have to introduce in advance the assumptions about the compactness of X_1, the regularity of the constraints, etc.

If the problem itself assures the boundedness of x_1 and y_1 then one can establish practically the same procedure as in the nonlinear case and the general case can be reduced to this one.

If we have a problem of the form

(12)
$$\begin{aligned} A_{11}\, x_1 + A_{12}\, x_2 &\geq 0 \\ A_{21}\, x_1 + A_{22}\, x_2 &\geq 0 \\ x_1, x_2 &\geq 0 \\ \min\,(c_1\, x_1 + c_2\, x_2) \end{aligned}$$

let us introduce a further constraint and a new variable to obtain the following one

(13)
$$\begin{aligned} - e\, x_1 \qquad\qquad &\geq -\beta \\ e\,\xi + A_{11}\, x_1 + A_{12}\, x_2 &\geq 0 \\ A_{21}\, x_1 + A_{22}\, x_2 &\geq 0 \\ \xi,\, x_1, x_2 &\geq 0 \\ \min\,(\gamma\xi + c_1\, x_1 + c_2\, x_2) \end{aligned}$$

where β and γ are some fixed positive numbers and e is a vector of appropriate dimension containing ones.

To apply the decomposition procedure given by (7), (8) and (9) one additional remark is necessary. If in some step the subproblem

$$\begin{aligned} A_{22}\, x_2 &\geq -A_{21}\, x_1^{n+1} \\ x_2 &\geq 0 \\ \min\,(c_2 - y_1^{n+1}\, A_{12})x_2 \end{aligned}$$

has no feasible solution [is not bounded] then there exists an extremal $y_2^{n+1} \geq 0$ $\left[x_2^{n+1} \geq 0\right]$ for which $y_2^{n+1}\, A_{22} \leq 0$ and $y_2^{n+1}(-A_{21}\, x_1^{n+1}) > 0$ $\left[A_{22}\, x_2^{n+1} \geq 0 \text{ and } (c_2 - y_1^{n+1}\, A_{12})\, x_2^{n+1} < 0\right]$ and one introduces a constraint $y_2^{n+1}(-A_{21}\, x_1) \leq 0$ $\left[(c_2 - y_1\, A_{12})\, x_2^{n+1} \geq 0\right]$ into the problem corresponding now to (7) [(8)] completing this iterative step. (I.e., the sets X_1 and Y_1 for which the assumptions implying the existence of $\varphi(x_1,y_1)$ are valid can be defined now by

$$X_1 = \{x_1 : 0 \le x_1 \le \beta,\ (-y_2 A_{21})\, x_1 \le 0 \quad \text{for all} \quad y_2 \in Y_2\},$$

$$Y_1 = \{y_1 : 0 \le y_1 \le \gamma,\ (c_2 - y_1 A_{12})\, x_2 \ge 0 \quad \text{for all} \quad x_2 \in X_2\}$$

where $X_2 = \{x_2 : A_{22}\, x_2 \le 0,\ x_2 \ge 0\}$ and $Y_2 = \{y_2 : y_2 A_{22} \ge 0,\ y_2 \ge 0\}$, but in the linear case they are generated by the procedure itself.)

Let now $\varepsilon > 0$ be arbitrarily fixed and let us consider two arbitrary monotonic increasing sequences $\{\beta_i\}$, $\{\gamma_i\}$, for which $\lim_{i\to\infty} \beta_i = \lim_{i\to\infty} \gamma_i = \infty$. Applying our procedure for solving problem (13) with $\beta = \beta_i$ and $\gamma = \gamma_i$ it follows, that after a finite number of steps we have one of the following possibilities: problem (13) has no feasible solution, the dual of problem (13) has no feasible solution, or we have feasible solutions to both of these problems and the difference of the objective values belonging to them is at most ε. Let us denote these solutions by (ξ_i, x_{1i}, x_{2i}) and (π_i, y_{1i}, y_{2i}) resp. and let be $\xi_i = \infty$ $\left[\pi_i = \infty\right]$ in the case of the first $\left[\text{second}\right]$ possibility.

Theorem 2. $\lim_{i\to\infty} \xi_i \left[\lim_{i\to\infty} \pi_i\right]$ always exists. (12) [the dual to (12)] has a feasible solution if and only if $\lim_{i\to\infty} \xi_i = 0$ $\left[\lim_{i\to\infty} \pi_i = 0\right]$. Furthermore in this case $\lim_{i\to\infty} |\gamma_i\, \xi_i - \varepsilon| = 0$ $\left[\lim_{i\to\infty} |\pi_i\, \beta_i - \varepsilon| = 0\right]$.

This means, that if (12) has an optimal solution then applying our procedure for solving problems (13) after a finite number of steps ξ_i and π_i become small and we can omit the additional constraint and variable. According to the last theorem one can stop the procedure now having an almost feasible and an almost optimal solution.

People, who are programming computers are not quite satisfied with this kind of results. We have no recommendation how to choose the sequences $\{\beta_i\}$ and $\{\gamma_i\}$ and the value of ε to obtain the most efficient computational scheme. On the other hand, if x_1 must belong to a compact set X_1, i.e. we have a constraint $A_{01}\, x_1 \le b_1$ describing a bounded set, then one does not need to introduce the variable ξ and the sequence $\{\gamma_i\}$. A procedure can be obtained by slightly modifying the original one which solves this case, and this modification does not bring any computational difficulties [5].

References

[1] Dantzig, G.B.: "Linear programming and extensions", pp. 476-477, Princeton University Press, Princeton, 1963.

[2] Fiacco, A.V.: "Sensitivity analysis for nonlinear programming using penalty methods", Technical paper, The George Washington University, Washington, 1973.

[3] Kronsjö, T.O.M.: "Decomposition of a nonlinear convex separable economic system in primal and dual directions". In: Optimization (ed. by R. Fletcher), pp. 85-97, Academic Press, London, 1969.

[4] Oettli, W.: "Eine allgemeine, symmetrische Formulierung des Dekompositionsprinzips für duale Paare nichtlinearer Minmax- und Maxmin-Probleme", Zeitschrift für Operations Research 18 (1974), 1-18.

[5] Stahl, J.: "Decomposition procedure for doubly coupled LP programs", INFELOR Közlemények 10, Budapest, 1975 (In Hungarian).

INFELOR
H-1137 Budapest
Ujpesti rkp. 8

Vol. 59: J. A. Hanson, Growth in Open Economies. V, 128 pages. 1971.

Vol. 60: H. Hauptmann, Schätz- und Kontrolltheorie in stetigen dynamischen Wirtschaftsmodellen. V, 104 Seiten. 1971.

Vol. 61: K. H. F. Meyer, Wartesysteme mit variabler Bearbeitungsrate. VII, 314 Seiten. 1971.

Vol. 62: W. Krelle u. G. Gabisch unter Mitarbeit von J. Burgermeister, Wachstumstheorie. VII, 223 Seiten. 1972.

Vol. 63: J. Kohlas, Monte Carlo Simulation im Operations Research. VI, 162 Seiten. 1972.

Vol. 64: P. Gessner u. K. Spremann, Optimierung in Funktionenräumen. IV, 120 Seiten. 1972.

Vol. 65: W. Everling, Exercises in Computer Systems Analysis. VIII, 184 pages. 1972.

Vol. 66: F. Bauer, P. Garabedian and D. Korn, Supercritical Wing Sections. V, 211 pages. 1972.

Vol. 67: I. V. Girsanov, Lectures on Mathematical Theory of Extremum Problems. V, 136 pages. 1972.

Vol. 68: J. Loeckx, Computability and Decidability. An Introduction for Students of Computer Science. VI, 76 pages. 1972.

Vol. 69: S. Ashour, Sequencing Theory. V, 133 pages. 1972.

Vol. 70: J. P. Brown, The Economic Effects of Floods. Investigations of a Stochastic Model of Rational Investment. Behavior in the Face of Floods. V, 87 pages. 1972.

Vol. 71: R. Henn und O. Opitz, Konsum- und Produktionstheorie II. V, 134 Seiten. 1972.

Vol. 72: T. P. Bagchi and J. G. C. Templeton, Numerical Methods in Markov Chains and Bulk Queues. XI, 89 pages. 1972.

Vol. 73: H. Kiendl, Suboptimale Regler mit abschnittweise linearer Struktur. VI, 146 Seiten. 1972.

Vol. 74: F. Pokropp, Aggregation von Produktionsfunktionen. VI, 107 Seiten. 1972.

Vol. 75: GI-Gesellschaft für Informatik e.V. Bericht Nr. 3. 1. Fachtagung über Programmiersprachen · München, 9.–11. März 1971. Herausgegeben im Auftrag der Gesellschaft für Informatik von H. Langmaack und M. Paul. VII, 280 Seiten. 1972.

Vol. 76: G. Fandel, Optimale Entscheidung bei mehrfacher Zielsetzung. II, 121 Seiten. 1972.

Vol. 77: A. Auslender, Problèmes de Minimax via l'Analyse Convexe et les Inégalités Variationelles: Théorie et Algorithmes. VII, 132 pages. 1972.

Vol. 78: GI-Gesellschaft für Informatik e.V. 2. Jahrestagung, Karlsruhe, 2.–4. Oktober 1972. Herausgegeben im Auftrag der Gesellschaft für Informatik von P. Deussen. XI, 576 Seiten. 1973.

Vol. 79: A. Berman, Cones, Matrices and Mathematical Programming. V, 96 pages. 1973.

Vol. 80: International Seminar on Trends in Mathematical Modelling, Venice, 13–18 December 1971. Edited by N. Hawkes. VI, 288 pages. 1973.

Vol. 81: Advanced Course on Software Engineering. Edited by F. L. Bauer. XII, 545 pages. 1973.

Vol. 82: R. Saeks, Resolution Space, Operators and Systems. X, 267 pages. 1973.

Vol. 83: NTG/GI-Gesellschaft für Informatik, Nachrichtentechnische Gesellschaft. Fachtagung „Cognitive Verfahren und Systeme", Hamburg, 11.–13. April 1973. Herausgegeben im Auftrag der NTG/GI von Th. Einsele, W. Giloi und H.-H. Nagel. VIII, 373 Seiten. 1973.

Vol. 84: A. V. Balakrishnan, Stochastic Differential Systems I. Filtering and Control. A Function Space Approach. V, 252 pages. 1973.

Vol. 85: T. Page, Economics of Involuntary Transfers: A Unified Approach to Pollution and Congestion Externalities. XI, 159 pages. 1973.

Vol. 86: Symposium on the Theory of Scheduling and its Applications. Edited by S. E. Elmaghraby. VIII, 437 pages. 1973.

Vol. 87: G. F. Newell, Approximate Stochastic Behavior of n-Server Service Systems with Large n. VII, 118 pages. 1973.

Vol. 88: H. Steckhan, Güterströme in Netzen. VII, 134 Seiten. 1973.

Vol. 89: J. P. Wallace and A. Sherret, Estimation of Product. Attributes and Their Importances. V, 94 pages. 1973.

Vol. 90: J.-F. Richard, Posterior and Predictive Densities for Simultaneous Equation Models. VI, 226 pages. 1973.

Vol. 91: Th. Marschak and R. Selten, General Equilibrium with Price-Making Firms. XI, 246 pages. 1974.

Vol. 92: E. Dierker, Topological Methods in Walrasian Economics. IV, 130 pages. 1974.

Vol. 93: 4th IFAC/IFIP International Conference on Digital Computer Applications to Process Control, Part I. Zürich/Switzerland, March 19–22, 1974. Edited by M. Mansour and W. Schaufelberger. XVIII, 544 pages. 1974.

Vol. 94: 4th IFAC/IFIP International Conference on Digital Computer Applications to Process Control, Part II. Zürich/Switzerland, March 19–22, 1974. Edited by M. Mansour and W. Schaufelberger. XVIII, 546 pages. 1974.

Vol. 95: M. Zeleny, Linear Multiobjective Programming. X, 220 pages. 1974.

Vol. 96: O. Moeschlin, Zur Theorie von Neumannscher Wachstumsmodelle. XI, 115 Seiten. 1974.

Vol. 97: G. Schmidt, Über die Stabilität des einfachen Bedienungskanals. VII, 147 Seiten. 1974.

Vol. 98: Mathematical Methods in Queueing Theory. Proceedings 1973. Edited by A. B. Clarke. VII, 374 pages. 1974.

Vol. 99: Production Theory. Edited by W. Eichhorn, R. Henn, O. Opitz, and R. W. Shephard. VIII, 386 pages. 1974.

Vol. 100: B. S. Duran and P. L. Odell, Cluster Analysis. A Survey. VI, 137 pages. 1974.

Vol. 101: W. M. Wonham, Linear Multivariable Control. A Geometric Approach. X, 344 pages. 1974.

Vol. 102: Analyse Convexe et Ses Applications. Comptes Rendus, Janvier 1974. Edited by J.-P. Aubin. IV, 244 pages. 1974.

Vol. 103: D. E. Boyce, A. Farhi, R. Weischedel, Optimal Subset Selection. Multiple Regression, Interdependence and Optimal Network Algorithms. XIII, 187 pages. 1974.

Vol. 104: S. Fujino, A Neo-Keynesian Theory of Inflation and Economic Growth. V, 96 pages. 1974.

Vol. 105: Optimal Control Theory and its Applications. Part I. Proceedings 1973. Edited by B. J. Kirby. VI, 425 pages. 1974.

Vol. 106: Optimal Control Theory and its Applications. Part II. Proceedings 1973. Edited by B. J. Kirby. VI, 403 pages. 1974.

Vol. 107: Control Theory, Numerical Methods and Computer Systems Modeling. International Symposium, Rocquencourt, June 17–21, 1974. Edited by A. Bensoussan and J. L. Lions. VIII, 757 pages. 1975.

Vol. 108: F. Bauer et al., Supercritical Wing Sections II. A Handbook. V, 296 pages. 1975.

Vol. 109: R. von Randow, Introduction to the Theory of Matroids. IX, 102 pages. 1975.

Vol. 110: C. Striebel, Optimal Control of Discrete Time Stochastic Systems. III. 208 pages. 1975.

Vol. 111: Variable Structure Systems with Application to Economics and Biology. Proceedings 1974. Edited by A. Ruberti and R. R. Mohler. VI, 321 pages. 1975.

Vol. 112: J. Wilhelm, Objectives and Multi-Objective Decision Making Under Uncertainty. IV, 111 pages. 1975.

Vol. 113: G. A. Aschinger, Stabilitätsaussagen über Klassen von Matrizen mit verschwindenden Zeilensummen. V, 102 Seiten. 1975.

Vol. 114: G. Uebe, Produktionstheorie. XVII, 301 Seiten. 1976.

Vol. 115: Anderson et al., Foundations of System Theory: Finitary and Infinitary Conditions. VII, 93 pages. 1976

Vol. 116: K. Miyazawa, Input-Output Analysis and the Structure of Income Distribution. IX, 135 pages. 1976.

Vol. 117: Optimization and Operations Research. Proceedings 1975. Edited by W. Oettli and K. Ritter. IV, 316 pages. 1976.